Heat Transfer in Polymer Composite Materials

Series Editor
Noël Challamel

Heat Transfer in Polymer Composite Materials

Forming Processes

Edited by
Nicolas Boyard

iSTE

WILEY

First published 2016 in Great Britain and the United States by ISTE Ltd and John Wiley & Sons, Inc.

ISTE Ltd
27-37 St George's Road
London SW19 4EU
UK

www.iste.co.uk

John Wiley & Sons, Inc.
111 River Street
Hoboken, NJ 07030
USA

www.wiley.com

Library of Congress Control Number: 2016931428

British Library Cataloguing-in-Publication Data
A CIP record for this book is available from the British Library
ISBN 978-1-84821-761-4

Contents

Preface xv

Chapter 1. Introduction to Heat Transfer During the Forming of Organic Matrix Composites 1
Didier DELAUNAY

1.1. Introduction 1
1.2. Examples of injection of short fiber reinforced composites 2
1.2.1. Heat transfer during the filling phase 2
1.2.2. Heat transfer during part consolidation 18
1.3. Injection on continuous fiber reinforcements 22
1.4. Conclusion: toward a controlled processing 24
1.5. Bibliography 25

Chapter 2. Experimental Determination and Modeling of Thermophysical Properties 29
Nicolas BOYARD and Didier DELAUNAY

2.1. Measurement of specific volume and shrinkage 29
2.1.1. Thermoplastic PvT diagram 30
2.1.2. Specific volume of thermosetting polymers 36
2.2. Determination of specific heat capacity of resin and composites 40
2.3. Thermal conductivity: a tricky task.... 43
2.3.1. A first assessment of experimental data 44
2.3.2. Overview of the main characterization techniques to measure the thermal conductivity of polymers and associated composites 46
2.3.3. Modeling of the thermal conductivity of composites 57

2.4. Conclusions 69
2.5. Bibliography 70

Chapter 3. Experimental Determination and Modeling of Transformation Kinetics 77
Nicolas BOYARD, Jean-Luc BAILLEUL and M'hamed BOUTAOUS

3.1. Introduction 78
3.2. What are the most suitable devices to analyze a reaction rate? 79
3.2.1. Conventional methods 79
3.2.2. Original methods 86
3.2.3. A first assessment of the current characterization methods 86
3.3. Modeling of the cure kinetics of thermosetting resins 88
3.3.1. Mechanistic models: complexity versus accuracy 89
3.3.2. Description of the kinetics with empirical models: the engineer approach 91
3.3.3. Modeling of the diffusion induced by vitrification 97
3.4. Overall crystallization kinetics of semi-crystalline thermoplastics 98
3.4.1. Most popular crystallization kinetics models for process simulations 100
3.4.2. Systems of differential equations 102
3.4.3. Measurements of crystallization kinetics and associated parameters 104
3.4.4. Models for specific crystallization phenomena and geometries 107
3.5. Concluding remarks 109
3.6. Bibliography 111

Chapter 4. Phase Change Kinetics within Process Conditions and Coupling with Heat Transfer 121
M'hamed BOUTAOUS, Matthieu ZINET, Nicolas BOYARD and Jean-Luc BAILLEUL

4.1. Introduction 121
4.2. Flow-induced crystallization: experimental observations 124
4.2.1. Relevant experimental techniques 124
4.2.2. Effect of flow on crystallization kinetics 125
4.2.3. Flow effect on crystalline morphology 131
4.2.4. Flow effect on crystalline growth rate 134

4.2.5. Effect of flow on rheological properties 134
4.2.6. Summary of experimental observations and guidelines for modeling 136
4.3. Flow-induced crystallization: modeling 138
4.3.1. Overall kinetics modeling 138
4.3.2. Explicit nucleation and growth modeling 139
4.3.3. Role of viscoelasticity 140
4.4. Effect of the composite components 143
4.4.1. Effect of nucleating agents 143
4.4.2. Effect of fibers 144
4.5. Concluding remarks 147
4.6. Bibliography 149

Chapter 5. From the Characterization and Modeling of Cure-Dependent Properties of Composite Materials to the Simulation of Residual Stresses 157
Yasir NAWAB and Frédéric JACQUEMIN

5.1. Introduction 157
5.2. Origin of residual stress 157
5.2.1. Mechanical levels of residual stress 158
5.2.2. Parameters contributing to the formation of residual stress 159
5.2.3. Problems generated by residual stress 161
5.3. Determination of composite properties 161
5.3.1. Modeling the mechanical properties of composites 162
5.3.2. Experimental determination of thermomechanical properties of composite 164
5.4. Modeling of residual stress 167
5.4.1. Linear approach or classical theory of laminates 168
5.4.2. Nonlinear approach 168
5.4.3. Minimization of energy approach 169
5.4.4. Application 170
5.5. Conclusion 171
5.6. Bibliography 172

Chapter 6. Heat Transfer in Composite Materials and Porous Media: Multiple-Scale Aspects and Effective Properties 175
Michel QUINTARD

6.1. Introduction 175
6.2. Effective thermal conductivity 177
6.2.1. Background on upscaling methods 178

6.2.2. A simple example: continuous thermal conductivity 179
6.2.3. Effective thermal conductivity: properties and bounds. 181
6.3. Local-equilibrium model and thermal dispersion. 184
6.4. Local equilibrium versus local non-equilibrium models 187
6.4.1. The two-equation model. 187
6.4.2. Further discussion 189
6.5. Various extensions 192
6.5.1. Effect of homogeneous and heterogeneous thermal sources 193
6.5.2. Interfacial thermal resistance 194
6.6. Conclusions. 195
6.7. Bibliography 196

Chapter 7. Thermal Optimization of Forming Processes 203
Vincent SOBOTKA

7.1. Context of optimization. 203
7.2. Heat transfer: optimization lever 204
7.3. Definition of the optimization criterion 206
7.4. Problem modeling. 207
7.4.1. Spatial scale 207
7.4.2. Time scale: process steps 207
7.4.3. Multi-physical aspects 207
7.5. Numerical optimization methods 208
7.5.1. The adjoint problem 210
7.5.2. Practical setting of the method 211
7.6. Example of process optimization: determination of optimal heat flux setpoint. 211
7.6.1. Experimental setup and constraint 212
7.6.2. Instrumentation of the mold and the preform 214
7.6.3. Thermal modeling 215
7.6.4. Experimental data from a composite part molding 218
7.6.5. Estimation of the thermal contact resistances 219
7.6.6. Determination of the optimal setpoint. 220
7.7. Optimal design of molds 222
7.7.1. OSOTO project 222
7.7.2. The considered thermoplastic part 222
7.7.3. General methodology 223
7.7.4. Conformal cooling approach 223

7.7.5. Heat transfer model in the process 224
7.7.6. Objective function 226
7.7.7. Minimization of the functional J 227
7.7.8. Cooling channel design 228
7.8. Conclusions and outlook 231
7.9. Bibliography 232

Chapter 8. Modeling of Thermoplastic Welding 235
Gilles REGNIER and Steven LE CORRE

8.1. Introduction 235
8.1.1. Polymer welding processes 235
8.1.2. Healing mechanisms of polymer interfaces 237
8.2. Physics of thermoplastic welding 237
8.2.1. Intimate contact at interface 237
8.2.2. Macromolecular diffusion 241
8.3. Linear viscoelasticity to quantify the macromolecular diffusion 246
8.4. Application to continuous welding of composite tape 248
8.4.1. Process description 248
8.4.2. Influence of processing conditions on interfacial strength 249
8.4.3. Modeling of macromolecular diffusion 249
8.4.4. Modeling of thermal aging 251
8.4.5. Thermal modeling of the process 253
8.4.6. Weldability prediction 254
8.5. Application to ultrasonic welding 255
8.5.1. Process description and time scale separation 255
8.5.2. Process modeling: necessity of a time homogenization framework 256
8.5.3. Numerical multi-physical model 257
8.5.4. Ultrasonic welding with energy directors: process analysis and optimization 259
8.6. Conclusion 262
8.7. Acknowledgments 263
8.8. Bibliography 263

Chapter 9. Multiphysics for Simulation of Forming Processes 269
Luisa SILVA, Patrice LAURE, Thierry COUPEZ and Hugues DIGONNET

9.1. Introduction 269
9.2. Multiscale, multiphysics and multidomain modeling 270
9.2.1. Flow equations 271

9.2.2. Thermal-rheological-kinetical coupling 273
9.2.3. Orientation and structure development during processing 275
9.3. Advanced numerical techniques and macroscale simulations 278
9.3.1. Implicit boundaries 280
9.3.2. Immersed subdomains and regularization 281
9.3.3. Multiphase flow and thermokinetical numerical resolution 282
9.3.4. Composite forming simulation illustrations 283
9.3.5. Parallel mesh adaptation and high-performance computing 286
9.4. Determination of equivalent properties and microscale simulations 289
9.4.1. Generation of representative numerical samples 289
9.4.2. Permeability of a composite 291
9.4.3. Stiffness tensor determination 294
9.5. Conclusions 294
9.6. Bibliography 296

Chapter 10. Thermal Instrumentation for the Control of Manufacturing Processes of Organic Matrix Composite Materials 301
Jean-Christophe BATSALE and Christophe PRADERE

10.1. Introduction 301
10.2. Methods based on contact measurement 302
10.2.1. Temperature sensors and fluxmeters 302
10.2.2. Heat flux estimation 304
10.2.3. Thermal probes for thermophysical property measurement in static conditions 308
10.3. Contactless heating 311
10.3.1. Photothermal methods with monosensors (under microscale characterization) 312
10.3.2. Thermal non-destructive evaluation: cracks or delamination detection in composite samples 313
10.3.3. Screening of chemical processes and microfluidic experiments 317
10.3.4. Principles for the factory global monitoring, example of conveyor belt parameter estimation 323
10.3.5. In a near future: the Big Data related to multiscale process survey 325

10.4. Conclusion 326
10.5. Bibliography 326

Chapter 11. Sensors for Heat Flux Measurement 333
Fabien CARA and Vincent SOBOTKA

11.1. Motivations: heat flux sensor 333
11.2. Principle of heat flux sensors. 335
11.2.1. Gradient heat flux sensor. 335
11.2.2. Inertial heat flux sensor 338
11.2.3. Inverse heat flux sensor 338
11.3. Main characteristics of HFS 340
11.3.1. Invasiveness 340
11.3.2. Time constant 340
11.3.3. Calibration 343
11.4. Type, positioning and use of heat flux sensors 344
11.4.1. Commercial sensors 345
11.4.2. Positioning of heat flux sensors. 347
11.4.3. Price. 348
11.5. Advantages and limitations of HFS compared to other *in situ* monitoring techniques 349
11.5.1. Advantages. 349
11.5.2. Limitation and care in using HFS. 349
11.6. Examples 349
11.6.1. Compression molding 350
11.6.2. Resin transfer molding flow front detection 352
11.6.3. Resin transfer molding: influence of mold temperature 353
11.6.4. Internal temperature prediction during infusion 354
11.6.5. Glass mat transfer, internal temperature monitoring 355
11.7. Conclusions 356
11.8. HFS suppliers. 356
11.9. Bibliography 357

Chapter 12. Thermal Radiative Properties of Polymers and Associated Composites 359
Benoit ROUSSEAU

12.1. Introduction 359
12.2. Fundamental requisites concerning thermal radiation 361
12.2.1. Spectral range of thermal radiation. 361

12.2.2. Radiant energy, radiant flux, radiative flux density and radiative intensity 363
12.2.3. Blackbody spectral emissive power 366
12.2.4. Radiative properties at interfaces 368
12.2.5. Radiative properties of semi-transparent slabs 372
12.3. Prediction of the radiative properties of homogeneous semi-transparent slabs: case of the isotactic polypropylene 373
12.3.1. Intrinsic optical properties for a homogeneous, isotropic and non-magnetic medium 374
12.3.2. Prediction of the radiative properties of polypropylene slabs at 20°C 376
12.4. Radiative properties of polymer composites with fiber structures 378
12.4.1. Normal spectral absorptance of Roving TWINTEX® 378
12.4.2. Normal spectral absorptance of sheet of PEEK with carbon fibers 380
12.5. Conclusion 381
12.6. Bibliography 381

Chapter 13. Infrared Radiation Applied to Polymer Processes 385
Yannick LE MAOULT and Fabrice SCHMIDT

13.1. Introduction 386
13.1.1. Why use infrared heating for polymers? 386
13.1.2. Application of radiative transfers in polymer processing 386
13.2. Infrared radiation characteristics 388
13.2.1. Radiative properties (basis and main definitions) 389
13.2.2. Infrared emitters: characterization 393
13.2.3. Infrared camera measurements 397
13.3. Modeling of infrared radiation 403
13.3.1. Opaque medium: surface to surface methods 403
13.3.2. Semi-transparent medium 405
13.3.3. Ray tracing method 407
13.4. Polymer processing applications 409
13.4.1. Optimization of preform temperature for the ISBM process 409
13.4.2. Optimal infrared composite curing 415
13.5. Future work 419

13.6. Acknowledgements 420
13.7. Bibliography 420

List of Authors 425

Index 427

Preface

This book is a recapitulation of the CNRS Summer School organized in June 2013 dedicated to heat transfer in organic matrix composite materials and their forming processes. The industry of composites has grown steadily in recent years due to the numerous advantages of these materials, such as their lightness and interesting mechanical properties, compared to aluminum and other metal-based alloys. To remain competitive, especially in a very strong international economic context, the quality of the produced parts must be fully controlled. This control requires an accurate knowledge of physical phenomena occurring during the various steps of their manufacturing process and in a context where the strong activity and the needs led to the emergence of new processes and increasingly fast production rates. The forming of composite materials has thus become a major topic of research in terms of experimentation, modeling and simulation, where several scientific disciplines must come together in order to achieve the control of manufactured parts and properties. We can notice that heat transfer is one of the main levers to control the forming processes and induced properties of the composite part. They have to be carefully analyzed during the manufacturing of these materials that also require a multidisciplinary approach. Thus, thermal sciences have to be coupled to other scientific fields such as mechanics and physical chemistry.

The first goal of this summer school was to bring together academic and industrial researchers from different disciplines within thermal sciences with transverse themes common to their activities. A second aim was also to provide the basis on heat transfer during polymer and composite processing as well as the latest methods and techniques from experimental, numerical and modeling points of view, useful to help in the solving of many issues. Therefore, the book takes this and gives theoretical and practical information to understand, measure and describe, in a relevant way, heat transfer during forming processes (in the tool as well as in the composite part) and introducing the required couplings. For this purpose, we relied on the experience of recognized French researchers.

This book is written in a comprehensive way for an audience that is already aware of the world of composites and associated processes: graduate students, researchers and people involved in R&D activities in industrial sectors. Our aim is to provide a tool, useful for the readers to start a study on composite processing where heat transfers are involved. Each chapter describes the concepts, techniques and/or models related to the developed topic and several examples are given for illustration purposes. A list of selected references is also given at the end of each chapter for a deeper complement of its content, which is necessary for more complex analyses and developments. Unfortunately, all topics and issues related to heat transfer in composite parts and processes cannot be addressed in a single book and a selection was made to cover a broad range of subjects and associated issues.

The introductive chapter presents heat transfer analyses and issues in polymer and composite processing through illustrative examples mostly from injection molding. Preconceived ideas, difficulties and simplified approaches are well highlighted. One key to success in heat transfer modeling is the accurate knowledge of thermophysical properties, phase change kinetics and their associated models for both thermosets and thermoplastics. Conventional as well as new methods to experimentally determine these properties and reaction rate parameters as a function of temperature are detailed in Chapters 2 and 3. Scientific and technical issues are also included. A comprehensive review of the effects of thermoplastics process conditions (shear and/or elongation induced by the flow) and the addition of other components (nucleating agents, fibers, etc.) on the transformation kinetics of the polymers, their rheological behaviors and final microstructures is detailed.

From all these data, the simulation of residual stresses developed during the matrix transformation and the cooling is discussed. For this purpose, thermokinetic and mechanical couplings are introduced and the prediction of cure-dependent mechanical properties is presented. In Chapter 6, modeling of heat transfer in multi-scale porous media, which can be encountered during the filling step of Resin Transfer Molding (RTM) mold, is discussed following a homogenization approach. The relationship between the physics at local-scale and the macroscale description is explained, also including the determination of effective properties.

The improvement of the quality of parts can be achieved by optimizing process parameters. Among them, the thermal control of the part is of strong importance and depends on the thermal control of the mold. Thus, optimization approach has to consider heat transfer in the tool and couplings to include contact conditions and transformation kinetics. Context, definition and methods of optimization are covered in this book and are illustrated with two detailed examples. Process modeling is introduced in Chapters 8 and 9. First, we discuss the peculiar case of thermoplastic welding, where no adding materials is required for assembling. The importance of intimate contact and macromolecular diffusion is emphasized from theoretical and practical points of view, including the strong temperature dependence. A simulation of forming processes is also addressed in a more general way. Several examples are proposed to present multiscale, multiphysics and multidomain modeling, which are representative of the complexity of forming processes.

Another important part in heat transfer analysis concerns the instrumentation for the thermal characterization and the control of manufacturing processes. From these data, it is possible to obtain information about the process, thermophysical properties and/or the matrix transformation (for example, using inverse method algorithms). An overview of the existing instrumentation (contact and contactless methods) is given in this book. A specific chapter has been specifically dedicated to heat flux sensors, since they provide relevant information to quantify heat transfer between the part and the tool. It is thus an important complement to temperature measurement. Available heat flux sensor technologies and their main characteristics are also mentioned and are completed with practical examples.

Toward the final part of this book, radiative heat transfer in polymer and composite forming are detailed. The processes using infrared heating are in development and the complexity of heat transfer analysis leads to several scientific issues. After a presentation of the basics to define thermal radiative properties, measurements are presented for classical semi-crystalline polymers and associated composites. Finally, after a description of infrared emitters and the temperature measurement using infrared camera, modeling of radiative heat transfer is introduced and polymer processing applications are included.

I would like to thank all my French colleagues who have done me the honor of participating to the CNRS Summer School in 2013 and then of accepting to contribute to this book project with their high-quality work. Special recognition goes to Didier Delaunay, CNRS senior researcher, for his scientific involvement and significant contribution to research in heat transfer in composites and forming processes. I hope that all readers, working in the broad field of polymer and composite processing, may find this book an interesting and valuable resource.

Nicolas BOYARD
January 2016

1

Introduction to Heat Transfer During the Forming of Organic Matrix Composites

In this chapter, we present some very illustrative examples of advances obtained in the analysis of the heat transfer in the forming processes of composites. These examples highlight at the same time the difficulties and scientific issues, as well as some simplified approaches to obtain, in an accessible way, a rapid estimation of the times of cooling or heating of a composite part. We also point out some preconceived ideas, in particular on the nature of transfers during the filling of molds in injection process, and for this we propose a new criterion to determine the transition between the thermal shock regime and the one of established convection, which is validated by experimental results. The selected processes are the injection of composites with short fibers (thermoplastic and thermosets), and the injection on a fabric. The examples are illustrated by results issued of more than 25 years of analysis of heat transfer in the processes, during thesis led within the framework of partnership programs with companies of the plastics processing industry. Finally, some directions for new developments are proposed.

1.1. Introduction

The mastery of composite forming processes raises a number of challenges on heat transfer and how to take them into account adequately. Indeed, the nature of these materials itself induces peculiarities. First, they have at least two components, which poses the problem of determining their effective properties according to those constituents. Composites are also multi-scale materials: the fibers are gathered into tows, which are woven to make a fabric. Reinforcements, but sometimes also the matrices, are anisotropic. The coupled phenomena introduce complex physics difficult to interpret without thorough knowledge of them and their interactions. Furthermore, the operating conditions in the process can be often considered as extreme. High cooling rates are frequently encountered, such as for the contact

Chapter written by Didier DELAUNAY.

between a cold mold and a hot composite. Shear rates can be very large at the wall of an injection mold, or in the micro channels between fibers (even if the flow rates are low). To compensate the shrinkages due to the cooling or the transformations, pressures are sometimes very high in the molding cavity (up to 200 MPa for injection). Low-conductive polymers and composites are in some cases subjected to overheating as a result of heat source release induced by transformation. An additional issue is the fact that the temperature measurement is very difficult because it is intrusive. It is not yet well known how to experimentally determine the temperature fields within a molded part or inside of the plies in a reinforcement stack. The size of thermocouples, which should be at least as small as the fibers, makes them very fragile, especially in a viscous fluid flow. To date, the precise determination of the inlet temperature in the injection channel of a mold is an open problem. Another particularly characteristic example is the location of the filling fronts and the saturation distribution in a composite made by liquid composite molding (LCM) since it is coupled to heat transfer. In addition, solving a heat transfer problem requires the accurate knowledge of the boundary conditions, which may be difficult. We can also give the example of the determination of the thermal contact resistance (TCR) to the wall of a molding cavity or between plies during the consolidation of a composite.

Temperatures are important to know, but the dynamics of a thermal system may be assessed only by measuring the heat flux. How to make a heat flux sensor non-intrusive and accurate in the environment of forming processes? The main question is ultimately whether a fine thermal knowledge is essential to achieve quality parts.

In this chapter, we will show, from a few illustrative examples, that the couplings involved at all levels must be adequately taken into account from the point of view of heat transfer, since they induce consequences on the quality of the final product appearance, size, shape and properties. The thermal scientific problem appears as inevitable, especially since productivity requires short cycles in mass production: there is indeed cycles of about 1 min for automotive parts. Everyone can easily understand that the heating and cooling of a part by varying its temperature sometimes several hundred degrees in very short times, with the objective to control the temperature fields and to obtain uniform final properties for complex shapes, requires a non-trivial strategy.

1.2. Examples of injection of short fiber reinforced composites

1.2.1. *Heat transfer during the filling phase*

1.2.1.1. *Case of semi-crystalline polymer matrices*

We will first discuss the injection of a polymer reinforced with glass fibers, taking the example of a widely distributed poly-aramid, whose trade name is IXEF

[PIN 09]. The scope of this study is a collaborative program ("FISH" program) involving LTN, IMP (Lyon), PIMM (ENSAM Paris) laboratories and Moldflow, Legrand, and Solvay companies.

An injection cycle is typically divided into four phases: the filling step (few seconds) is short compared to the total cycle time, during which high shear rates may occur. The packing phase consists of applying a pressure on the polymer/composite to compensate the thermal and crystallization shrinkages. The third step is the isochoric cooling under pressure after the gel of the injection gate, preventing the entry or exit of polymer from the molding cavity and finally the cooling at atmospheric pressure after the possible unsticking of the part. At the end of the cooling, the solid polymer part is ejected and the new cycle can begin.

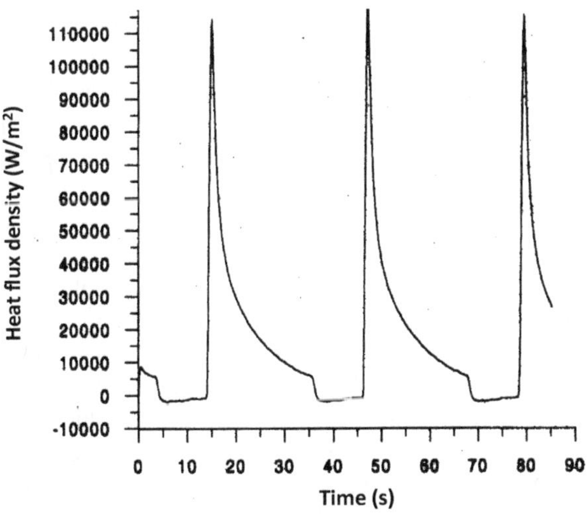

Figure 1.1. *Heat flux to the wall of a molding cavity in injection process*

The thermal behavior of the mold is periodic: the heat flux exchanged between the part and the mold is very large at the beginning of the cycle, decreases and finally is negative at the ejection, as shown in Figure 1.1. This particularity has to be taken into account since an established periodic state is required for constant quality parts. The typical example of heat flux at the wall of a molding cavity displayed in Figure 1.1 highlights a dramatic and very quick decrease. How can we interpret this behavior? Let us consider the flow in the channel formed by the molding cavity. If we assume legitimately that the forced convection is in the steady state, it is possible under these conditions to evaluate a local Nusselt number at the distance x from the entry of the molding cavity using a conventional correlation for a prescribed wall temperature [PIN 09], for a shear thinning fluid with a rheological power law type:

$$Nu(x) = 1.16((3n + 1)/4n)^{1/3} (PeD_h/x)^{1/3} \quad [1.1]$$

The Peclet number is defined as $Pe = VD_h/a$; D_h is the hydraulic diameter of the channel, here twice its thickness, *a* is the diffusivity of melt, *n* is the index of the power law viscosity. For the considered instrumented molding cavity [LEG 06] (incomplete part is shown in Figure 1.2 and instrumented cavity in Figure 1.10), we obtain the position of the sensor near the gate *a* value of 20 for the Nusselt number, which corresponds to a constant heat flux during the filling phase close to 1.10^5 W/m^2. The heat flux obtained as such (see detailed calculations in section 1.2.1.2 applied for bulk molding compound (BMC) processing) is the good order of magnitude (see Figure 1.1) but the experimental one decreases very quickly and we do not observe a constant value even during a short time. The analysis is thus invalidated: the evolution of the heat flux does not correspond to a regime of established convective exchange between the polymer and the wall of the molding cavity. Here is a preconceived idea, which constitutes an approach nevertheless classic but erroneous.

Let us test then the hypothesis that the heat flux decrease is due to the coupling with the conduction in the mold. There is an analytical solution [CAR 59] to the problem of a flowing fluid, which is suddenly put into contact with a wall. In this solution, exchanges by convection are based on a constant convective heat transfer coefficient h. This latter, calculated from the equations [1.1] and [1.11], is in this case approximately equal to 660 $W/m^2/K$ (we take the value $n = 0.308$). The heat flux density is given by $\Phi(t) = h\, F(t/\tau)\, (T_m - T_i)$. T_m is the average temperature of the melt, and T_i is the initial temperature of the mold. F is a decreasing function of time [CAR 59]. τ is given by $\tau = \lambda^2/h^2 a$, where λ is the mold thermal conductivity and *a* is the melt diffusivity. For the molding cavity studied, we find $\tau = 35586$ s, which corresponds to about 10 h. The rapid decrease in the observed heat flux is completely incompatible with this law since for $t = \tau$ the value of the function F is approximately 0.4. The result of this analysis using *a reductio ad absurdum* that convective heat transfer during the filling phase is not the key to analyze the heat transfer with the mold.

Let us consider the incomplete part shown in Figure 1.2. The energy equation for the filling of the molding cavity may under certain simplifying assumptions (constant thermophysical properties in particular) be written as:

$$\rho C_p DT/Dt = \beta T DP/Dt + \eta\dot{\gamma}^2 + \lambda\, \Delta T + Q \quad [1.2]$$

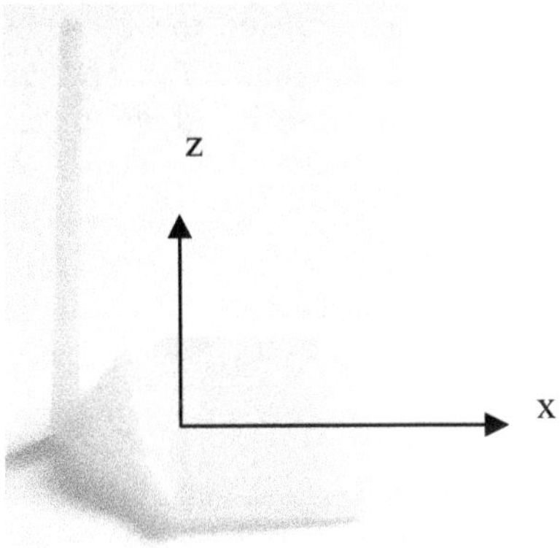

Figure 1.2. *Incomplete part obtained with the SWIM mold [LEG 00]*

In this equation, D/Dt is the operator ($\partial/\partial t$ + V.grad). The left term represents the thermal inertia. The first term on the right is due to the compressibility. The analysis of its order of magnitude shows that in first approach, it can be neglected. The second term is due to the viscous dissipation. Although it is not always negligible, it is also an order of magnitude lower than the other terms, especially early in the filling phase. The third term is the thermal diffusion, which occurs mainly in the z direction. Q is the possible source due to the phase change of the polymer (in the case of semi-crystalline thermoplastics, for example). A commonly accepted hypothesis is that transfers are one-dimensional (1D) along z, so that we can neglect viscous dissipation and the effect of the compressibility. Equation [1.2] can be thus simplified:

$$\rho C_p(\partial T/\partial t + v_x\,\partial T/\partial x) = \lambda\,\partial^2 T/\partial z^2 \qquad [1.3]$$

The second term of the left side equation [1.3] is, at the beginning of the filling phase, of several orders of magnitude, lower than the first one. Indeed, the term $\partial T/\partial t$ is very large and takes precedence over the other. We can then neglect this one and the equation is reduced to the classical 1D Fourier equation:

$$\rho C_p \partial T/\partial t = \lambda\,\partial^2 T/\partial z^2 \qquad [1.4]$$

The part is then subjected to a heat shock: for a short time (fast filling), it can be assumed that during this stage only the surface layers of the part and the mold in the vicinity of the wall of the molding cavity are involved in heat exchanges. The temperature field is thus the one observed in two semi-infinite media suddenly put into contact. At this time, the temperature in the polymer can be assumed to be uniform and equal to the injection temperature T_{inj}. The temperature in the vicinity of the molding cavity wall is quasi-uniform and equal to the temperature of the mold surface at the time of the beginning of the filling phase. We denote it by T_{mi}. The solution of the contact problem is well known. In the case of an amorphous polymer, if we neglect the contact resistance between the mold and the polymer, the mold surface temperature T_c at the contact time is given by equation [1.5]

$$T_c = (T_{inj}b_p + T_{mi}b_m)/b_p + b_m) \quad [1.5]$$

In this equation, b_p and b_m denote, respectively, the effusivity of the polymer and the mold, given by the square root of the product thermal conductivity λ, volumic mass ρ and specific heat C_p: $b = (\lambda\rho C_p)^{0.5}$. Let us apply this result on the example presented in Figure 1.3, where the time evolution of the mold surface temperature is plotted for the injection of an ABS in a steel mold. The thermal properties of this steel are $\lambda = 36$ W.m^{-1}.K^{-1}, $\rho = 7850$ kg.m^{-3}, $C_p = 460$ J.kg^{-1}.K^{-1}.

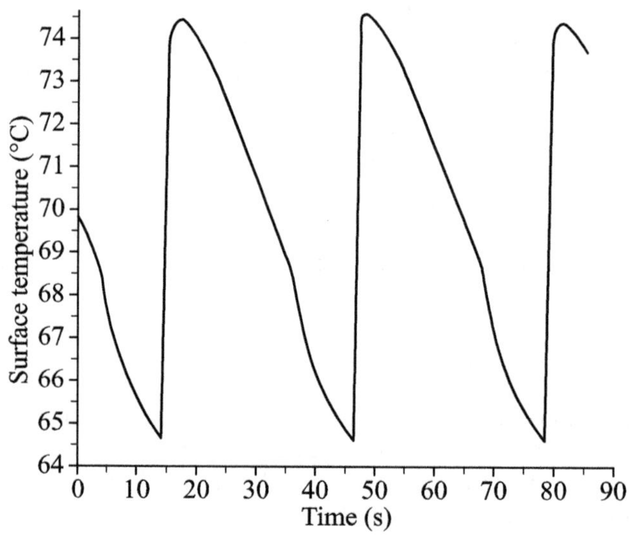

Figure 1.3. *Surface temperature of a mold during injection of an amorphous polymer (ABS)*

The injection temperature of the molten polymer is 240°C. In a periodic regime, we can see that the experimental temperature of the injection mold, which is the minimum value of the curve, is about 64.7°C. The effusivity of the mold steel is 11,474 $J.K^{-1}.m^{-2}.s^{-1/2}$. The ABS used for this test had the following properties: thermal conductivity $\lambda = 0.2\ W.m^{-1}.K^{-1}$, density $\rho = 961\ kg.m^{-3}$, specific heat $C_p = 2{,}300\ J.kg^{-1}.K^{-1}$. Its effusivity is, therefore, 665 $J.K^{-1}.m^{-2}.s^{-1/2}$. From equation [1.5], the contact temperature is equal to $T_c = 74.3°C$. This value is obviously very close to the recorded maximum temperature reached by the molding cavity surface, which corresponds to the time of contact of the mold with the hot polymer. This first result is in line with the validation of the heat shock phenomenon governed by the conduction in the part thickness. We note a temperature range of 10 K for the increase in the wall temperature. This result is classic. Soon after, the temperature decreases, because the condition of semi-infinite medium is only valid in the first times of the cycle for a dimensionless characteristic time $\tau = at/e^2 < 0.05$ (empirical result). In the expression of τ, a is the thermal diffusivity of the polymer and e is the half-thickness of the part. In our case, $e = 1.5.10^{-3}$ m and the diffusivity of the ABS is $9.10^{-8} m^2.s^{-1}$. The corresponding time below which the assumption of infinite medium is valid is 1.25 s. From the end of the filling that lasts 1 s, the core temperature of the part decreases and, as a result, the surface temperature.

The heat flux exchanged between the part and the mold associated with equation [1.5] is decreasing and is given by:

$$\Phi(t) = \lambda\ (T_{inj} - T_c)/\sqrt{\pi a t} \qquad [1.6]$$

This expression of the heat flux, inversely proportional to the square root of the time, is consistent with the experimental observation. It was compared with the experimental data recorded during the injection of an isotactic polypropylene [LEB 98]. Figure 1.4 illustrates the good description of the experimental heat flux by this law during the filling phase, within the first second after contact, when the polypropylene is in the molten (and thus amorphous) state. Then, the beginning of the crystallization is observed, since it induces an increase in the heat flux. At this time, the model deviates from the experiment since it cannot take into account the effect of the phase change. This result completes the demonstration that in the early time of the injection cycle, conductive thermal shock allows us to interpret perfectly the heat exchanges between the part and the mold. Is it possible to transpose this result to a charged polymer, additionally for a semi-crystalline one?

The first approach is to consider the neat matrix. Figure 1.5 displays a typical example of the wall temperature of the molding cavity for the injection of a semi-crystalline polymer [SOB 13].

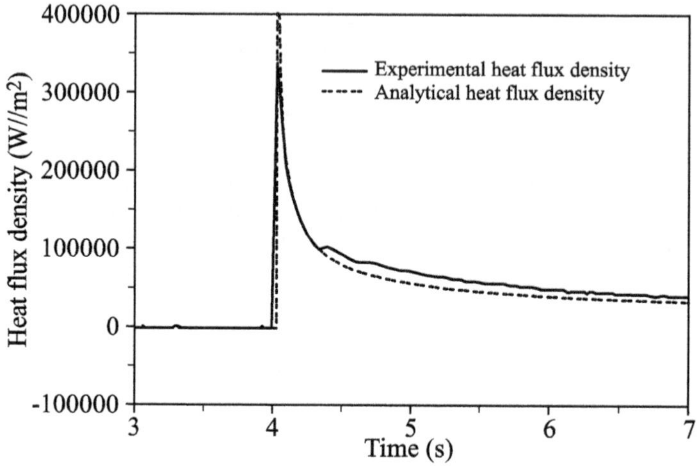

Figure 1.4. *Comparison between the heat flux given by equation [1.6] (dotted line) and the measurement (solid line) in thermal shock regime during the injection of an isotactic polypropylene [LEB 98]*

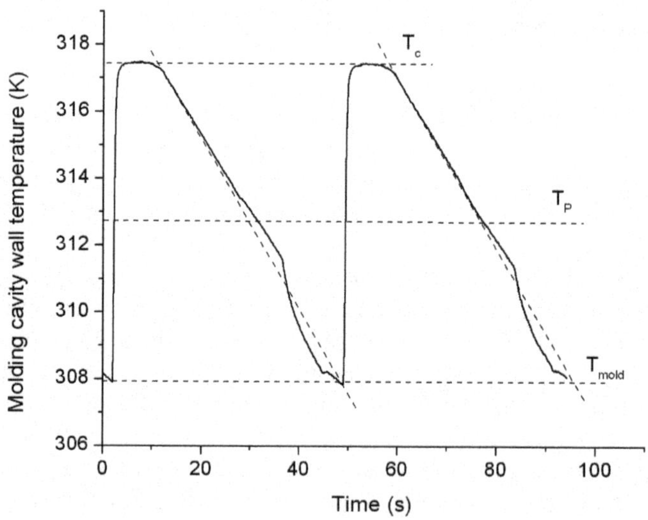

Figure 1.5. *Wall temperature in the molding cavity during the injection of isotactic polypropylene [SOB 13]*

Unlike amorphous polymers, we observe that the temperature curve shows a net plateau during the first seconds before the decreasing due to cooling. This is due to the latent heat of crystallization. The solution of a thermal shock problem with phase change (Stefan problem [SOB 13]) induces a change (in comparison with the classical case) of the contact temperature equation [LOU 97]:

$$T_c = \frac{b_s T_F + b_m erf(\xi/v) T_{mi}}{b_s + b_m erf(\xi/v)} \quad [1.7]$$

In this relation, $v = \sqrt{a_s/a_l}$ where a_s and a_l are the solid and liquid (i.e. molten) phases diffusivity, respectively. ξ is the solution of the transcendental equation

$$Ste_s v \frac{e^{-\xi^2/v^2}}{erf(\xi/v)} - \frac{Ste_l e^{-\xi^2}}{erfc(\xi)} = \sqrt{\pi}\xi \; ; \; t > 0 \quad [1.8]$$

where $Ste_l = Cp_l (T_{inj} - T_F)/L$, $Ste_s = Cp_s (T_F - T_c)/L$ are, respectively, the liquid and solid Stefan numbers. T_F is defined as the temperature of the crystallization quasi-plateau that appears in the core of the polymer upon cooling [SOB 13]. This is a new thermophysical characteristic, which depends on the pressure conditions. It can be determined experimentally or by simulation taking into account the kinetics of crystallization and its changes under pressure. The iterative solution of equations [1.7] and [1.8] allows calculating ξ and T_c. The quasi-plateau observed in the core of the part is related to the heat released during the crystallization, maintaining T_F until the complete solidification. Therefore, the constant half-thickness temperature is imposed, which has the effect to extend the semi-infinite character of the phenomenon, hiding what happens "behind" the isotherm T_F. This plateau, therefore, lasts until complete solidification of the part. What happens if we add short fibers to the semi-crystalline polymer?

Figure 1.6 shows a typical example of the evolution of the mold surface temperature during the injection of a poly-aramide reinforced with short glass fibers. A temperature plateau is clearly observed and its time is even longer than the reinforcement ratio is low. This is related to the solidification time since the latent heat decreases when the reinforcing ratio increases. The plateau temperature is also even higher than the reinforcing ratio increases. It can be explained by the increase in the contact temperature induced by both the increase in ξ, when latent heat decreases (the solidification is faster) and the increase in the effusivity of the composite when the glass fiber weight ratio is higher. The accurate calculation of T_c confirms these effects and the experimental results. 1D heat transfer in the composite through the thickness of the part has been demonstrated in several studies

[DEL 11, LEG 10, LEG 11], the fibers being oriented in the part in a core-skin structure.

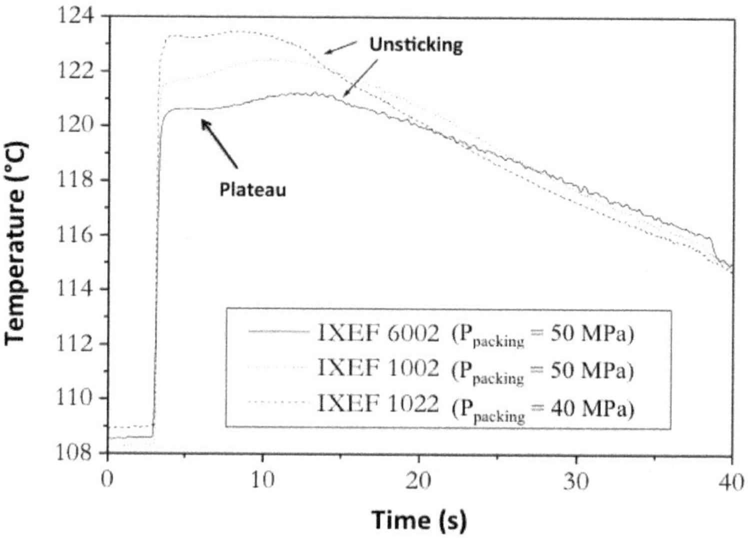

Figure 1.6. *Molding cavity wall temperature during the injection of poly-aramide composites [LEG 06]: no fiber (IXEF 6002), 30 wt% fiber (IXEF 1002), 50 wt% (IXEF 1022)*

This characteristic behavior has the advantage of applying the method described in [SOB 13] to easily estimate the cooling time of the composite part and thus to choose the optimum molding parameters. Indeed, observing Figure 1.6, we note that the molding cavity surface temperature may be less than 1 K with accuracy schematically described by successively a plateau (ordinate T_c) and the negative slope line connecting the point corresponding to the solidification time with ordinate T_c (120.5°C for IXEF 6002) to the point corresponding to the part ejection (39 s, 116.5°C for IXEF 6002). The solidification time may be calculated analytically [SOB 13]. It is then possible to analytically calculate the temperature changes in the part and therefore the time for which an ejection criterion is reached. Needless to say that such a simplified approach is not a substitute for professional software that continuously become more accurate, as Moldflow© for example, but it provides the expert with a relatively precise and very quick estimate. This reduced model is also an advantage for optimization software that uses intensively direct problems for solving.

1.2.1.2. *Bulk molding compound injection molding*

This section is dedicated to research work done as part of a program with the ultimate goal of controlling surface aspect of parts made of BMC, involving several laboratories: CRMD of Orleans, the Department of Polymers and Composites Technology of the School of Mines of Douai, LTN laboratory and Menzolit company (now known as IDI Composites), material supplier. Our specific goal in this study that covered several cooperative programs between 1997 and 2003 was to better control heat transfer in the injection of these composites constituted of unsaturated polyester resin, thermoplastic additive (to compensate the shrinkage induced by the crosslinking), mineral fillers and short glass fibers. The results are detailed in [MIL 01]. An experimental mold equipped with pressure and heat flux sensors was designed, as shown in Figure 1.7.

Figure 1.7. *Picture of the instrumented mold for the heat transfer study of the injection of a BMC plate*

This mold was mounted onto a vertical press (200 tons clamping force). The injection unit is horizontal. It is equipped with heating elements to heat the material up to 40°C in order to reduce its viscosity, which facilitates its injection. A water-cooled nozzle avoids the warming induced by the hot mold. Material gelation in the nozzle is then avoided. The injected part had a size enough large to be representative of the process: 0.4 m × 0.15 m × 3 mm. The heating channels of the mold are such that the thermal gradient along the walls of the molding cavity is as small as possible. An oil regulator unit is used to heat the mold up to 180°C. Two pressure sensors and four thermal ones (heat flux and temperatures) have been placed into the mold according to the positions indicated in Figure 1.8. The pressure sensors are Kistler type. Data acquisition is done by a Tektronix recorder (200 measures/s). The heat flux sensors conception and their manufacturing have been done in our laboratory [QUI 98] and are composed of three small thermocouples (25 µm diameter). Each sensor is made with the same steel (Z40CDV3) as the mold so that it is fully non-intrusive from a thermal point of view. A numerical treatment of the data (using the 1D sequential inverse method of Beck [SOM 12]) gives the temperature and the flux density at the sensor surface. Practically speaking, the thermocouple placed far (4.9 mm in our case) from the surface defines the boundary condition; the two others are used for the criterion calculation. This criterion is defined as the quadratic difference between the calculated temperatures (for the position of each thermocouple) and the measured ones. The minimization of the criterion gives the surface heat flux density and the temperature evolution with time. Theoretically, two thermocouples are sufficient for calculation. However, the analysis with three thermocouples gives a better confidence in the results. The thermophysical properties of the steel have been carefully measured due to classical methodologies. Note that short fibers are all oriented in the plane of the part, so that heat transfer arises mainly in the direction of the thickness of the sample. The heat flux sensors have a time constant lower than 5 ms. Indeed, the thermocouples are semi-intrinsic with separated junctions. The metal of the sensor itself ensures the electric continuity. The sensor thermocouples are K type (the thermoelectric power of about 40 µV/°C). This low voltage generates acquisition difficulties and amplification of the signal is necessary.

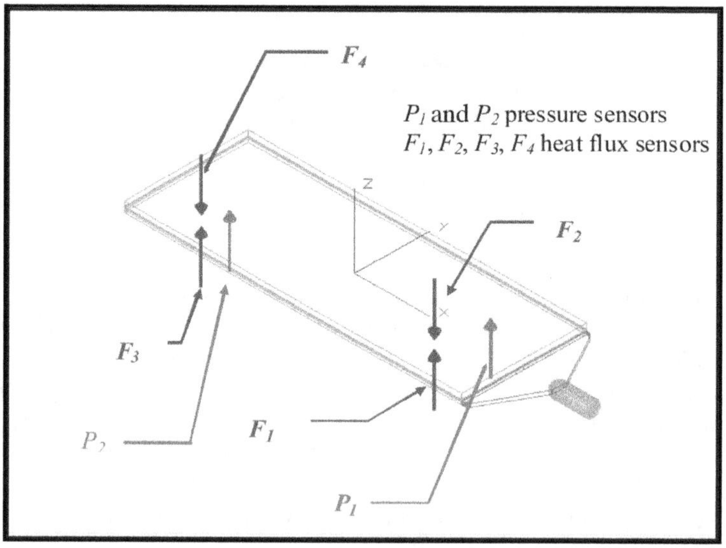

Figure 1.8. *View of the BMC part indicating the locations of the sensors*

A characteristic result of heat flux and pressures during the filling phase is shown in Figure 1.9 and we intend to use it to illustrate the power of relevant thermal analysis. The mold temperature before the injection is set to 140°C, whereas the BMC is maintained at 40°C in the injection unit. During the injection, when the material arrives in front of the heat flux sensors F1 and F2, the heat flux decreases sharply, nevertheless increasing in absolute value. We note that between 0.40 and 0.45 s, the sensors facing each other (one on top of the mold and the other on the lower part) have different responses. This is due to a particularity of the BMC injection, which is also observed in some cases of fiber-reinforced thermoplastics: the existence of a "non-sticking front" ahead of the front. BMC is a yield stress fluid, implying that it has a "piston-type" flow until the velocity and duration of shear is not high enough (with respect to the yield stress). It does not adhere to the wall of the molding cavity. Therefore, there is no contact on the sensor F2 that does not see the front, while there is a poor contact on the sensor of the lower mold part (i.e. F1). Both curves of heat flux F1 and F2 converge after 0.05 s to a maximum absolute value before rising up to an injection time equal to 0.6 s. This initial change is characteristic of a heat shock regime with imperfect contact between the "cold" BMC and the hot mold. From t = 0.6 s, the heat flux given by the sensors F1 and F2 stabilizes at almost the same value. The mold is relatively long and, unlike the case of thermoplastics, there is time to reach an established convective regime.

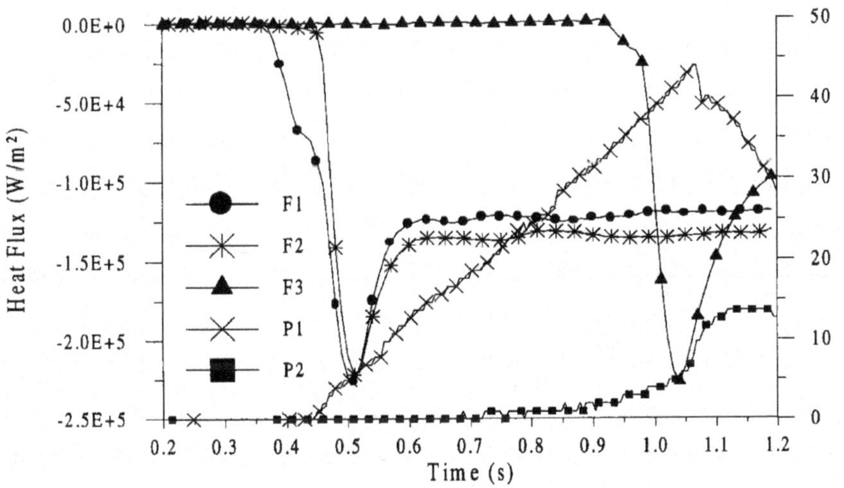

Figure 1.9. *Characteristic record of the heat flux and pressure during the filling of the mold*

The same behavior is observed on the F3 sensor at the bottom of the molding cavity, the maximum magnitude of the heat flux being the same as on the sensors near the injection gate. The composite was not heated significantly during its flow in the mold. This hypothesis is confirmed by the observation of the given pressure P1 by the sensor: the linear evolution during the filling is correlated to the increase in the pressure drop between the sensor position and the BMC front, the filling taking place with a constant flow rate. The fact that the slope is constant reflects that the variation of the viscosity, and therefore the composite average temperature, are almost negligible.

Therefore, heat flux sensors are excellent detectors of events and these data may be used to initially test the rheology of the composite. Indeed, while with the pressure sensors it is difficult to detect the passage front of the material (particularly with P2 sensor), we can estimate the filling velocity by identifying, for example, the minimum flux on the sensors F2 and F3. The distance between these sensors is 0.273 m, and the peaks are separated by 0.55 s. The result is an average velocity $\overline{V}$ = 0.496 m.s^{-1}.

Rheological measurements [MIL 01] demonstrate that the rheological behavior can be modeled with an Oswald–de Waelde-type shear thinning model $\eta = K(T)\,\dot{\gamma}^{m-1}$ with m = 0.43 and K(T) = 7600.4 exp(–0.025T). This is consistent with what is proposed in [LE 07]. Let us consider the filling of the molding cavity between the sensor locations, defining a channel of width W considered as infinite, with a length L = 0.273 m, and a height h = 3.10^{-3} m. The flow is therefore in this two-dimensional (2D) channel. The volume flow rate is expressed by:

$$Q = \overline{V} h W \qquad [1.9]$$

Given the rheological law of the BMC, the volumetric flow rate is classically given by:

$$\frac{m}{2(2m+1)}\left(\frac{1}{2K}\frac{\Delta P}{L}\right)^{1/m} h^{\frac{1+2m}{m}}\, W = Q \qquad [1.10]$$

From equations [1.9]–[1.10], taking into account values of m and h, $\overline{V}$ can be inferred from the value of ΔP and the apparent value of the consistency index K(T): the mold can be viewed as a sort of rheometer. As shown in Figure 1.9, the pressure increases from 5 to 40 bars between 0.5 and 1.05 s, i.e. a pressure variation equal to 3.5 10^{6} Pa. We thus determine at the average temperature the fluid composite consistency index: $K(T_{inj})$ = 8.54 10^{2} Pa.s. Using the classical expression of the apparent shear rate $\dot{\gamma}_a = 6Q/Wh^2$, we obtain a value $\dot{\gamma}_a$= 1,000 s^{-1}. Using the expression of the viscosity given above, the apparent viscosity of the uncured composite would be η_a = 16.65 Pa.s. The laminar 2D flow in the mold is an advantage to characterize *in situ* the composite and to compare the results with laboratory techniques, which are unfortunately not always appropriate. The aging of the polymer can thus be identified by changes in viscosity.

Let us go further into the analysis and try to quantify the heat flux at the molding cavity wall, in the steady convective regime (in the plateau between 0.6 and 1.2 s). For this purpose, we use the classical correlation given in equation [1.1]. The Peclet number value is $Pe = \overline{V}D_h/a$ where a is the thermal diffusivity. We noted the fiber orientation is in the flow plane (in this plane, they exhibit an orientation distribution in x and y directions indicated in Figure 1.8). The transverse thermal conductivity, measured in guarded hot plate, was found independent of temperature and equal to $\lambda = 0.6\ W.m^{-1}.K^{-1}$ whereas the density is 2,000 $kg.m^{-3}$ and the specific heat is 1,200 $J.Kg^{-1}.K^{-1}$. BMC diffusivity is then $2.5.10^{-7}\ m^2s^{-1}$. The hydraulic diameter D_h being equal to 2 h, so we deduce $Pe = 1.2.\ 10^4$. The F1 and F2 sensors are located 67 mm from the injection gate. This singularity is thus the origin of the thermal boundary layer. We obtain by applying the formula given in equation [1.1], taking the origin of x at the injection gate, the value of the Nusselt number to the position of the sensor $Nu(x = 67.10^{-3}m) = 12.77$. Then, we can deduce the heat transfer coefficient by equation [1.11]:

$$h\,(x = 67.10^{-3}m) = Nu\,\lambda\,/D_h \qquad [1.11]$$

We obtain the value $h = 1277\ W.m^{-2}.K^{-1}$. The heat flux density is then given by:

$$\Phi = h\,(\,T_p - T_{inj}\,) \qquad [1.12]$$

where T_p represents the average temperature of the molding cavity wall during the filling step; T_p = 132°C between 0.5 and 1.2 s at 132°C for the sensor F2 and T_p = 130°C for the sensor F1. The corresponding heat flux is in absolute value $1.18.10^5\ W.m^2.K^{-1}$ for sensor F2 and 1.15 $W.m^2.K^{-1}$ for sensor F1. Experimentally, the sensors record the entering heat flux in the mold as negative because the hot mold transfers the heat to the cold composite (T_{inj} = 40°C). Experimental absolute values given in Figure 1.9 are very close to the values estimated by equations [1.1], [1.11]–[1.12] (better than 10% accuracy for the predicted heating). Calculations also indicate that the heat flux given by F2 is greater than F1 in absolute value. The heat flux estimation is of great interest because conversely, it allows calculating the average temperature in the part thickness from a heat balance during the residence time in the mold. As such, we can build simplified models for heat transfer.

Nevertheless, a question arises on the transition into the mold between the heat shock regime and the convective flow regime. An attempt to estimate this time can be to find the time for which the conductive heat flux due to thermal shock decreasing in $1/\sqrt{t}$ (equation [1.6]) becomes equal to the steady regime of convective heat flux given by equations [1.1], [1.11–1.12].

Conductive heat flux during a thermal shock is given by equation [1.6]. Equations [1.11]–[1.12] combined with the expression of the Nusselt number in a plane channel (equation [1.1]) are used to calculate the convective flux. By identifying these two expressions, and by dividing both sides of the equality by λ ($T_{inj} - T_c$), we obtain equation [1.13] where x is the distance between the gate and the position of the sensor:

$$\frac{1}{\sqrt{\pi a t}} = \frac{1.276}{D_h}(\frac{3m+1}{4m})^{\frac{1}{3}}(\frac{PeD_h}{x})^{\frac{1}{3}} \qquad [1.13]$$

For the BMC, the index of the power law being m = 0.43, this previous expression becomes:

$$\frac{at}{D_h^2} = 0.1955(\frac{x}{Pe.D_h})^{\frac{2}{3}} \qquad [1.14]$$

Knowing that D_h is equal to 2 h, by defining a Fourier number Fo = at / h^2, we obtain the time t for which both heat fluxes are equal. This is the time t such that:

$$\text{Fo} = 0.78(\frac{x}{Pe.D_h})^{\frac{2}{3}} \qquad [1.15]$$

Let us apply this relationship to the results displayed in Figure 1.9. We remind that the Peclet number is Pe = 1.2. 10^4, D_h = 2h = 6.10^{-3}m, and the sensor F1 is at x = 67 mm from the gate. The diffusivity is $a = 2.5.10^{-7}$ $m^2.s^{-1}$. This gives a time t = 0.27 s. If it is added to the time of passage in front of the sensors, we obtain the end of the shock regime between 0.65 and 0.7 s. It is seen in Figure 1.9 that this time corresponds remarkably well with the beginning of the plateau.

If we perform the same exercise on the SWIM mold of section 1.2.1, we also obtain very interesting results. Figure 1.10 presents the position of the sensors on the wall of the SWIM molding cavity. The closest one is at x = 17 mm to the gate. Figure 1.11 (extracted from [LEB 98]) depicts the heat flux measured during the injection of a polypropylene in the SWIM mold. We note again that the amplitudes of the heat flux close to the gate and the cavity bottom are almost identical. The filling velocity is estimated from the maximum values of the flux to 0.1 $m.s^{-1}$. If we assume an Oswald–de Waelde-type rheological behavior for the PP, the power law index is close to 0.3 (0.308 for the studied PP). The diffusivity is $8.7.10^{-8}$ $m^2.s^{-1}$. The Peclet number is thus equal to 6,900. We apply the same approach as for the BMC, noting that the expression $(\frac{3m+1}{4m})^{\frac{1}{3}}$ now has the value 1.16. A value close to the one obtained for the BMC, which differs only by 5%. This parameter has little importance on the transition between the two regimes. The calculation gives a time t = 0.454s for equality between the two terms of the heat flux, and therefore the transition between regimes. We see in Figure 1.11 that the filling is finished at this time. We conclude that the whole filling occurs during the heat shock regime.

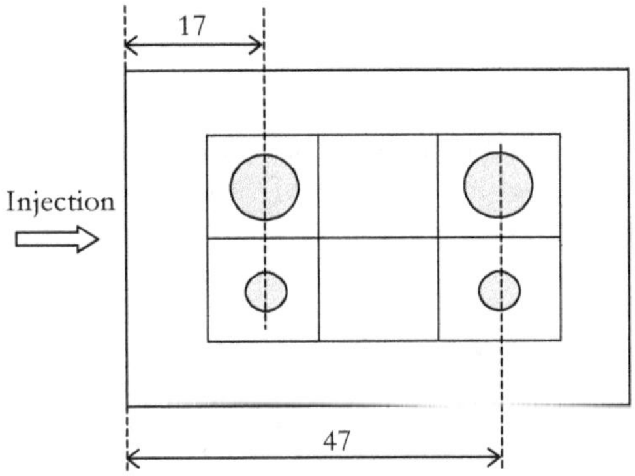

Figure 1.10. *Position of the sensors on the mold SWIM*

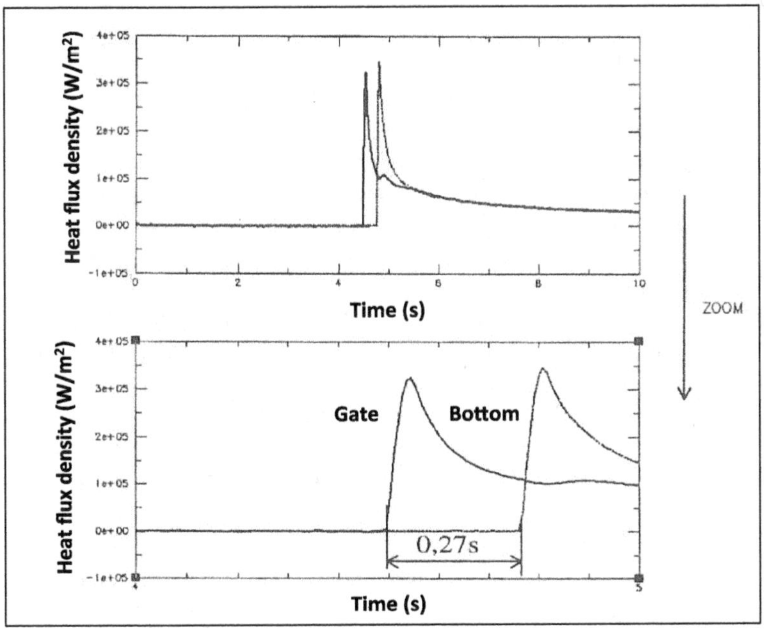

Figure 1.11. *Heat flux density during the passage of the molten PP in front of the sensors*

Finally, considering a domain of variation of the power law index including most of the injected polymers, we are able to propose a criterion determining the transition between the regimes. If a Fourier number Fo, built on the thickness of the piece and estimated at a distance x from the molding cavity gate, satisfies:

$$\mathrm{Fo} < 0.6\,\left(\frac{x}{Pe.D_h}\right)^{\frac{2}{3}} \qquad [1.16]$$

Then, it can be considered that the regime of heat shock prevails at this location. This equation takes into account the range of variation of the power law index of the polymer-based pastes.

1.2.2. *Heat transfer during part consolidation*

1.2.2.1. *Consolidation of the thermoplastic parts*

During the consolidation phase, new phenomena must be taken into account. A very strong coupling exists between the heat transfer and the phase change in the case of semi-crystalline. The coupling is initially induced by a strong dependence of thermophysical properties on temperature, pressure and the physical state, i.e. liquid or solid (including the crystalline fraction). These properties are the specific volume, the specific heat and the thermal conductivity. We will discuss this in Chapter 2, but it can be said that these data are accessible with suitable apparatuses. The term of compressibility may be non-negligible, especially when the pressure varies very quickly at the commutation time in injection molding (temperature rise of few degrees). Viscous dissipation is generally negligible because of the low flow velocities in these phases. The source term must be of course taken into account adequately for semi-crystalline polymers. It is related to the kinetics of crystallization and its release participates in the coupling between the phase change and heat transfer. Provided that the homogenization is possible in the composite, the source is weighted by the weight fraction of the matrix. This question of homogenization is not trivial and research works remain to be undertaken to indicate in which circumstances it is possible. The coupling between the heat transfer and crystallization is controlled by the ratio of two characteristic times called thermal Deborah number *De*. This is the ratio between the characteristic time of heat transfer $\tau = e^2/a$, where e is a characteristic length of the part and a is the thermal diffusivity (typically 1.10^{-7} $m^2.s^{-1}$) and the characteristic time of the transformation t_t at the temperature where it occurs in the core of the part. We can choose the transformation half-time t_c, which is accessible experimentally. Let us take the example of a PP part with a half-thickness equal to 2 mm. Crystallization occurs at 100°C, the crystallization half-time is 1 s, which is small compared to τ = 40 s.

Depending on the value of *De*, one can fall in one of the three situations illustrated in Figure 1.12.

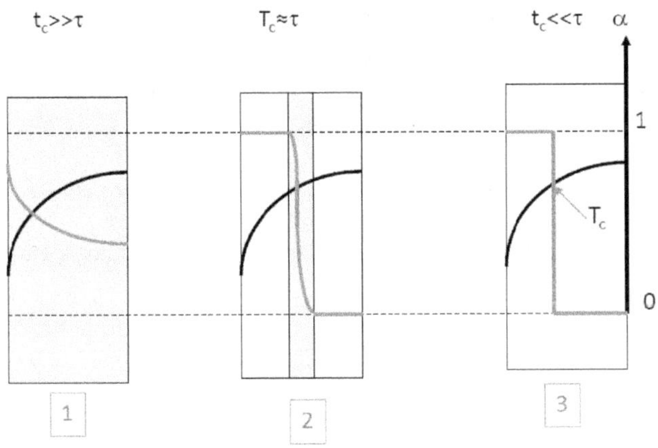

Figure 1.12. *Coupling between the temperature and phase change fields. The temperature field is in black and the transformation field is in red. For a color version of the figure, see www.iste.co.uk/boyard/heat.zip*

Of course, solid and liquid Stefan numbers defined in section 1.2.1.1 play a role in the coupling. If these numbers are large, the heat source is negligible and heat transfers are the same as those in a part made with anamorphous thermoplastic. If Stefan numbers are small, the thermal behavior will depend on the characteristics time ratio. If *De* is very large, the polymer behaves like two phases with a localized (sharp) interface, and the change of state occurs at T_c (case 3, in Figure 1.12). A plateau is observed at T_c on the temperature evolution in the centre of the part. This is almost true for the solidification example given above for a PP, for which De = 40. If $De \approx 1$, there is a quasi-plateau of phase change in the core of the part and the transformation zone has limited thickness compared to the part one (case 2, Figure 1.12). A good approximation of solidification and cooling times is possible with simplified models already mentioned in section 1.2.1.1 (see [SOB 13]). In the case where the phase change time is large compared to the thermal time constant (case 1 in Figure 1.12), *De* is small and the thermal history imposes locally the phase change. Heat transfer and phase change can be uncoupled. We are then in a case similar to the forming of an amorphous polymer, for which the cooling is easy to model, noting that the time evolution of molding cavity surface temperature shown in Figure 1.3 may be represented by a triangle. The average temperature of the molding cavity surface during the cycle $\overline{T}_s$ can then easily be determined and used in a conventional analytical solution to calculate the cooling time.

Kinetic laws, which provide the relationship between the rate of transformation $d\alpha/dt$ to the temperature and the transformed fraction α (see Chapter 3), are used to make the coupling between the energy conservation equation and the equation which describes the evolution of the transformed fraction. Indeed, the source in the energy equation, related to the phase change, is:

$$Q = \rho\,(P, T, \alpha)\,\Delta H\,\partial\alpha/\partial t \qquad [1.17]$$

In this expression, ρ is the density and ΔH is the phase change enthalpy.

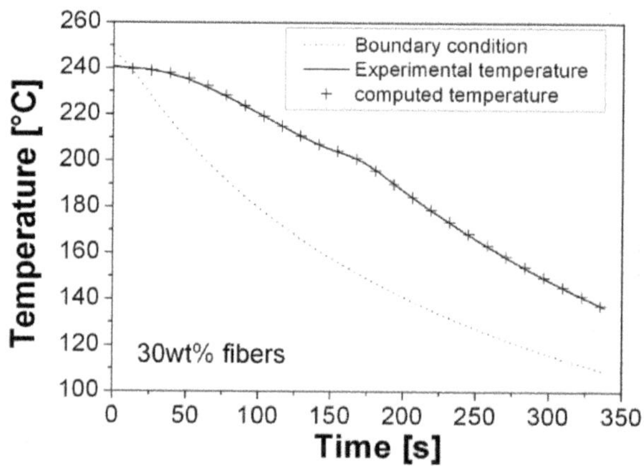

Figure 1.13. *Measured and calculated temperature during the solidification of a reinforced poly-aramide (IXEF 1002)*

The kinetics is piezo-dependent: the increase in pressure moves the crystallization toward higher temperatures. The heat source is thus released at temperatures, which depend on the pressure and this has a strong impact on the temperature fields. Many formulations are available for the source and the readers will find a detailed presentation later in this book (Chapter 3). Nevertheless, it is observed that the temperature difference called supercooling: $\Delta T = T_m^0 - T$ appears in the expression of the crystallization kinetics. The temperature T_m^0 refers to the thermodynamic melting temperature. The effect of pressure on the kinetics is related to the increase in the thermodynamic melting temperature when the pressure P increases, which has to be taken into account in the expression for the kinetics [LEG 10, FUL 01]. Figure 1.13 shows the excellent agreement between simulated and measured temperatures in the core of the part [LEG 11]. The modeling of heat transfer includes crystallization heat source, including the pressure effect. Similar results are obtained for a composite polyamide/UD glass fiber [FAR 15]. The cooling rate decreases (quasi-plateau) during the crystallization. It should also be noted that this level of accuracy requires a good modeling of the thermal properties, in particular depending on the phase change, which will be discussed in Chapter 2.

1.2.2.2. *Part consolidation with a thermosetting matrix*

The problem is identical to the previous one for the thermosetting matrix composites. It can be estimated that the calculation of the temperature fields coupled to the source of polymerization (cross-linking) is now performed correctly, subject to a number of precautions, especially related to a good estimate of the thermal properties and their changes according to the degree of conversion [DEL 14a, DEL 14b]. An example of the level of accuracy in the prediction of coupled fields is shown in Figure 1.14 [BAI 98].

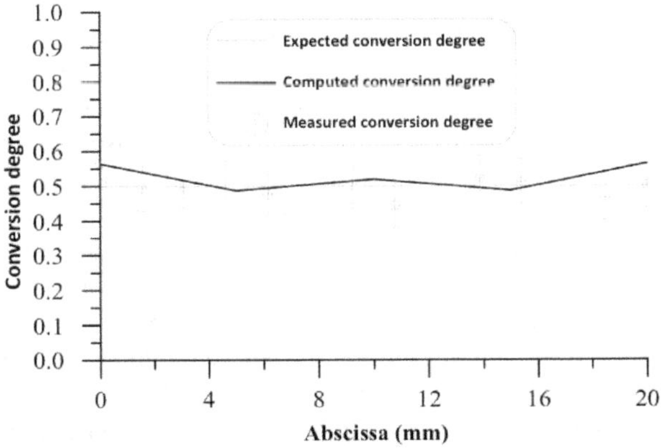

Figure 1.14. *Comparison between the measured and calculated conversion degree in an epoxy/UD glass fiber part in which partial and uniform polymerization was carried out through the thickness [BAI 98]*

Indeed, the solved problem included an inverse method to compute the optimized temperature cycle to impose on the surfaces of a 10 mm thick plate of composite epoxy/glass fiber to obtain a uniform and partial conversion degree in the thickness of the plate. Of course, the accurate knowledge of the cross-linking kinetics is mandatory since it is necessary to avoid uncontrolled heating in the centre of the part, linked to the release of the heat due to the chemical reaction in such an insulating medium. The conversion degree was verified by cutting small strips in the thickness of the plate and then measuring the residual enthalpy by DSC. It is seen that the control of the transformation with an optimized temperature cycle, which is very difficult to achieve, is efficient and a partial predetermined state of cure can be reached with a good accuracy.

1.3. Injection on continuous fiber reinforcements

This problem is particular. In this section, we are discussing about process of the type liquid composite molding (LCM), for which continuous reinforcement is placed in the mold, which is closed, and then the resin is injected to impregnate the preform. An alternative is to partially close the mold, such as in C-RTM, to impregnate the reinforcement with a transverse flow in the thickness. This is, first of all, multi-scale medium: the microscopic scale is associated with the fibers and meshes (10–100 μm), the mesoscopic scale is related to the ply (millimeter scale) and finally the macroscopic scale is associated with the part (meter is the order of magnitude). Chapter 6 will deal with homogenized transfer equations in such a medium. During the consolidation of parts, problems are identical to those presented in the previous section. A new generation of composite arises now, based on thermoplastics. This is of great interest, because of their ecological potential (recycling, bio-based polymers), the possible reduction of the forming cycle, the lack of post-curing, their potential welding and post-forming and their increased mechanical properties (impact resistance). The high viscosity of thermoplastic polymers raises the difficulty of impregnating continuous reinforcement with a viscous polymer. Two solutions are thus possible to overcome this difficulty, either coupling the RTM process and polymer chemistry by injecting a monomer onto the reinforcement, polymerization (to create macromolecules) and crystallization are then produced in the mold, or injecting low viscosity polymers which then crystallize in the mold. Both ways are developed by polymer suppliers, raising many scientific and technological obstacles. Indeed, the challenges are numerous for parts manufacturers: to achieve as fully as possible the polymerization in the case of reactive injection, to achieve an optimum crystallinity, to obtain a composite with controlled porosity (air/water), to control shrinkage induced by transformations, to get a good surface aspect, to ensure reasonable cycle times depending on the application (including automotive) and finally to develop an "industrial" process. Scientifically, the challenges are no less pithy: the first one is to understand and control the coupling between crystallization and polymerization: from a certain conversion, the polymer will crystallize, hindering the macromolecule growth. There is a competition between both mechanisms that have to be controlled by an accurate temperature control.

Obtaining a composite with controlled porosity supposes an understanding of how the double scale flow evolves, within and between the meshes of a fluid whose chemo-rheological properties vary with the time and the temperature changes due to heat transfer. The lack of knowledge on mechanisms provokes a bad impregnation and voids caused by air, reaction gas and/or water bubbles entrapment. The situation is generally observed and analyzed in [VIL 15, NOR 12, RUI 12] for example, and is also schematically shown in Figure 1.15.

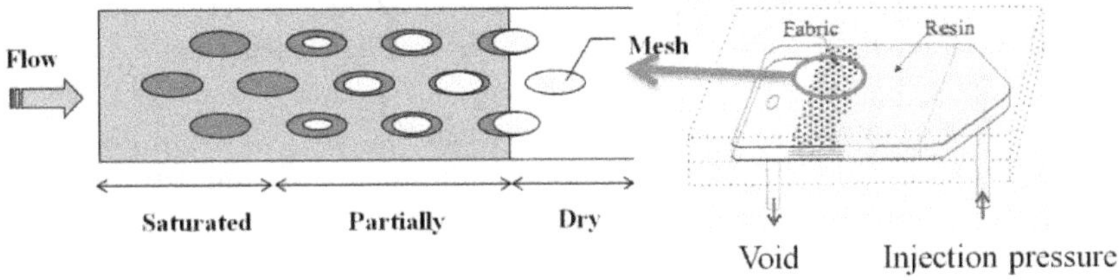

Figure 1.15. *Saturated and unsaturated flows into the fabric of a composite part during the impregnation of a continuous reinforcement (fluid is in yellow). For a color version of the figure, see www.iste.co.uk/boyard/heat.zip*

a)

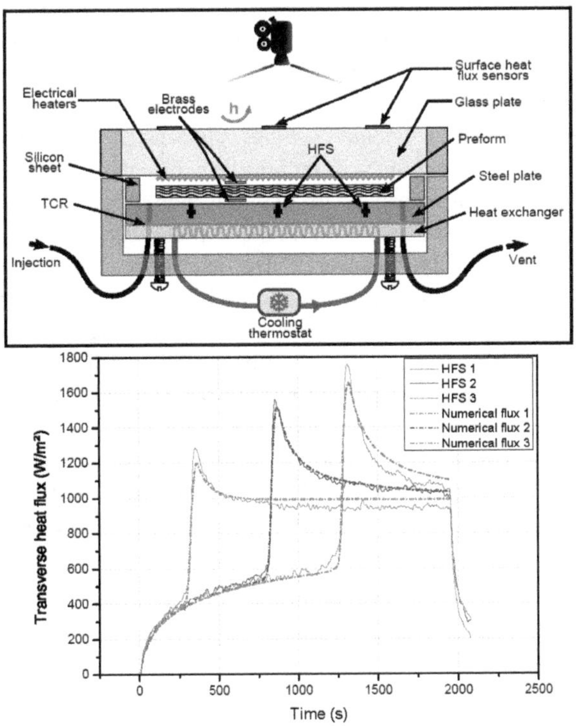

b)

Figure 1.16. *a) Experimental bench and b) recorded and calculated heat flux at the sensor locations [VIL 15] . For a color version of the figure, see www.iste.co.uk/boyard/heat.zip*

According to the respective velocity of impregnation between the meshes and in the meshes, air bubbles can be trapped in the composite, therefore degrading its quality. Modeling heat transfer in and through the structure during the impregnation has been addressed by the heat transfer community and is an important scientific challenge. It should be noted that the thermal conductivity has a high sensitivity to the saturation [VIL 15], but it varies in a differentiated manner depending on the intra-mesh or inter-mesh saturation. Nevertheless, it has been shown that the heat transfer modeling can be performed with a good accuracy. This is illustrated in Figure 1.16(b) that shows the comparison between the simulated and measured heat flux when a known heat flux is dissipated by a thin heater on the wall of a mold, in which of a fluid impregnates the reinforcement, as shown schematically in Figure 1.16(a).

When the flow front reaches the sensor position, the heat flux sharply increases and then decreases until it reaches a value corresponding to the established convective regime. We can see in this figure the very good agreement between the measured and calculated heat flux, the latter taking into account both the heat transported by the flow and that passing through the composite thickness during the saturation. Another phenomenon must be taken into account and is known as thermal dispersion. It appears when the temperature and flow rates at the microscopic scale are different from their mean values. Physically speaking, local convection phenomenon between meshes increases the effective thermal conductivity. It is then necessary to include the effect of tortuosity. During the tortuous flow between the meshes, in the presence of a transverse temperature gradient in particular, heat is transported in a complex way, resulting in an increased apparent thermal conductivity. Lecointe [LEC 99] shows that we must introduce a dispersive thermal conductivity λ_d, which depends on the Peclet number built on the hydraulic diameter of the inter-meshes channels. However, some problems remain open today in this field: the thermal homogenized conditions of entry and the modeling of a coupled heat transfer-unsaturated multi-scale flow. On the contrary, the consolidation phase is solved, it is identical to the case described in the previous section, provided that a correct coupling between the polymerization reactions (identical to the case of thermosets) and the crystallization exists, which can be influenced by the shear in the meshes and the confinement between the fibers (transcrytallinity).

1.4. Conclusion: toward a controlled processing

As can be seen in this chapter, many problems related to heat transfer during composite processing were solved during numerous studies dedicated to this topic. Heat transfers during the filling of short fiber composites are well described by the models, provided that the orientation tensor is, of course, correctly estimated. The consolidation is also described satisfactorily, the coupling with the transformation kinetics (polymerization or crystallization) allowing us to accurately estimate the overheating due to the source release during the transformations. Many challenges are still to be met. Without being exhaustive, these challenges include the problem of modeling the non-isothermal saturation of the reinforcing fabrics, taking into account the influence of shear on the crystallization at the various scales of these reinforcements, the influence of shear on the thermal conductivity (effects of anisotropy by the orientation of the macromolecules or the crystalline structures), the contact resistance between a part and a tool, or between plies, the adequate consideration of radioactive effects in semi-transparent medium (for example, composite PA/glass fibers). Finally, advances in thermal metrology in the composite processing to validate the models, especially to measure composite surface temperatures, temperatures inside the composite without intrusivity or also 2D heat flux fields. Once these many challenges are confronted, we can extend the works to thermomechanical coupling to optimize the processing parameters to meet criteria on the residual stresses and war pages.

1.5. Bibliography

[BAI 98] BAILLEUL J.L., DELAUNAY D., JARNY Y., "Optimal thermal processing of composite materials, an inverse algorithm and its experimental validation", *11th International Heat Transfer Conference*, pp. 87–92, 1998.

[CAR 59] CARSLAW H.S., JEAGER J.C., *Conduction of Heat in Solids*, 2nd ed., Oxford University Press, London, 1959.

[DEL 11] DELAUNAY D., "Thermal conductivity of an injected polymer and short glass fibers composite part: measurement and model", *Keynote in 27th world Congress of the polymer Processing Society*, Marrakech, pp. KN–15–517, 2011.

[DEL 14a] DELAUNAY D., SOBOTKA V., THOMAS S. *et al.*, "Thermal properties", in THOMAS C., SINTUREL, THOMAS R. (eds), *Micro and Nanostructured Epoxy/Rubber Blends*, Wiley-VCH, 2014.

[DEL 14b] DELAUNAY D., BOYARD N., "Pressure-volume-temperature (PVT) analysis", in THOMAS S., SINTUREL C., THOMAS R. (eds), *Micro and Nanostructured Epoxy/Rubber Blend*, Wiley-VCH, 2014.

[FAR 15] FARAJ J., PIGNON B., BAILLEUL J.L. *et al.*, "Heat transfer and crystallization modeling during compression molding of thermoplastic composite parts", *Key Engineering Materials*, vols. 651–653, pp. 1507–1512, 2015.

[FUL 01] FULCHIRON R., KOSCHER E., POUTOT G. *et al.*, "Analysis of the pressure effect on the crystallization kinetics : dilatometric measurements and thermal gradient modelling", *Journal of Macromolecular Science: Physics*, vol. B40, pp. 297–394, 2001.

[LE 07] LE T.H., ORGEAS L., FAVIER D. *et al.*, Rhéologie des BMC lors de leur injection, Compte-rendu du 18ème Congrès de Mécanique, 2007.

[LEB 98] LE BOT P., Comportement thermique des semi-cristallins injectés. Application à la prédiction des retraits, PhD Thesis, University of Nantes, 1998.

[LEC 99] LECOINTE D., Caractérisation et simulation des processus de transferts lors d'injection de résine pour le procédé RTM, PhD Thesis, University of Nantes, 1999.

[LEG 06] LE GOFF R., Etude et modélisation des transferts thermiques lors de la solidification de pièces injectées en polymère semi-cristallin chargé de fibres, PhD Thesis, University of Nantes, 2006.

[LEG 10] LE GOFF R., BOYARD N., SOBOTKA V. *et al.*, "A heat transfer analysis of injection molding of fiber-reinforced poly (M-xylylene Adipamide)", *International Journal of Material Forming*, vol. 3, no. 1, pp. 805–808, 2010.

[LEG 11] LE GOFF R., BOYARD N., SOBOTKA V. *et al.*, "Inverse estimation of the crystallization kinetic function of semi-crystalline polymers and short fiber reinforced composites in moderate cooling conditions", *Polymer Testing*, vol. 30, pp. 678–687, 2011.

[LOU 97] LOULOU T., DELAUNAY D., "The interface temperature of two suddenly contacting bodies, one of them undergoing phase change", *International Journal of Heat and Mass Transfer*, vol. 40, no. 7, pp. 1713–1716, 1997.

[PIN 09] PINHO F.T., COELHO P.M., *Non-Newtonian Heat transfer in Rheology*, Encyclopedia of Life Support Systems, Eolss Publishers, Oxford, UK, 2009.

[MIL 01] MILLISCHER A., DELAUNAY D., "Experimental and numerical analysis of heat transfer in bulk molding compound injection process", *Journal of Reinforced Plastics and Composites*, vol. 20, no. 6, pp. 495–512, 2001.

[NOR 12] NORDLUND M., MICHAUD V., "Dynamic saturation curve measurement for resin flow in glass fibre reinforcement", *Composites Part A*, vol. 43, no. 3, pp. 333–343, 2012.

[QUI 98] QUILLIET S., Transferts thermiques à l'interface polymère métal dans le procédé d'injection des thermoplastiques, PhD Thesis, University of Nantes, 1998.

[RUI 12] RUIZ E., LEBEL F., TROCHU F., "Experimental study of capillary flows, voids formation and void migration in LCM manufacturing", *11th Flow Processes in Composite Materials*, pp 136–143, 2012.

[SOB 13] SOBOTKA V., AGAZZI A., BOYARD N. *et al.*, "Parametric model for the analytical determination of the solidification and cooling times of semi-crystalline polymers", *Applied Thermal Engineering*, vol. 50, pp. 416–421, 2013.

[SOM 12] SOMÉ S.C., GAUDEFROY V., DELAUNAY D., "Estimation of bonding quality between bitumen and aggregate under asphalt mixture manufacturing condition by thermal contact resistance measurement", *International Journal of Heat and Mass Transfer*, vol. 55, pp. 6854–6863, 2012.

[VIL 15] VILLIERE M., GUEROULT S., SOBOTKA V. *et al.*, "Dynamic saturation curve measurement in liquid composite molding by heat transfer analysis", *Composites Part A*, vol. 69, pp. 255–265, 2015.

2

Experimental Determination and Modeling of Thermophysical Properties

In all industrial sectors using composite materials, an accurate description of heat transfer is a key point to succeeding in simulation and optimization of a process, part quality being strongly linked to the thermal history during the molding. Indeed, phase change (due to crystallization or cross-linking) is very sensitive to temperature. Therefore, process simulation requires full and accurate knowledge of thermophysical properties to provide coherent results. A complete characterization has to be done under well controlled conditions that are as close as possible to the process. This includes thermal conductivity, specific heat and specific volume. The modeling also requires detailed knowledge of the reaction rate, which will be detailed in the next chapter. It is essential to predict temperature and conversion degree fields in the molded part. After a brief presentation of the context and the issues associated with the characterization of thermophysical properties, we present the main classical equipment, the principle of measurements and the methodologies to determine these properties for resins and composites. We will detail advantages and drawbacks, the conditions in which they work (temperature range, etc.) and technical characteristics. Examples of results will give orders of magnitude of these properties. Finally, several modeling approaches are proposed.

2.1. Measurement of specific volume and shrinkage

The specific volume, or its inverse the specific mass, is absolutely indispensable to correctly simulate the thermal as well as mechanical behavior of a material, notably during its transformation (cross-linking and crystallization): it is a fundamental parameter to determine the shrinkage, and to calculate the residual stresses. It is an essential element for the engineer, in order to master the surface

---

Chapter written by Nicolas BOYARD and Didier DELAUNAY.

aspect of the parts, as well as their dimensions. In polymer sciences, the PvT measures permit us to follow several phenomena: glass transition, phase change and volume thermal expansion.

2.1.1. *Thermoplastic PvT diagram*

2.1.1.1. *PvT apparatus*

Two main techniques exist to measure the PvT data: the piston-die technique and the confining-fluid technique. The first method is represented schematically in Figure 2.1(a) extracted from [CHA 02]. This technique is also used by the society Thermo Haake GmbH and was commercialized for a long time as Thermo-Haake PvT100, used in numerous references. The principle is based on the compression of a sample by two pistons inside a cylinder. A displacement sensor allows measuring the sample volume during the imposed temperature cycle through the piston displacement.

The main drawback is that the stresses on the specimen in the solid phase are not hydrostatic and the material must be in conditions such that the friction of the sample surface with the interior cylinder surface remains negligible. The shrinkage and the deformation of the sample surfaces in contact with the pistons are compensated by the use of a ductile joint (Teflon® or elastomer).

The second method, the confining-fluid technique, consists of immersing the polymer in a fluid (mainly mercury), on which the pressure is applied (Figure 2.1(b)). The advantage is that the pressure is hydrostatic (applied by a bellows piezometer), in both solid and fluid states. However, heating and cooling rates are limited (few degrees per minute) due to the thermal inertia of the high-pressure vessel. This type of device, besides the use of mercury forbidden by the legislation of some countries, imposes the use of samples with a geometry which is not well defined, making it difficult to take into account the temperature and transformation gradients, this point being essential for samples of several grams with heating or cooling rates representative of the matrices conditions of use.

PvT devices, such as Gnomix, have the advantage of working over large pressure and temperature ranges (up to 200 MPa and 400°C, respectively). It is then used to characterize high-performance thermoplastics (e.g. PEEK [ZOL 89], Figure 2.2(a)) using isothermal compressive runs and isobaric runs associated with heating or cooling cycles. The PvT 100 apparatus (piston-die technique) is the most used in the literature [LUY 00, WON 01, VAN 06] but is not manufactured anymore.

Figure 2.2(b) provides an example of a PvT diagram (unfilled squares) for polypropylene. It also highlights the effect of temperature gradients on the measured specific volume, which can be viewed as an apparent one.

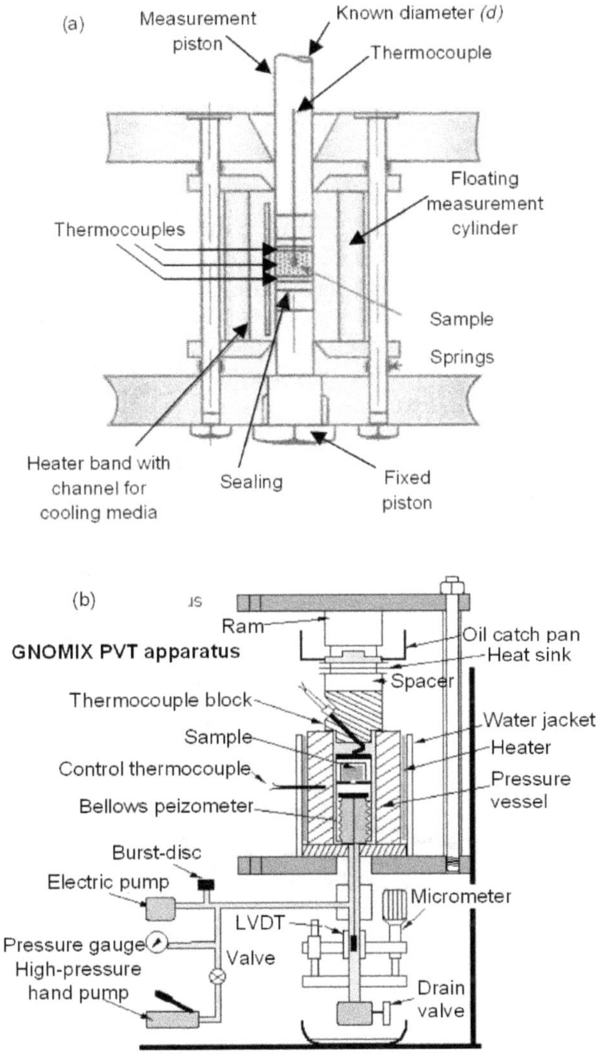

Figure 2.1. *a) The piston-die configuration [CHA 02] and b) Gnomix apparatus [ZOL 76]*

Considering the needs of PvT data with respect to available techniques, new non-standard devices are developed in some research laboratories with operating conditions more representative of process ones. Chakravorty [CHA 02] develops an apparatus to reach high cooling rates up to 300 K/min using thin and annular specimens. Other authors [VAN 06, FOR 09, VAN 05, FOR 09] work to quantify the effect of pressure and shear simultaneously, combining piston-die and Couetterheometer.

a)

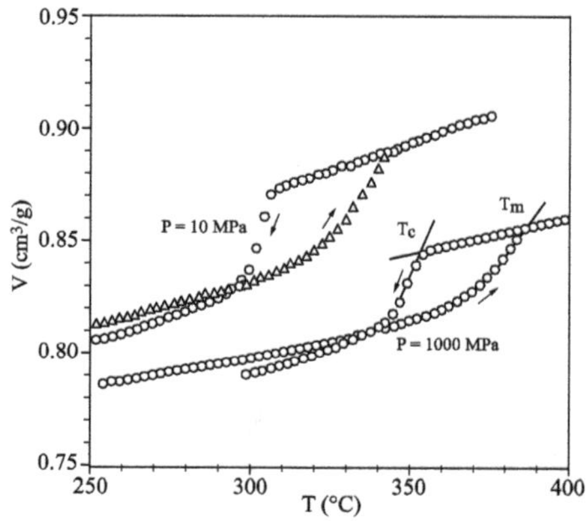

b)

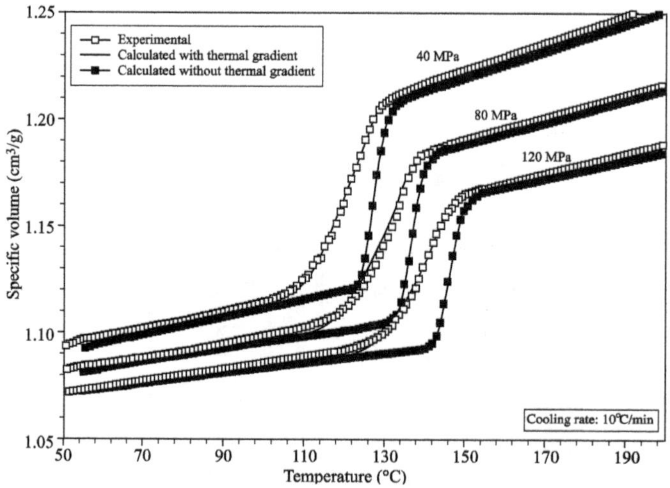

Figure 2.2. *a) PEEK PvT using Gnomix device [ZOL 89], b) PP PvT from PvT 100 apparatus; influence of the thermal gradient on specific volume measurement [LUY 00]*

Recently, a new version (PvT-xTmold) [TAR 12, PIG 15] was developed to make measurements at high temperature (up to 400°C), in a large pressure range (up to 200 MPa). It is also instrumented with a heat flux sensor and heat transfer is controlled to be in the radial direction of the cylindrical sample. Another interest is to be able to measure the crystallization enthalpy and to identify the crystallization kinetics under pressure from specific volume measurements, taking into account thermal gradients.

Table 2.1 gathers the main characteristics of PvT used and developed in the literature. Data in brackets are the maximal ones whereas others are those actually used in papers.

Authors	*PvT name*	*Technology*	*Sample shape*	*Pressure [MPa]*	*Temperature [°C]*	*Rate [K/min]*	*Shear [s^{-1}]*
Standard devices							
Luyé [LUY 00]	PvT-100	Piston	Cylindrical	120	220	30	no
Won [WON 01]	PvT-100	Piston	Cylindrical	250 (250)	310 (420)	15	no
Zoller [ZOL 89]	Gnomix	Fluid		200	400	2.5	no
Non-standard devices							
Millischer [MIL 01]	PvTα	Piston	Cylindrical	10	210	10	no
Tardif [TAR 12, PIG 15]	PvT-xT	Piston	Cylindrical	(200)	(400)	(100)	no
Watanabe, 2003 [WAT 03]	SF-PvT	Piston	Cylindrical	20	200	4	0.5
Van der Beek [VAN 06, VAN 05]	Pirouette	Piston	Annular	60 (100)	210 (260)	2,088 (6,000)	78 (80)
Forstner [FOR 09, FOR 09]	Pirouette	Piston	Annular	60 (100)	220 (300)	5,040 (6,000)	67.5 (200)
Chakravorty [CHA 02]	NPL PvT	Piston	Annular	160 (250)	220 (420)	200 (300)	no

Table 2.1. *Existing PvT devices for specific volume measurements of thermoplastics*

2.1.1.2. *Semi-empirical models*

Parameters of theoretical models being hard to estimate, semi-empirical models were developed and widely used to describe PvT diagrams. Two domains are considered to take into account amorphous and semi-crystalline states.

2.1.1.2.1. Tait model

Tait original equation [TAI 88, DYM 88] allows describing the evolution of the specific volume of a liquid from two parameters A and π to estimate:

$$\frac{v_0 - v}{p v_0} = \frac{A}{\pi + p} \tag{2.1}$$

This equation is thus modified by Tammann [TAM 95] to get the following expression:

$$v = v_0(T)\left[1 - C ln\left(1 + \frac{P}{B(T)}\right)\right] \tag{2.2}$$

where $v_0(T)$ corresponds to the specific volume at atmospheric pressure, $B(T)$ is a function to identify with respect to temperature, C is a constant equal to 0.0894 and P is the pressure.

Hieber [HIE 02] demonstrates that it is possible to describe a PvT diagram of an amorphous material considering 10 parameters and a linear evolution of the glass transition temperature with pressure:

$$T^* = b_5 + b_6 P$$

$$v_0(T) = \begin{cases} b_{1s} + b_{2s}(T - b_5) & if\ T < T^* \\ b_{1l} + b_{2l}(T - b_5) & if\ T > T^* \end{cases}$$

$$B(T) = \begin{cases} b_{3s}\exp\left(-b_{4s}(T - b_5)\right) & if\ T < T^* \\ b_{3l}\exp\left(-b_{4l}(T - b_5)\right) & if\ T > T^* \end{cases} \tag{2.3}$$

This model was then applied empirically to describe PvT diagrams of semi-crystalline polymers, assuming that the crystallization temperature evolves linearly with pressure. However, this model does not include the influence of the cooling rate on this transition temperature T*.

2.1.1.2.2. IKV model (modified Spencer and Gilmore model)

The IKV model (used, for example, in [LUY 00] and adapted from the one developed by Spencer and Gilmore [SPE 50]) is composed of two domains. The transition from one domain to the other is characterized by a temperature T* (defining glass transition or crystallization), as presented in the previous section:

$$T^* = b_5 + b_6P \qquad [2.4]$$

$$v = \frac{a_1}{a_4+P} + \frac{a_2T}{a_3+P} + a_5e^{a_6T-a_7P} \qquad if\ T < T^* \qquad [2.5]$$

$$v = \frac{b_1}{b_4+P} + \frac{b_2T}{b_3+P} \qquad if\ T > T^* \qquad [2.6]$$

2.1.1.2.3. Linear model

Luyé [LUY 00] proposes a linear model to describe the specific volume as a function of temperature, when the polymer is in the rubbery state (T>Tg).

$$v_a = A_1(P) + A_2(P)T$$

$$v_{sc} = A_3(P) + A_4(P)T \qquad [2.7]$$

where v_a and v_{sc} are the specific volumes in the amorphous and the semi-crystalline states, respectively. A_i are the parameters to identify and depend on pressure.

During the crystallization, the specific volume is modeled according to a mixing law, which is a function of the relative crystallinity α and the absolute crystallinity X_∞.

$$v = \alpha X_\infty v_{sc} + (1 - \alpha X_\infty)v_a \qquad [2.8]$$

Note that this approach is the best one. Indeed, it is not appropriate to model the specific volume in the crystallization domain by an empirical law, which depends only on the temperature. The measured specific volume is associated with a given absolute crystallinity, meaning that we should consider $PvTX_\infty$ diagram instead of PvT. Moreover, it is almost impossible that, for any thermal history, a unique link between the temperature and the crystallinity exists. Therefore, the use of a unique PvT diagram made at a low cooling rate is not correct for process simulations.

2.1.1.3. *Which experimental protocol for PvT diagram?*

PvT data can be obtained following several protocols (see below). However, the question is to know what the best one is and if some of them should be prohibited to avoid getting erroneous results.

1) *Isothermal compression while increasing the temperature level*: for a given temperature, measurements are made for several pressure levels. Then, the specimen is heated up to the next temperature step;

2) *Isothermal compression while decreasing the temperature level:* same procedure as the first one but the temperature steps are done from cooling;

3) *Isobaric heating*: the measurement is done at constant pressure and the sample is heated using a constant heating rate. For another pressure level, the temperature cycle has to be applied again;

4) *Isobaric cooling*: same procedure as the previous one but a cooling rate is imposed.

It is first interesting to note that methods 1 and 3 are not adapted when we need data for processes where polymer/composite parts are manufactured by cooling. These methods only highlight the evolution of the specific volume during the melting, but not associated with the crystallization. However, the associated characteristic temperatures are different (several tens of Celsius degrees) due to the well-known supercooling phenomenon (see Figure 2.2(a) as an example).

It is known that transition temperatures (melting and crystallization as well as glass transition) increase with pressure (from 0.2 to 0.5°C/MPa). The PvT diagram can be thus depend on the experimental protocol. More specifically, methods 1 and 2 exhibit a transition temperature (melting for method 1 and solidification for method 2), which is independent of the pressure level. It is obviously an artifact. In method 1, the melting occurs for the lowest pressure level and when the pressure is increased, the polymer does not have time to recrystallize before the new measurement. In method 2, the crystallization occurs for the highest pressure. As a result, the next measurement (at a lower temperature and the lowest pressure) is made on a specimen already solidified (or at least partially solidified) and which cannot remelt since the melting temperature is higher than the crystallization one.

Finally, the best measurement protocol is the isobaric cooling because it enables us to observe the effect of the cooling rate at a given pressure on the crystallization. The main drawback is the temperature gradient within the specimen. Luyé *et al.* [LUY 00] demonstrate that even a rather small thermal gradient can lead to a non-negligible bias (Figure 2.2(b) in the measurement of the specific volume, including a crystallization gradient). Therefore, these gradients have to be taken into account during the data treatment or to make the measurement at low cooling rate.

2.1.2. *Specific volume of thermosetting polymers*

The PvT diagram is quite difficult to measure since the cross-linking is irreversible, leading to the jamming of the molding cavity. However, several publications deal with the evolution of the specific volume of such polymers using the above-mentioned apparatus.

Goyanes and co-workers [GOY 04] made some measurements on cured epoxy resins considering different hardeners using a PvT 100 device by Haake. The tests were carried from 20 in 80 MPa, starting from room temperature to 200°C at 5 K/min, in isobaric heating mode.

Several authors followed the curing of epoxy resins including, or not, additives while using a PvT device, often coupled with DSC measurements. The resin can be sealed in a thin metallic leaf (nickel) so that it does not stick to the PvT 100 device or it can be confined in mercury by the Gnomix device.

Ramos *et al.* [RAM 05] study the influence of the pressure on the curing kinetics of a pure epoxy resin. They use a SWO/Haake PVT 100 in isobaric and isothermal mode, to follow the evolution of the specific volume. Data clearly show the shrinkage induced by the crosslinking reaction. One question is the accuracy of the measurement if the sample loses the contact with the molding cavity. The conversion degree is then estimated for a time t at a given temperature T and pressure P, from the specific volume evolution. The authors show that the kinetics depends strongly on the temperature, but it is much less sensitive to the pressure. The behavior is obviously dependent on whether the resin is above or below the glass transition temperature Tg. The pressure increases the curing rate, but it decreases it after the glass transition, in the diffusion regime.

Muller *et al.* [MUL 07] also use the PvT measurements obtained with a Gnomix device to follow the crosslinking of an ABS-modified epoxy resin, for a ratio of ABS up to 20%, considering an isothermal and isobaric mode. Assuming a linear relationship between the formation of the polymer network and the volume shrinkage, they note that the blend with the highest modifier ratio shows the weakest shrinkage, as well as faster kinetics.

More advances in this domain were made in the last 12 years due to the development of a specific device named PvTα [ABO 09, BOY 03, BOY 04, NAW 12, NAW 14]. This mold (Figure 2.3) consists of a piston, which can move in a cylindrical stainless steel cavity having a 61 mm internal diameter. This mold is designed so that heat transfers are one-dimensional (1D) through the sample thickness. The heating is ensured by three heaters located at the top and bottom to heat the piston and cavity, respectively. The mold is placed between two platens of an electric press. The position of the piston is adjusted in real time to keep the pressure constant. In this instrument, the experiments can be done on bulk samples.

Before starting the PvT–α experiment, neat resin or the associated composites were sealed in a low modulus elastomer capsule, which is selected to avoid the jamming of the mold. Impregnated composite can be done by infusion of resin in the capsule using vacuum assistance to remove air bubbles or directly by stacking preimpregnated plies. Capsule was then placed in PVT–α cavity under the piston, which moves following the variations of the sample volume to get a constant pressure. These variations were recorded by LVDT-type displacement sensor with a precision of 1 μm and a limit of 10 mm. The heat flux was measured by two heat flux sensors placed in the bottom of cavity and in the piston. Since the composite sample was placed in the deformable but incompressible elastomer cavity, the pressure is assumed to be hydrostatic and the surface of the elastomer, on which the pressure was applied, does not vary. As a result, the measured thickness is directly related to the sample volume and its variations.

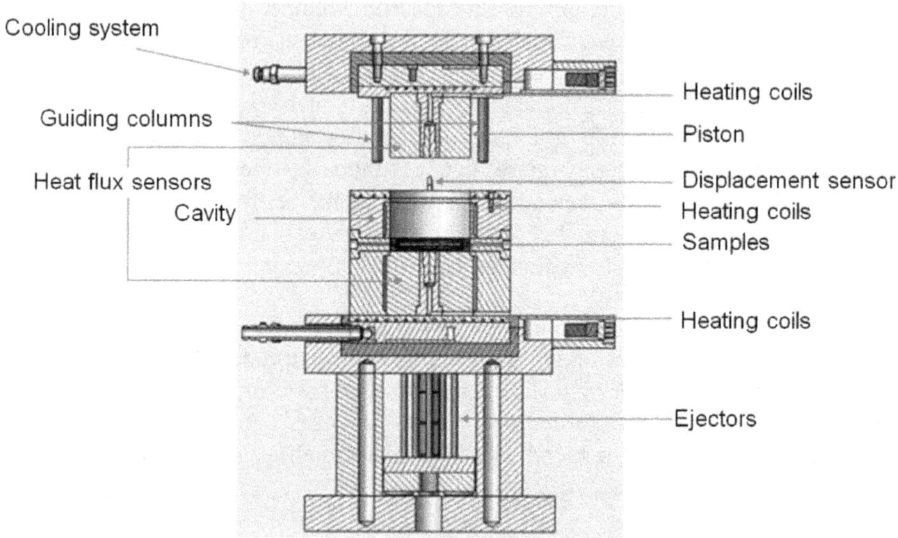

Figure 2.3. *PvTα apparatus*

A typical evolution of the resin thickness (representative of volume) during the curing cycle, sample surface temperature and the associated degree of cure α (calculated from heat flux density data) are plotted versus time in Figure 2.4 [NAW 12]. During the heating cycle, this thickness variation is the result of thermal expansion and chemical shrinkage, while the only prominent phenomenon during the cooling cycle was the thermal contraction.

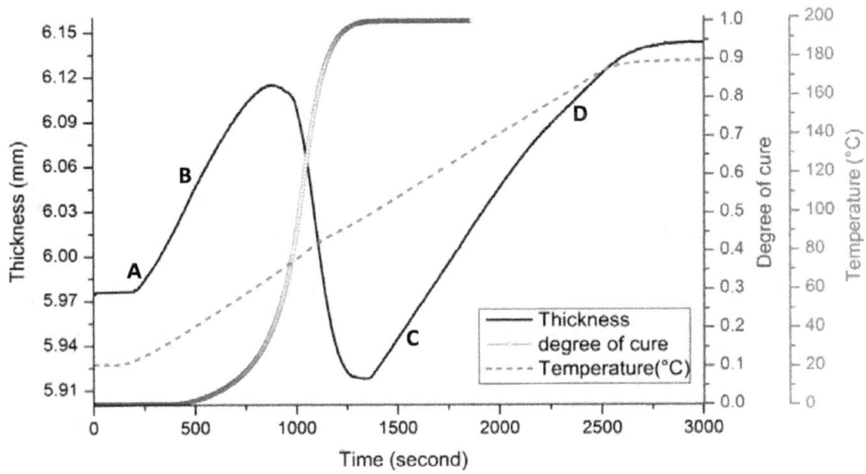

Figure 2.4. *Thickness, temperature and degree of cure of neat resin sample versus time for a vinylester resin [NAW 12] . For a color version of the figure, see www.iste.co.uk/boyard/heat.zip*

Linear (A–B) part of the thickness curve represents the thermal expansion of neat resin. As the reaction starts at the low temperature, this region is very small and it is difficult to determine coefficient of thermal expansion (CTE) of raw resin accurately. Between the points B and C (crosslinking zone), the chemical shrinkage of resin also appeared in addition to thermal expansion, and tended to decrease the sample thickness up to point C where the resin was close to being fully cured (as $\alpha = 1$). A linear thickness curve from point C to D shows that the only prominent effect was again the thermal expansion. When the thickness is plotted versus temperature, slopes of parts AB and CD of that curve give the CTE of raw and cured rubbery resin, respectively.

In a first time, the evolution of the volume shrinkage is determined by subtracting a baseline and assuming that the whole sample is isothermal and isoconversional at each time step. However, it is not true, especially during the resin crosslinking. So, it is necessary to take into account thermal and conversion degree gradients in the specimen so as to deduce the exact contribution of thermal expansion and chemical shrinkage to the experimental displacement curve, as demonstrated in [NAW 12]. For this resin, the volume shrinkage is found to be equal to 7.1%. The variation of the volume shrinkage according to the conversion degree is almost linear during the whole crosslinking, when considering the mean degree of cure, with a slope (defined as the shrinkage coefficient) equal to CS = 0.06.

2.2. Determination of specific heat capacity of resin and composites

The specific heat capacity corresponds to the amount of heat required to raise the temperature of 1 g of a substance by 1 Kelvin. In the case of an isochoric evolution at variable pressure, the specific heat capacity $c_v = (\partial u/\partial T)_v$ is used, u being the internal energy for the unit of mass. If the material encounters an isobaric evolution, the specific heat at constant pressure $c_p = (\partial h/\partial T)_p$ must be used, h being the enthalpy for the unit of mass.

The specific heat capacity at constant pressure is generally determined by using a calorimeter, and especially a differential scanning calorimeter (DSC). This device consists of heating a sample with a linear temperature ramp and measuring the heat transfer in this sample relative to a reference. The dimension of the sample must be small enough to assume a uniform temperature at each instant, but large enough so that its properties are representative of the homogeneous material. The size of the sample is, therefore, the result of a compromise. In practice, the mass of the sample is about 10 mg.

In the case of a power compensation DSC, the sample and the reference are placed in two separate ovens. The difference dQ/dt between the powers required for heating the sample of mass m_s and the reference of mass m_r so that they follow the same temperature evolution allows us to determine at constant heating rate (r) the difference between the specific heat capacities of the sample c_s and the reference c_r, provided that the heat losses on the sides of the sample and on the reference are strictly identical (case of a power compensation DSC):

$$dQ/dt = r(mscs\text{-}mrcr) \qquad [2.9]$$

Another type of DSC consists of placing the sample and the reference in the same oven. They receive the same power with, as a consequence, a difference of temperature between them, because of their different properties. In this case, the specific heat capacity is deduced from the difference of the temperature. Interested readers can refer to the book edited by J.D. Menczel and R.B. Prime [MEN 09], which provides a complete review of DSC technologies and data analysis.

In any case, the time constant of the device $\tau = RC$ (for the sample, R is the total thermal contact resistance including the thermal resistance between the pan and the calorimetric block as well as between the sample itself and the pan) is an important parameter and is often found between 30 and 60 s. Therefore, conventional DSCs cannot follow rapid transitions. Measurements of the specific heat capacity obtained in transformation zones associated with melting, crosslinking, crystallization, etc. are only apparent results, without any physical sense. This is especially true for transitions of melting or crystallization as illustrated in [LEB 08] for indium. Such a property has to be measured only outside such zones.

The experimental protocols developed for c_p measurements are detailed in [MEN 09] for conventional devices. At least two runs with the same temperature cycle are required: a blank run (also called baseline) and a sample run:

$$c_{p,s} = \frac{\phi_s(T) - \phi_{blank}(T)}{m_s \frac{dT}{dt}} \qquad [2.10]$$

where ϕ is the measured heat flux.

For higher accuracy, a third one with a sapphire specimen allows for calculating an instrument calibration constant K for different temperatures:

$$c_{p,s}(T) = K(T) * \frac{\phi_s(T) - \phi_{blank}(T)}{m_s \frac{dT}{dt}} \qquad [2.11]$$

$$K(T) = c_{p,sapphire}(T) \frac{m_{sapphire} \frac{dT}{dt}}{\phi_{sapphire}(T) - \phi_{blank}(T)} \qquad [2.12]$$

If the DSC can also work on the modulated mode (modulated temperature signal is imposed), c_p is easily obtained from the reversing signal. Figure 2.5(a) shows a characteristic example of the evolution of the specific heat capacity according to the temperature of an epoxy resin before and after crosslinking [BAI 96]. The evolution will remain practically the same as in the case of an epoxy/rubber blend. The values of the specific heat are different before and after crosslinking. However, for each state, a strong variation of specific heat capacity according to the temperature is observed. The glass transition (T_g) can also be observed in this figure; it corresponds to the strong increase in c_p between 120 C and 140°C for the crosslinked resin. This leads to a global increase of 50% between 40 and 180°C for the crosslinked resin. In the glassy (under T_g) and rubbery (above T_g) states, the variations are linear according to the temperature. In the zone of transition, the values depend on the cooling rate, and as explained previously, are only apparent values. We can, therefore, postulate a value of T_g and then extrapolate the values of c_p up to this temperature.

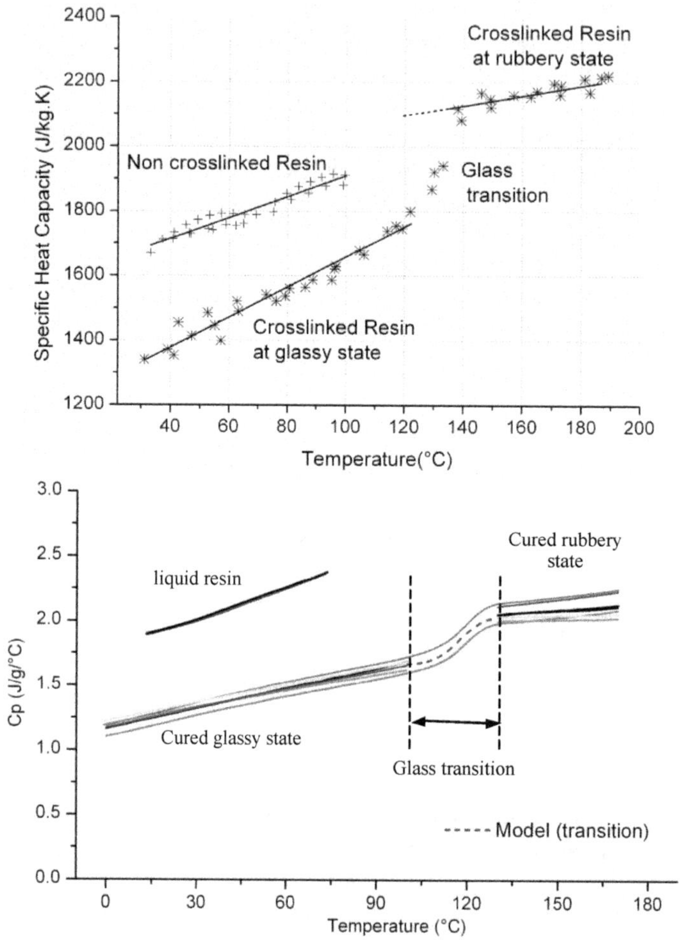

Figure 2.5. *Specific heat capacity of a) an epoxy resin (from [BAI 96]), b) a vinylester resin [NAW 12]. For a color version of the figure, see www.iste.co.uk/boyard/heat.zip*

For the neat resin, the upper limit of the temperature domain in which the measure is possible corresponds to the beginning of the crosslinking (100°C for the measure in Figure 2.1, in the heating conditions used). The rapid evolution of the properties associated with the source released by the chemical reaction makes the measure impossible during the crosslinking. An extrapolation for higher temperatures is then necessary. The evolution of the specific heat during the crosslinking is modeled by a mixture law (equation [2.13]) that weights the specific heat capacity of the resin in both states by the crosslinking degree.

$$c_p(T, \alpha) = (1 - \alpha)\, c_p(T, 0) + \alpha c_p(T, 1) \qquad [2.13]$$

In equation [2.13], the expressions $c_p(T, 0)$ and $c_p(T, 1)$ stand, respectively, for the evolution of the specific heat according to temperature for $\alpha = 0$ and $\alpha = 1$ in the rubbery state. An important remark must be introduced here concerning the values used in the law of mixture: during the crosslinking, the resin remains in the rubbery state. A classical error consists of weighting c_p by the raw values found on the curves (for example, at 100°C to weight the values of c_p = 1,900 J/(kg.K) for the non-crosslinked resin and c_p = 1,650 J/(kg.K) for the crosslinked resin). Two different states would be weighted: the rubbery state for the uncured resin and the glassy state for the crosslinked resin. However, during the crosslinking, the resin generally remains in the rubbery state. It is then necessary to extrapolate linearly and at temperatures lower than Tg, the behavior of the crosslinked resin in the rubbery state. This corresponds to the dashed straight line in Figure 2.5(a). The value at 100°C is then of 2,050 J/(kg.K). The error reaches 10% for $\alpha = 0.5$ and 20% for $\alpha = 1$, while using a mixture law on α. The excellent agreement between the temperatures measured inside the resin during the crosslinking and the predicted values using this relationship validates the assumption of using a law of mixture during the crosslinking to model the specific heat capacity.

Figure 2.5(b) displays another example of results for another thermoset polymer (vinylester resin). The results are in the same order of magnitude and it is easy to differentiate the liquid from the solid state and to observe the glass transition. The dotted line represents a modeling of the transition from the glassy to the rubbery state of the cured vinylester resin.

2.3. Thermal conductivity: a tricky task...

Heat transfer modeling in composite parts during their processing requires an accurate knowledge of the thermophysical properties. Any uncertainty on these parameters has a direct impact on the accuracy of the calculated heat flux and temperature fields in the composite structure. It is thus of crucial importance to have reliable and optimized methods at our disposal. Among these thermal properties, thermal conductivity is clearly among the most difficult to measure as well as to model. We recall that it is this property that provides an indication of the rate at which energy is transferred by heat diffusion.

Since the composite materials are heterogeneous and their fibrous reinforcement can have several orientations, thermal conductivity has an anisotropic character, which imposes evaluating a tensor, not a single value. Measurements have thus to be made in several directions (they should be the main ones), leading to time-consuming and tedious characterization because almost no device is able to estimate the components of the thermal conductivity tensor with a minimum of experiments and samples. Moreover, it depends on the physical state (solid/liquid) of the polymer and the temperature.

The determination of the thermal conductivity can be done following two complementary approaches. The first is experimental and aims to identify the effective anisotropic properties of representative specimens from measurements carried out in stationary or transient methods. The second consists of using analytical models and/or homogenization theories to predict the effective thermal property from the thermal properties of the elementary constituents of the material and from information on the composite microstructure, also including the effect of the polymer transformation.

2.3.1. *A first assessment of experimental data*

A cross-bench was organized in 2005 by Airbus Group (France) and the French Thermal Society (SFT). The aim of this study was to compare the results provided by different measurement methods on a single composite material at room temperature. The composite material was a pure unidirectional carbon/epoxy composite: a stack of 27 layers of AS4/8552 RC34 AW196 UD tape (fiber volume content: Vf = 56 vol%). Coupons were manufactured by Airbus (Getafe – Airbus Spain) using the same manufacturing process: an autoclave curing with a vacuum bag process. As the measurement methods required different sample shapes and sizes, a large composite plate (500 mm × 500 mm) of 5 mm thick was sent to each involved laboratory, which was in charge of the cutting of the plate so as to extract the desired sample. The following coordinate system has been chosen: (Ox) corresponds to the fibers' direction and (Oz) represents the thickness direction. UD composite has a high anisotropic behavior. Thermal conductivity or diffusivity must be measured in each main direction. The laboratories involved in this cross-bench used several techniques such as hot guarded plate, hot wire, transient plane source, periodic and laser flash methods.

Among all the results, it is interesting to note that diffusivity values are low scattered (<±5%), whatever the longitudinal or the transverse ones. However, it is obviously not the case for the thermal conductivities (Figures 2.6(a) and (b)): the scattering reaches ±15% around the mean value (when extremes are not considered). For example, the values of the transverse thermal conductivity range from 0.59 to 0.76 W/mK at room temperature.

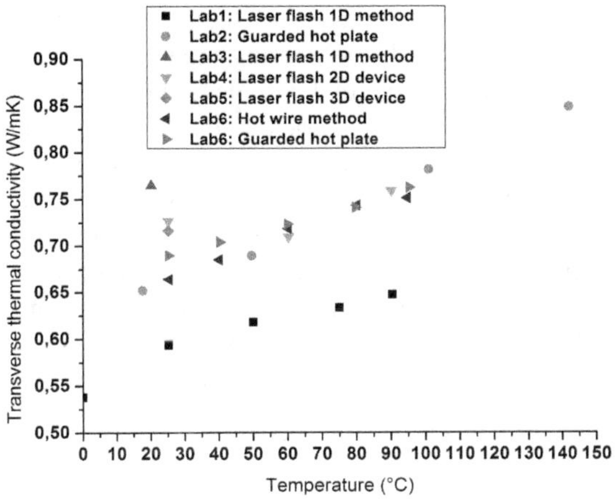

a)

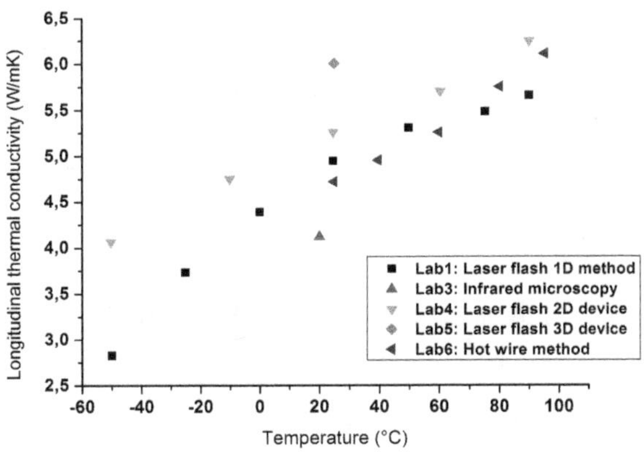

b)

Figure 2.6. *Comparison of the measured thermal conductivities within the framework of the cross-bench: a) transverse conductivity (along (Oz); b) longitudinal conductivity (along Ox: fiber direction). For a color version of the figure, see www.iste.co.uk/boyard/heat.zip*

Due to the high anisotropic character of the carbon fiber conductivity, the differences between the experimental results along the (Ox) direction (i.e. fiber direction) and the (Oy) and (Oz) directions are consistent. *A contrario*, high thermal contrast is not expected between the directions (Oy) and (Oz) but some results tend to show a difference. Thomas *et al.* [THO 08] show that existing resin-rich interplies induced by the composite manufacturing can explain such experimental data.

Finally, this cross-bench reveals that the measurement of the thermal conductivity is not straightforward, considering the scattering of the results given at room temperature and for an assumed simple structure. It also underlines that many experiments are necessary to characterize the whole thermal conductivities of composites, which is not efficient from cost and accuracy points of view. A third remark is about the directions along which the measurements are made: nobody knows if these directions are the main ones or not and can lead to errors. From a practical point of view, we also have to keep in mind that thermal conductivity data for dry and impregnated reinforcements are necessary to simulate processes, not only for consolidated composites. It is not addressed in this benchmark but there is almost no routine device dedicated to such materials, which is a current big issue.

2.3.2. *Overview of the main characterization techniques to measure the thermal conductivity of polymers and associated composites*

Experiments are carried out to determine the components of the effective thermal conductivity tensor when the studied material is a composite. These values represent the macroscopic thermal properties of an equivalent homogeneous material. Fourier's law can be written from averaged heat flux and temperature gradient:

$$\langle \varphi \rangle = -\bar{\bar{k}}_e \langle \nabla T \rangle \qquad [2.14]$$

where the operator $\langle \ \rangle$ is an average and $\bar{\bar{k}}_e$(W.m^{-1}.K^{-1}) is the effective thermal conductivity tensor.

Generally speaking, the experimental methods consist of thermally exciting a sample (with heater, heat flux pulse, laser, etc.) and measuring its thermal response (with thermocouples, IR camera, flux meter, etc.) so as to estimate its heat transport properties. The methods detailed in this section are all based on conductive heat transfer. As for specific heat, to avoid the coupling with another equation, the characterization methods are conducted, most of the time, outside of the phase change. Note that complete reviews and comparisons of the existing techniques are given in [DEG 94, RID 09]. These methods fall within two categories: thermal steady-state methods and transient ones.

2.3.2.1. *Stationary methods*

This first class of methods is based on the use of the Fourier's law to determine the thermal conductivity. The basic principle is to measure simultaneously the heat flux crossing a sample and a temperature difference. The most well-known and standardized method [STA 91] is the guarded hot plate. It consists of placing a specimen between two plates at two different temperatures T_1 and T_2. Considering φ as the heat flux crossing a sample, the thermal conductivity of the sample (in the direction of the temperature gradient), for one temperature level, is basically obtained with:

$$k = \frac{\varphi e}{(T_1 - T_2)} \qquad [2.15]$$

where e is the sample thickness and φ (W/m^2) is the heat flux density.

Many devices were developed so as to provide the most accurate results. The geometry of the specimen is such that the thickness (few millimeters) is small compared to the in-plane dimensions. A thermal guard often made with the same material to the one to be characterized limits the lateral heat losses. This method can be used over a temperature range (between 20 and 500°C).

The main drawbacks of this method are the long amount of time necessary to reach the steady state (especially for low diffusivity materials), the lack of convenience for measurement on liquid material, notably for sealing reasons and the heat transfer quality at the sample interfaces. To reduce the thermal contact resistance strongly, thermal conductive grease is often used.

Guarded hot plate is a very popular method but it can only measure the transverse thermal conductivity of a material. However, measurements of in-plane conductivity remain possible by cutting and then assembling the small parts such that the direction of interest is the one of the heat flux dissipated by the device. The heat flow meter is a variant of the guarded hot plate technique. A heat flux sensor is used to measure the heat flux through the specimen placed between the hot and cold plates.

2.3.2.2. *Transient methods*

The following methods are time-dependent and provide either directly to the thermal conductivity, or to the thermal diffusivity ($a = k/(\rho Cp)$), where ρ is the density.

2.3.2.2.1. Flash method

Flash method is widely used to estimate the diffusivity of homogeneous and isotropic materials. It was then extended to composite materials. In its first version, this method is built upon the analysis of the back face temperature rise of a specimen, whose front face is submitted to an energy pulse of short time compared to the observed phenomenon. The speed of the temperature propagation through the sample is directly linked to the thermal diffusivity of the material and to its geometrical characteristics. The specimens are typically 10–20 mm in diameter and 1–3 mm thick. In the case of transparent specimens, they had to be coated prior to testing to ensure absorption and emission at the front and rear faces, respectively.

Following the method and assumptions (perfectly insulated sample, the pulse is infinitesimally short and geometrically small) given by Parker [PAR 61], it is possible to deduce the transverse thermal diffusivity from the expression $a = \frac{0.139e^2}{t_{1/2}}$, where e is the thickness and $t_{1/2}$ is the half-time rise of the temperature on the back face. Then, more complex models were developed to take into account convection and radiative heat losses [CAP 63], laser beam shape, etc.

This method is then extended to anisotropic media. Among others, we can cite the approach proposed by Donaldson [DON 75] to simultaneously estimate the transverse and radial diffusivity. It is based on a heterogeneous heat flux pulse on the front face to create a thermal gradient in the sample plane. Other works [DEG 86, DEG 96, DEM 97, LAC 91, KRA 04] show that it is possible to estimate the whole set of components of the diffusivity tensor. The flash method uncertainty factors are mainly due to the experimental conditions that are difficult to realize, e.g. the convective and radiative exchanges on front and back faces, the duration and inhomogeneity of the incident heat flux or the nonlinear effects.

We have to keep in mind that flash methods do not lead directly to the thermal conductivity tensor but to the thermal diffusivity tensor. To obtain the thermal conductivities, two strategies can be used. On the one hand, we can calibrate the incident heat flux on the sample. A technique [PLA 04] consists of using a reference material of known thermal properties, and submitting it to the energy pulse in order to determine, according to its temperature rise, the incident heat flux. Then, assuming the repeatability of the device and the good confidence in the knowledge of the thermal properties of the reference material at measurement temperature, the experiment can be realized on the sample with unknown thermal conductivities.

On the other hand, a second strategy consists of combining the flash method with dedicated measurements of the specific heat and the density. In both cases, the estimation of the thermal conductivity with flash methods induces additional uncertainties compared to the thermal diffusivities.

For composites, it is important to avoid 2D or 3D effects, which can be induced by fibers, so that experiment times should not be too long. Moreover, thin samples also have to be avoided otherwise they could not be considered as representative of bulk material and equivalent to a homogeneous one. This can lead to large errors on the estimated diffusivity. Finally, this method is mainly dedicated to solid specimens but new sample holders have now been developed to carry out measurement on paste, liquids (many precautions have to be taken) and fibers over a large temperature range (from –90 to 500°C, even more).

2.3.2.2.2. Hot wire method

Initially, the hot wire method consists of radially dissipating a known heat flux via a resistant wire of low section located inside the homogeneous and isotropic material. The measurement of the induced temperature rise can be realized from the wire electrical resistance evolution or with a thermocouple placed as close as possible to the wire. After a certain time, this temperature is a linear function of the natural logarithm of time so that the thermal conductivity in the plane orthogonal to the wire can be determined by using the equation:

$$k = \frac{q_0}{4\pi P} \qquad [2.16]$$

where P is the slope of the linear part of the temperature variation when plotted versus the natural logarithm of time, and q_0 is the dissipated heat flux per length unit.

Based on the assumption of infinite media, it is necessary to take care that the measurement time is not too long to avoid edge effects. However, this time has to be long enough to observe the asymptotic transient regime where equation [2.16] is valid. An optimal time window should be found.

These transient measurement methods were then adapted to more complex issues, such as orthotropic materials. Jarny [JAR 01] uses an inverse method to estimate the two thermal conductivities orthogonal to the wire direction as well as ρc_p. This method consists of a symmetrical assembly in which a resistant wire is sandwiched between two pair of samples. The assembly is insulated on lateral faces, and microthermocouples are located in the two directions perpendicular to the wire. The latter is assumed to have an infinite length and negligible radius, and the hypothesis of an infinite sample is considered (the boundary effects are thus neglected). Initially, the assembly is isothermal, the temperature being imposed with thermoregulated plates. At t > 0, the wire aligned along (Oy) direction is powered and dissipates a known constant power q_0. Due to the anisotropic character of the composite samples, isotherms are elliptic in the plane (Oxz). A least-square criterion based on the difference between experimental ($T_{experimental}$) and computed ($T_{computed}$) temperatures is minimized to identify the conductivities and the heat capacity. Numerical temperatures are computed by solving a 2D heat conduction problem.

Some limitations could be the inertia of the hot wire as well as the possible thermal contact resistance between the wire and the sample. As these parameters only have an influence on the temperature increase for short times, they do not influence the estimation of the conductivities that are performed after a certain time. Nevertheless, the specific heat is mainly sensitive to short time. Precautions thereby have to be taken. Nonetheless, the main drawback of this method is the instrumentation since the wire and the micro thermocouples have to be placed on the sample. This step is first of all time-consuming. Moreover, it induces results and uncertainties linked to the uncertainty of thermocouple locations.

2.3.2.2.3. Shock probe

The principle of the shock probe (or thermal probe) is the same as the hot wire. The difference lies in compact cylindrical or spherical geometries. The temperature is measured from the electrical resistance of the heating element. This resistance being much higher compared to the hot wire, it is thus easier to measure. The shock probe is often used to determine the thermal conductivity of liquids or homogeneous porous materials. For example, it was included in the PvT 100 device. Theoretically, a perfect contact is assumed between the specimen and the probe, which is almost impossible in the case of solids. This method is also not adapted to anisotropic media.

2.3.2.2.4. Periodic methods

The development of periodic methods for the determination of thermal diffusivity is based on a theory originally stated by Ångström [ANG 63]. The principle consists of periodically heating a sample, then the temperature along the sample also varies with the same period, but with an amplitude that decays exponentially. Moreover, as the temperature wave travels along the sample with finite velocity, there is a varying phase relationship. The measurement of phase lag between periodic input and thermal response of the material can be directly related to the thermal diffusivity.

An example of application is the local excitation with a periodic laser flux creating a homogeneous isotropic media a thermal wave. Then the periodic temperature increase at a distance x is given by:

$$\Delta T(x,t) = \frac{Q}{4\pi k} exp\left(-\frac{x}{\mu}\right) cos\left(\omega t - \frac{x}{\mu}\right) \quad [2.17]$$

where $\mu = \sqrt{\frac{a}{\pi f}}$ is the thermal diffusion length, f is the modulation frequency of the harmonic heat source, Q is the heat source power and x is the distance from the excitation.

It can be seen that the phase shift $\varphi = -x/\mu$ between the periodic temperature rise at the surface of the medium and the modulation of the heat source linearly decreases as x increases. The slope of the phase curve along x is constant and equal to $-1/\mu$. Thus, the recording of this curve enables us to evaluate the thermal diffusivity by a linear fit using equation [2.17].

Many other models and experimental devices [AUT 05] are based on Ångström's method so as to measure heat transfer coefficients [GRI 99], thermal diffusivity of anisotropic media [SER 05] or thermal properties on small dimension media (e.g. thermal properties of fibers contained in composite materials [PRA 95, ROC 05]).

2.3.2.2.5. Transient plane source

Transient plane source (TPS) methods derived from works realized by Gustafsson [GUS 91] who developed a technique able to simultaneously estimate the thermal conductivity and diffusivity of homogeneous and isotropic solids. TPS methods consist of a flat heating element in the shape of a strip, strip pattern, disk, or disk pattern, which can be sandwiched either between two identical pieces of the sample (double-sided configuration) or applied to only one sample piece (single-sided configuration).When powered, the known heat generated by Joules effect is dissipated through the sample. Then, from the measurement of the temperature increase in the probe, thermal properties can be calculated. Different formats of sensor were developed to accommodate a large variety of samples. Later, developments were made so as to estimate the in-plane $k_{xx}= k_{yy}$ and the transverse k_{zz} thermal properties of orthotropic materials [GOB 04], provided that ρc_p is known.

The main advantage of these methods lies in the fact that both thermal conductivity and diffusivity are estimated. Moreover, the probe used generally constitutes both heat source and temperature sensor (thermocouples can be placed inside the probe), which is convenient from different aspects: ease of application, rapid instrumentation time or uncertainty on thermocouples locations. "Hot Disk" and C-Therm probes are commercialized systems based on this method. The C-Therm probe can be adapted to solids, pastes and liquids in a temperature ranging from room temperature to 200°C, but is limited to isotropic and homogeneous materials.

2.3.2.2.6. Inverse methods

Inverse methods are quite general and, for example, allow the identification of the thermal conductivity as a function of temperature of polymers (thermosets and thermoplastics) and composites in several states. These methods associate experimental data and a numerical model coupled to a minimization algorithm to identify the parameters. In our case of interest, a typical experiment consists of measuring temperatures with thermocouples at several locations through the thickness of a sample, which is placed between two heating platens, assuming 1D heat transfer. The upper and lower thermocouples are used as Dirichlet boundary conditions for the solving of the heat conduction equation. The inverse method aims to minimize a least-square criterion (equation [2.18]) based on the difference between experimental ($T_{experimental}$) and computed ($T_{computed}$) temperatures.

$$J(\lambda(T)) = \int_t T_{computed} - T_{experimental})^2 \, dt \qquad [2.18]$$

Practically, the conjugate gradient algorithm is often used, where the gradient of the criterion is determined by considering a set of adjoint equations [JAR 03]. Note that this method requires us to already know the specific heat and density of the studied material. As already mentioned, such a class of methods has a general character. As an example, this approach has been successfully applied to estimate several thermophysical properties from several experiments on specific multifunctional devices [PIG 15, NAW 12, TAR 12].

As a brief summary, Table 2.2 collects the main characteristics of the discussed methods. It is far from exhaustive but gives an overall view of the state of the art.

Examples of results from these abovementioned devices are given in Figures 2.7 and 2.8 for polypropylene and thermoset resins, respectively. Figure 2.7 shows a scattering of the data (at atmospheric pressure), demonstrating the difficulty in determining the thermal conductivity of thermoplastics, both in the molten and solid states. Rides *et al.* [RID 09] reach a similar observation for PMMA (a morphous polymer). Such variations of the measured properties can have large consequences in the heat transfer predictions.

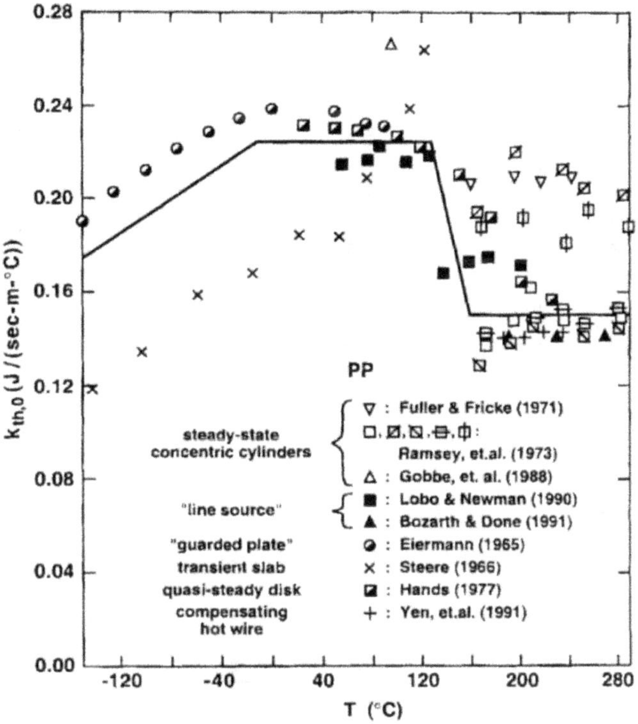

Figure 2.7. *Comparison of thermal conductivity data for an isotactic polypropylene [HIE 02]*

This scattering can be explained mainly either by the measurement method, or by the forming conditions of the sample (cooling rate and pressure). This second point is particularly important for semi-crystalline polymers.

Two characteristic examples of the evolution of the thermal conductivity versus the temperature of an epoxy/glass fiber composite [BAI 96] and a vinylester resin [NAW 12], before and after crosslinking, are presented in Figures 2.8(a) and (b). All these results are obtained from an inverse identification method. For both materials, the values of the thermal conductivity are higher after crosslinking. For example at 60°C, the thermal conductivity of the composite is close to 0.25W/m.K in the uncured state and to 0.58W/m.K in the fully crosslinked state (but below the glass transition). However, the thermal dependency is slightly more pronounced for this uncured composite. It is also interesting to note the authors' success in measuring the thermal conductivity for a partially cured resin.

	Variation	*Thermal property*	*Thermal conductivity range ($W.m^{-1}.K^{-1}$)*	*Temperature range (°C)*	*Measurement time*	*Sample state*	*Anisotropy*	*Limits*
Guarded hot plate		k_z	0.05-5	20 to 500	Several hours	solid	In-plane conductivity possible by cutting	Measurement time
Flash method		a_z	0.1-2000	-120 to 400	Several seconds	solid (liquid if adapted sample holder)	In-plane conductivity possible	Difficulty to master radiative heat losses
Hot wire	12	k (ρCp andtensor under conditions)	0.1-10	20 to 1500	Several minutes		Possible	Thermal contact resistance between wire and sample Long and difficult instrumentation
Shock probe		a	0.01-10	-20 to 90	Lower than 500s	solid, liquid, paste	no	Optimal time range Possible bad contact between probe and solid sample
Periodic methods		a_x			From several minutes to hours, depending on the sample	solid	no	Long and difficult instrumentation
Transient plane source	Hot Disk	a (ork_z+$k_x$in orthotropic mode)	0.03-100	-220 to 1200	Several minutes	solid	Orthotropic material only	Sample geometry has to respect specific conditions
	C-Therm	a	0-120	-50 to 200	Several seconds	solid, paste or liquid	no	Contact between probe and sample
	PIMS method [56]	k_x, k_y, $k_z$andρCp	<10	20 to 120	Several minutes	solid	yes	Time for inverse method estimation

Table 2.2. *Summary of the characterization methods dedicated to the thermal conductivity of polymer and composites*

The measurements for a neat vinylester resin are also interesting since Figure 2.8(b) exhibits first an increase in the thermal conductivity of the cured resin up to 90°C and then a decrease beyond this point. This behavior is a characteristic of the passage from the glassy state to the rubbery state, meaning that 90°C can be considered as the glass transition temperature. In the uncured state, the thermal conductivity remains almost constant.

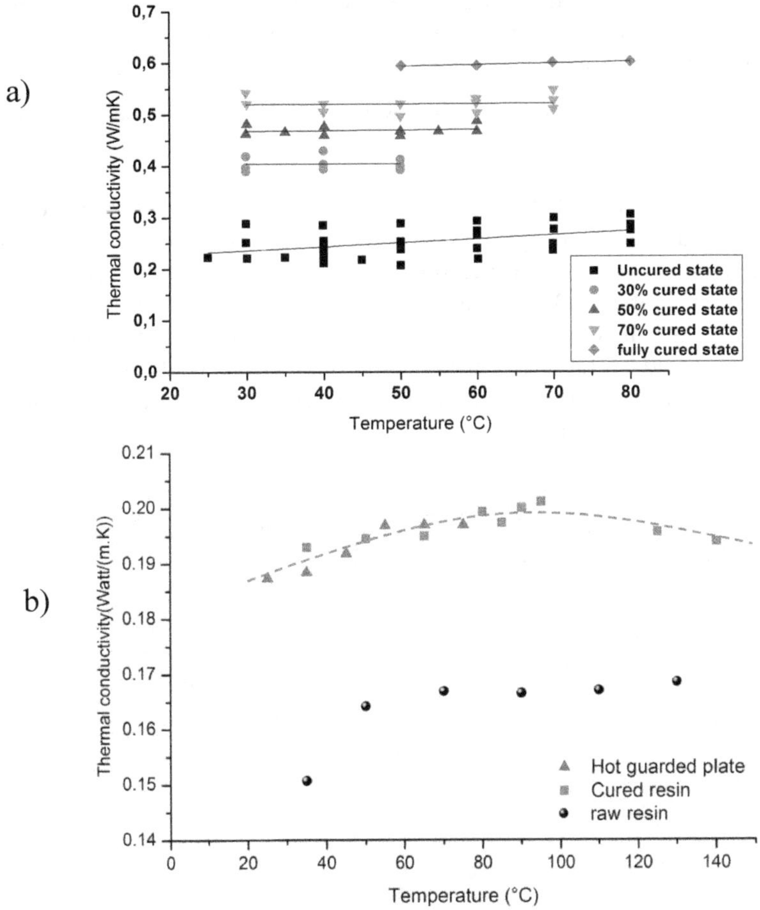

Figure 2.8. *Thermal conductivity measurements a) for an epoxy/glass fiber composite [BAI 96], b) for a vinylester resin [NAW 12]*

This technical review dedicated to the thermal conductivity measurements leads to several conclusions:

1) existing devices are mainly adapted to measurements on solid materials;

2) most of the developed methods propose to determine the thermal conductivity in only one direction (often the transverse one);

3) in the specific case of the polymer and composite forming, thermal properties have to be known in the solid and liquid states as well as for dry and impregnated reinforcements, over a large temperature range. Moreover, these data represent, for composite materials, the components of a tensor;

4) A new "all in one" method without any intrusive instrumentation and an easy handling is a dream.

Obviously, there is a gap between the needs and the existing methods. However, specific devices and methods [THO 10, VIL 12] are developed in specialized research laboratories to overcome such problems, even if some technical difficulties currently remain. Table 2.3 provides a typical example of in-plane thermal conductivity results, which can be obtained with such non-standard methods, i.e. dry preform, impregnated preform and cured composite. The considered materials are representative of those used in the aeronautics industry to manufacture parts with resin transfer molding (RTM) process.

State	Part	Experimental conductivity ($W\,m^{-1}\,K^{-1}$)	Predicted conductivity (Maxwell) ($W\,m^{-1}\,K^{-1}$)	Predicted conductivity (Charles & Wilson) ($W\,m^{-1}\,K^{-1}$)
Dry-preform	1	2.15	2.13	2.13
	2	2.70	2.81	2.81
	3	3.03	3.19	3.19
Uncured	1	2.33	2.32	2.29
	2	2.87	3.07	3.03
	3	3.40	3.47	3.41
Cured	1	2.62	2.50	2.46
	2	(2.91)	3.28	3.23
	3	3.73	3.68	3.62

Table 2.3. *Measured in-plane thermal conductivities for all states of aeronautic composite. Calculated conductivity from Maxwell and Charles & Wilson models is also indicated for comparison purposes [VIL 12]*

Note that transverse thermal conductivity data are also obtained for the same material and with respect to the fiber volume fraction V_f. The results show a logical increase in the identified conductivity with V_f. Furthermore, the increase between uncured and cured composites is about 25% to reach 0.755 W/(m.K) for $V_f = 55.3$ vol%. However, the influence of temperature is negligible in this case.

2.3.3. *Modeling of the thermal conductivity of composites*

2.3.3.1. *Definition of bounds*

A first approach consists of admitting the lack of information on the microstructure or from the difficulty of considering the microstructural information in the models, defining a range in which the result is guaranteed. The philosophy of these methods is to define the upper and lower bounds of the effective properties of heterogeneous materials. Historically, the Voigt and Reuss models are the first and the simplest models to provide the upper and lower bound of effective properties of heterogeneous materials. In fact, these bounds correspond to the well-known rules of mixtures. These basic models consist of considering the phases as serial (Reuss) or parallel (Voigt) thermal paths.

$$k_{Reuss} = \frac{k_f k_m}{V_f k_m + (1-V_f)k_f} \leq k_\perp^c \leq V_f k_{f,\perp} + (1-V_f)k_m = k_{Voigt} \quad [2.19]$$

where k_f and k_m are the thermal conductivity of the fiber and the matrix, respectively, and V_f is the fiber volume fraction. These bounds, which only depend on the inclusion volume fraction, are valid for any kind of inclusion arrangement, shapes or orientations, leading to a rather wide conductivity range.

Hashin and Shtrikman (HS) reduce this range using the following equations:

$$k_{HS^-} = k_m + \frac{V_f}{\dfrac{1}{(k_{f,\perp} - k_m)} + \dfrac{(1-V_f)}{2k_m}} \leq k_\perp^c \leq k_{f,\perp} + \frac{(1-V_f)}{\dfrac{1}{(k_m - k_{f,\perp})} + \dfrac{V_f}{2k_{f,\perp}}} = k_{HS^+} \quad [2.20]$$

In order to give an idea on the predictions realized using some of the presented models, results of thermal conductivity are presented as a function of the inclusion volume fraction (Figure 2.9(a)).

First, let us consider the case of a matrix reinforced with infinitively long fibers along (Ox) direction. No assumption is performed on fiber arrangement, which is only supposed to be homogeneous. All phases are assumed to be isotropic and the thermal contrast p (ratio between the thermal conductivities of inclusion and matrix) equals 10 since k_m= 0.2 W/m.K and k_f= 2 W/m.K. Models are used to predict the isotropic thermal conductivity in the transverse direction of fibers. As expected,

Figure 2.9(a) shows that the range of possible effective material properties described by the HS bounds is significantly closer than that given by the Voigt–Reuss bounds.

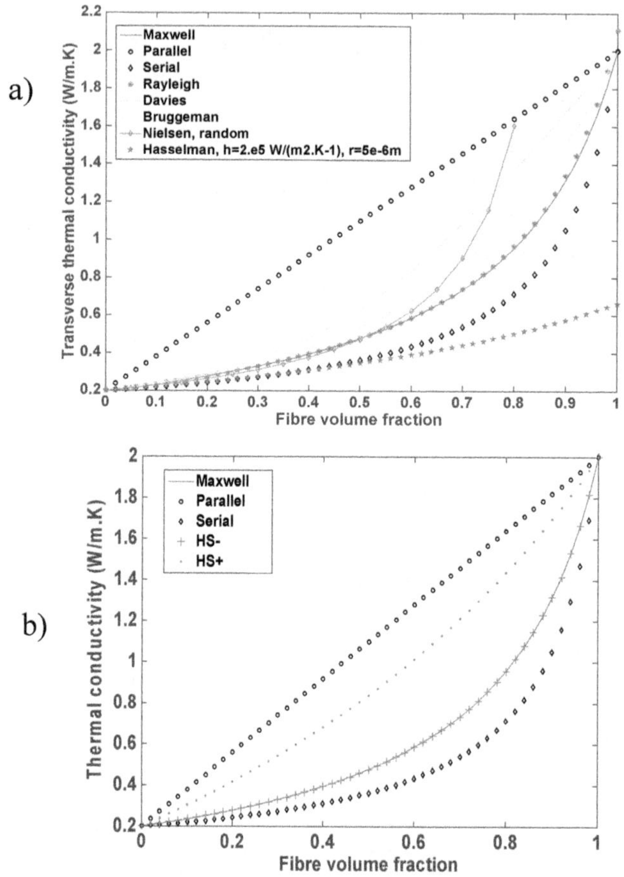

Figure 2.9. *Evolution of a composite transverse thermal conductivity versus the fiber volume fraction V_f with k_m = 0,2 W.m^{-1}.K^{-1} and a contrast A = k_f/k_m = 10. a) Illustration of the conductivity bounds. b) Comparison of several predictive models. For a color version of the figure, see www.iste.co.uk/boyard/heat.zip*

2.3.3.2. *Case of periodic microstructures*

First, microstructures are periodic arrangements (cubic, hexagonal, etc.) of unidirectional (UD) fibers. In the following, we distinguish the transverse thermal conductivity $k_\perp$ and the longitudinal one $k_\parallel$ (with respect to fibers).

The longitudinal conductivity is given by the parallel model and often provides satisfactory results:

$$k_{\parallel} = V_m k_m + V_f k_{f,\parallel} \qquad [2.21]$$

where V_m is the matrix volume fraction ($V_m = 1 - V_f$).

The prediction of the transverse thermal conductivity is more controversial. Mainly based on the analogy of electric potential, many models are available in the literature. The simplest one (for an arrangement of parallel fibers) is the serial model:

$$\frac{1}{k_\perp} = \frac{V_f}{k_f^\perp} + \frac{(1-V_f)}{k_m} \qquad [2.22]$$

Other classical expressions could be cited. A famous one was proposed by Lord Rayleigh for a cubic arrangement of cylinders of spheres (only low interactions are considered).

$$\frac{k_{\perp,Rayleigh}}{k_m} = 1 - \frac{2V_f}{\upsilon + V_f - \frac{C_1}{\upsilon}V_f^4 - \frac{C_2}{\upsilon}V_f^8}, \text{ with } \upsilon = \frac{1+\frac{k_{f,\perp}}{k_m}}{1-\frac{k_{f,\perp}}{k_m}} \text{ and } \begin{matrix} C_1 = 0.3058 \\ C_2 = 0.0134 \end{matrix} \qquad [2.23]$$

We can also talk about the models by Bruggeman (more dedicated to higher fiber volume fraction compared to Rayleigh's model):

$$k_\perp^c = \frac{1}{2}\lambda_m\left[(1-V_f)\left(\frac{k_f^\perp}{k_m}-1\right)^2 + 2\frac{k_f^\perp}{k_m} - \sqrt{\left[(1-V_f)\left(\frac{k_f^\perp}{k_m}-1\right)^2 + 2\frac{k_f^\perp}{k_m}\right]^2 - 4\left(\frac{k_f^\perp}{k_m}\right)^2}\right] \qquad [2.24]$$

McPhedran (square arrangement of cylinders in contact) [PHE 86] or Nielsen (square arrangement of UD fibers with maximum compactness Ψ_1, hexagonal arrangement with compactness Ψ_2 or random distribution with compactness Ψ_3):

$$k_{\perp,Nielsen} = \frac{1+A\beta V_f}{1-\Psi\beta V_f}, \text{ with } \beta = \frac{k_f/k_m - 1}{k_f/k_m + A} \text{ and } \psi = 1 + \frac{(1-\phi_{max})V_f}{\phi_{max}^2} \qquad [2.25]$$

where A is a coefficient which depends on the shape of the reinforcement (A = 0.5 for UD fibers and transversal conductivity) and ϕ_{max} is the maximum packing fraction (examples: square arrangement of UD fibers: ϕ_{max} = 0.785, hexagonal arrangement: ϕ_{max} = 0.907, random distribution ϕ_{max} = 0.822).

The model developed by Halpin and Tsaï [HAL 67] is also based on a square arrangement of cylinders with elliptical section. Other references can be found in [PRO 76, LOR 07, THO 08].

2.3.3.3. *Case of random microstructures*

This second simplified microstructure tends to be closer to the actual ones. The first model on two-phase systems goes back to Maxwell (1873) who derived an expression for the specific resistance of a dilute suspension of spheres in an infinite isotropic conductor [MAX 73]. Accurate rational models of two-phase composites with spherical or infinitely long cylindrical inclusions have been developed then for predicting elastic, thermal, transport and other properties. For infinitely long cylindrical inclusions, and when V_f is very small, Maxwell shown the effective thermal conductivity is given by

$$\frac{k_{\perp,Maxwell}}{k_m} = \frac{A+1+(A-1)V_f}{A+1-(A-1)V_f}, \text{ with } A = \frac{k_{f,\perp}}{k_m} \quad [2.26]$$

For illustration purposes, we present a comparison of several models, considering two thermal contrast "A". For low V_f (< 40 vol%), all models are almost superimposed (except the parallel and serial models which can be seen as bounds) but above this point, the models provide more scattered data.

Thus, Hasselman [HAS 87] extends this model to include the effect of a thermal contact resistance (TCR) between the fibers and the matrix:

$$k_\perp^c = k_m \frac{\left(\frac{k_f}{k_m} - \frac{k_f}{r.h} - 1\right)V_f + \frac{k_f}{k_m} + \frac{Bk_f}{r.h} + 1}{\left(1 - \frac{k_f}{k_m} + \frac{k_f}{r.h}\right)V_f + \frac{k_f}{k_m} + \frac{Bk_f}{r.h} + 1} \text{ with B=1 (for fiber) or 2 (sphere)} \quad [2.27]$$

where r is the radius and h = 1/TCR. The thermal contrast and the adimensional parameter Bk_f/rh are the key parameters of the model. Note that when h is high, this equation is equivalent to Maxwell's model. In Figure 2.9(b), h has to be higher than $2e^6$ W/(m^{2}K) (i.e. already low TCR) to be close to Maxwell's and Rayleigh's models.

Finally, the model developed by Charles and Wilson [CHA 81], also based on the analogy of the electrical potential, provides good predictions.

$$k_{\perp,CW} = k_m \left[\frac{k_{f,\perp}(1+V_f)+k_m(1-V_f)}{k_{f,\perp}(1-V_f)+k_m(1+V_f)}\right] \quad [2.28]$$

A direct application of all these models is given in Figure 2.9(b).

2.3.3.4. *Mean-field models*

Starting from the fact that a representative volume element (RVE) is equivalent to a macroscopic domain Ω, mean-field models are classical approaches to estimate the effective properties of heterogeneous materials. They consist of approximating the local fields of heat flux $\phi(\boldsymbol{x})$and temperature gradient $\boldsymbol{\nabla}_{\mathbf{x}}T$ by the following averages:

$$\langle\phi(\boldsymbol{x})\rangle^i = \frac{1}{\Omega^i}\int_{\Omega^i}\phi(\boldsymbol{x})d\Omega \quad [2.29]$$

$$\langle\boldsymbol{\nabla}_{\mathbf{x}}T\rangle^i = \frac{1}{\Omega^i}\int_{\Omega^i}\boldsymbol{\nabla}_{\mathbf{x}}T(\boldsymbol{x})d\Omega \quad [2.30]$$

where i stands for the matrix m and the fiber f domains.

The effective (macroscopic) thermal conductivity tensor is thus written:

$$K^* = \frac{\langle\phi(x)\rangle}{\langle\boldsymbol{\nabla}_{\mathbf{x}}T\rangle} \quad [2.31]$$

The mean-field approach contains a large number of different models. The basic philosophy underlying most of these developments is to focus on one peculiar scatterer and to replace the surrounding random medium by an effective homogeneous one. Two types of effective-medium approximation are distinguished: the matrix-inclusion type models, where the particulate phases are represented as inclusions embedded in a supposed infinite matrix phase, and the self-consistent (SC) ones, where all the phases are modeled as inclusions enclosed in a yet unknown effective medium supposed to be infinite.

The most important analytical solution is probably developed by Eshelby [ESH 57]. Numerous models have been developed on this base for various inclusion sizes and shapes. However, it is important to note that this model does not take into account possible interactions between inclusions. Effective thermal conductivity is then linearly dependent on V_f, which limits the use of this model to very low volume fraction of heterogeneities. Mori–Tanaka methods were introduced in order to generalize the dilute model of Eshelby to the non-dilute case. To develop their model, Mori and Tanaka [MOR 73] carry on Eshelby's works but considering interactions between inclusions. For short-fiber reinforced composites, Hatta and Taya [HAT 85] propose a model extended from the Mori–Tanaka model, considering the orientation distribution using an orientation tensor. It was successfully applied for reinforced thermoplastic parts made by injection molding [LEG 11].

Self-consistent models are also well-known models to estimate the effective elastic properties of composite materials. They appear to have originated with the works of both Budiansky [BUD 65] and Hill [HIL 65]. Contrary to Mori–Tanaka model, which incorporates fibers interactions in the influence of the matrix, self-consistent models incorporate them in a field that symbolizes the composite we want to characterize. More clearly, these models consist of embedding an inclusion in a fictitious homogeneous material of infinite dimensions possessing the composite's unknown properties, considering homogeneous boundary conditions [AUR 09].

2.3.3.5. *Laminated composites*

For higher structure complexity, such as a UD composite laminate with a given stacking sequence (Figure 2.10), orientations have to be taken into account. Let us define θ the angle between the fiber of a ply and the axis (Ox_1). The thermal conductivity tensor of a ply *i* in the composite coordinate system (O x_1, x_2, x_3) is given by:

$$K^i_{(Ox_1x_2x_3)} = \begin{pmatrix} k_{\parallel}\, cos^2\theta_i + k_{\perp}\, sin^2\theta_i & (k_{\parallel} - k_{\perp})cos\theta_i . sin\theta_i & 0 \\ (k_{\parallel} - k_{\perp})cos\theta_i . sin\theta_i & k_{\parallel}\, sin^2\theta_i + k_{\perp}\, cos^2\theta_i & 0 \\ 0 & 0 & k_{\perp} \end{pmatrix} \quad [2.32]$$

where $k_{\parallel}$ and $k_{\perp}$ are the longitudinal and transverse thermal conductivities of a UD ply, respectively. Possible models are discussed in the previous sections.

For the N plies of the composite, parallel and serial models can be applied to calculate the in-plane components and the transverse components, respectively (equation [2.33]).

The effective thermal conductivity tensor of the composite in the main directions is obtained by diagonalizing K^C matrix. It also enables us to know the angle existing between the chosen coordinate system (O x_1, x_2, x_3) and the one associated with the main directions.

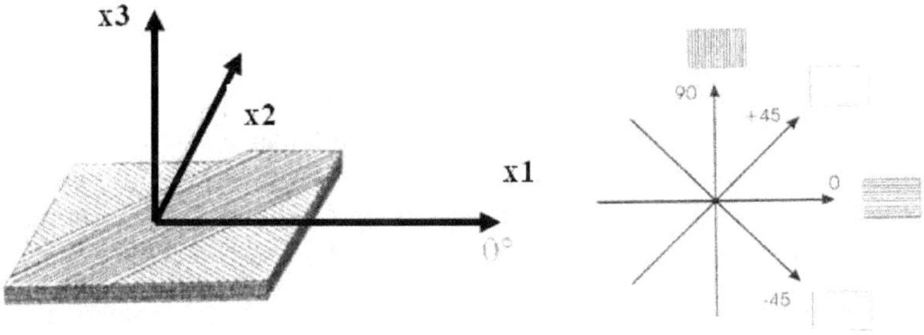

Figure 2.10. *Definition of the coordinate system of a composite composed of a stack of UD oriented plies*

Finally, in order to calculate in-plane thermal conductivities of a ply of fabric reinforcement, the latter can be broken down into a sequence of unidirectional plies. As an example, a twill-weave fabric made up of tows at θ = 0° and θ = 90° will be split into two unidirectional plies having as much as fibers in both directions [VIL 12, KUL 97].

$$K^{C}_{(Ox_1x_2x_3)} = \begin{pmatrix} \frac{1}{N}\left(\sum_{i=1}^{N} k^{i}_{11}\right) & \frac{1}{N}\left(\sum_{i=1}^{N} k^{i}_{12}\right) & 0 \\ \frac{1}{N}\left(\sum_{i=1}^{N} k^{i}_{21}\right) & \frac{1}{N}\left(\sum_{i=1}^{N} k^{i}_{22}\right) & 0 \\ 0 & 0 & \frac{1}{N}\left(\sum_{i=1}^{N} \frac{1}{k^{i}_{33}}\right)^{-1} \end{pmatrix} \qquad [2.33]$$

We assume that all plies have identical thickness and fiber fraction. Applying the previous equations where the twill-weave fabric is constituted by 2 UD plies (N = 2, where $\theta_1 = 0°$ and $\theta_2 = 90°$), we obtain:

$$k_{x,y}^{fabric\,ply} = \frac{k_{\perp} + k_{\parallel}}{2} \qquad [2.34]$$

$$k_{z}^{fabric\,ply} = k_{\perp} \qquad [2.35]$$

Note that the transverse thermal conductivity is not affected. In the specific case where weft and warp yarns are not orthogonal, the previous equations have included the β angle between these yarns:

$$k_{x}^{tissé} = \frac{k_{\perp}cos\beta + k_{\parallel}sin\beta}{2} \qquad [2.36]$$

$$k_y^{tissé} = \frac{k_\perp sin\beta + k_\parallel cos\beta}{2} \quad [2.37]$$

For the stacking of fabric plies, we use the same approach as the one presented for the lay-up of UD plies.

Fiber waviness is inherent to many reinforcing fabrics, unlike unidirectional composites, for which all fibers are nearly perpendicular to the z-axis. In fact, fiber crimping can be observed in the case of woven reinforcements. In such situations, each fiber bends over or under another and its axis deviates from a straight line and follows a simple or complex wavy path. A model that takes account of this effect is thus proposed in [VIL 12]. It brings a correction to the previous estimated conductivities.

2.3.3.6. *Mathematical homogenization*

Beyond the previous models, it is of prime importance to recall the concept of mathematical homogenization [BEN 78] following the asymptotic expansion method. Commonly employed in mechanics, it remains rarely used for thermal applications [ANA 10, ELG 06 MAT 13] but is a very interesting answer to the thermal conductivity prediction of composites.

The mathematical formulation of the physical problem has led to the homogenization theory and the exact prediction of an effective tensor of periodic media as well as the macroscopic evolution of the temperature and heat flux fields. The asymptotic development is an efficient method to obtain the homogenized problem [BAK 89, SAN 92] and can also be applied for random materials if the RVE is determined. The scale separation is mandatory for the geometry and the studied properties and is defined as follows:

$$\frac{l}{L} = \varepsilon \ll 1 \quad [2.38]$$

where l is the characteristic length of the unit cell (i.e. the RVE) and L is the characteristic length at the macroscopic scale (with a coordinate system $x = (x_1, x_2, x_3)$).

These two characteristic lengths then introduce a second space variable $y = (y_1, y_2, y_3)$ with $y = x/\varepsilon$. x is the macroscopic space variable associated with the composite (Ω domain) and y is the microscopic space variable associated with the RVE, also called the periodic cell (Figure 2.11). The asymptotic study of a problem containing the small parameter ε consists of describing the properties of the solution when ε tends to zero. It is also important to know that the resulting equation of this problem can be computed with finite elements. Such a tool provides the exact effective property of any arrangement of inclusions with periodic conditions.

It is assumed that the unknown temperature and heat flux fields can be written following the asymptotic expansions:

$$T_\varepsilon(x,y) = T_0(x,y) + \varepsilon T_1(x,y) + \varepsilon^2 T_2(x,y) + O(\varepsilon^2) \qquad [2.39]$$

where T_k is a function at the order k and is assumed to be a function of both space variables and periodic on **y**.

The expression of T_ε is thus introduced in the heterogeneous heat transfer equation to solve (for example, the stationary heat conduction equation), leading to specific equations (no details are given here but the interested readers can refer to [MAT 13, SAN 92]).

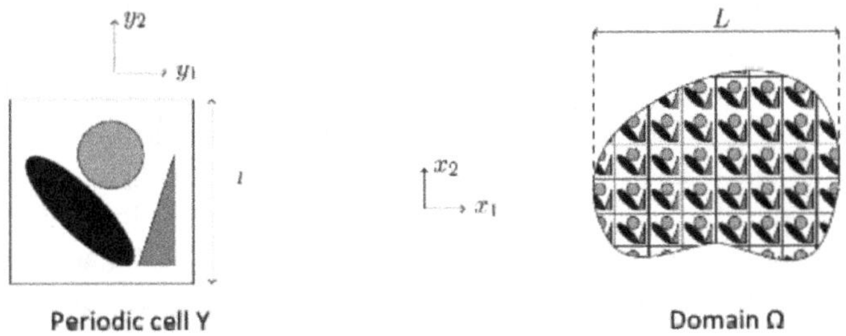

Figure 2.11. *Description of the scales used in asymptotic expansion method*

It is then possible to demonstrate that the mean heat flux density at the order ε=0 can be written:

$$\langle \phi_0 \rangle(\boldsymbol{x}) = \boldsymbol{K}^* \nabla_x T_0(\boldsymbol{x}) \text{in } \Omega \qquad [2.40]$$

With

$$\langle . \rangle = \frac{1}{|Y|} \int_Y . dx$$

where the components of the homogenized (effective) thermal conductivity tensor $\boldsymbol{K}^*$ are given by:

$$K_{ij}^* = \frac{1}{|Y|} \int_Y \left[\boldsymbol{K}(\boldsymbol{y})(\boldsymbol{e}_i + \nabla_y \boldsymbol{w}_i(\boldsymbol{y}))\right]^t \boldsymbol{e}_j d\boldsymbol{y} \qquad [2.41]$$

where w_i are the solution vectors of the elementary problems on the periodic cell Y:

$$div_y\left(\boldsymbol{K}(\boldsymbol{y})\left(\boldsymbol{e}_i + \boldsymbol{\nabla}_y \boldsymbol{w}_i(\boldsymbol{y})\right)\right) = 0 \quad \text{in Y} \qquad [2.42]$$

$$\boldsymbol{w}_i \text{ periodic on } \partial \text{Y} \qquad [2.43]$$

2.3.3.7. *Thermal conductivity estimation from real microstructures*

Although the analytical models are attractive, any attempt to extend their applications to complex geometries such as those we can find in real composites invariably leads to mathematical intractability. However, the accessibility of fast computers has allowed the development of advanced computational methods able to realize calculations on discrete microstructure. The main advantage of these methods is their flexibility since they can be applied to any kind of arrangement, and especially on real composite microstructures.

The natural following question lies in the way the information contained in composite microstructure images is integrated into a finite element computational model. In the literature, two distinct approaches are employed. In the first one, dedicated algorithms were developed so as to build some numerical generic phase arrangements that are statistically equivalent to the real microstructure [TOR 98, ZEM 03]: a statistical analysis of the microstructure is realized, then statistical representations of the real microstructure are numerically generated, and numerical calculations are then performed on them to estimate the effective properties. One of the advantages of this method is that the geometry that supports the finite element simulation can be easily built from basic geometrical object implemented into the used finite element tool. The second approach aims to make calculations directly on real microstructure images (2D or 3D) obtained from serial cuts or tomography. Generally, the selected numerical method is the finite element one.

This introductory paragraph underlines the requirement to consider three steps: (1) to obtain a geometry compatible with finite element calculations, (2) to define a RVE and (3) to choose a homogenization method.

The discretization of the geometry is of paramount importance and clearly influences the estimation quality. Different methods may be considered. A first method consists of applying a threshold of the gray value so as to clearly distinguish the phase (inclusion or matrix) for each pixel (or voxel). Then, finite element meshes from square elements are automatically associated with the corresponding phase. It is important to note that with such an approach, the discretization of the microstructure directly depends on the image quality and resolution.

A second approach is based on the detection of the contours of the inclusions with an adapted image processing tool. Then, each phase is identified and a classic discretization of the geometry can then be performed.

A third approach can be used in the peculiar case of elliptical or circular inclusion shapes. Mathematical morphology tools or algorithms such as the Hough transform can be used for shape detection. Information such as inclusion barycenter positions, aspect ratios or equivalent radii results from the use of such image processing tool, and if they are an important source of (statistical) information on the microstructure, they also enable us to build numerical image or finite element geometry extremely close to the real microstructure.

Admitting the existence of the RVE, the natural following question to be raised is its size. In the literature, quantitative RVE sizes can be found for various materials. It is difficult to compare them since RVE size is related to a given property, and depends on other parameters such as the inclusions volume fraction, the ratio between the inclusion and matrix property (i.e. the thermal contrast in our case) and the desired relative precision for the estimation.

A convergence analysis is a classic method to define both RVE size and effective properties in the same time. The idea is to perform calculations on growing window length and to study the convergence of the computed property. A limit to the convergence, ξ, then has to be defined. The effective property and the associated RVE size are thus estimated for a given ξ. Different effective properties and RVE can therefore be found for different ξ. The main drawback of this method lies in the fact that it may require computation on a large image and is then limited by computational capacity. RVE has to be representative of the whole microstructure. It may be important to check that the computed thermal conductivity is really independent of the location of the RVE. However, generally, such precautions are not taken. More details are given in [THO 08].

On the other hand, computing apparent properties on a small volume element (smaller than the VER) is efficient but leads to local properties instead of an effective property. However, if the number of realizations is high enough, a distribution of the apparent property is obtained and the mean value is relatively close to the value computed on a large image. Means, bounds and scattering of overall response of the inhomogeneous material can therefore be estimated. Note that a good ratio has to be found between the small window size and the number of realizations. Moreover, calculations performed on windows are subjected to some limitations due to the boundary perturbations. These limitations decrease with growing window size.

Once the geometry and the associated mesh are obtained, finite element calculation can be realized. In the literature, two common approaches are used in order to determine the effective properties. A simple and classical method is the unit cell method.

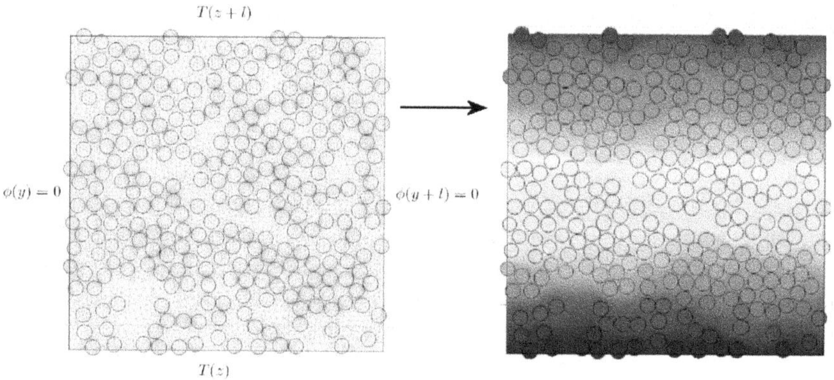

Figure 2.12. *Unit cell method. Example of meshing and associated thermal response of a real matrix-inclusion composite with Dirichlet boundary conditions [THO 08]. For a color version of the figure, see www.iste.co.uk/boyard/heat.zip*

Figure 2.12 illustrates the principle of the unit cell approach. It is defined as a parallelepipedic (or cubic for 3D analyses) area (of size l) representing the phase arrangement. Once the geometry meshed and the thermal conductivity tensors are allocated to each phase of the composite, a unitary transverse thermal gradient is imposed and adiabatic conditions are imposed on the other boundaries. The thermal response of the cell is computed from the solving of the conduction heat transfer equation in the stationary regime. Note that this method is equivalent to the guarded hot plate principle. The resulting heat flux density is integrated on a boundary ($z = 0$ or $z = l$) so as to calculate the apparent transverse thermal conductivity:

$$k^*_{zz} = \frac{\frac{1}{l}\int_0^l \phi(z)dz}{\left(\frac{\partial T}{\partial z}\right)} \quad [2.44]$$

The apparent thermal conductivity along y direction is then obviously obtained by applying the thermal gradient in this direction.

This methodology was successfully applied on an epoxy/UD carbon fiber composite [THO 08]. It was then possible to demonstrate that the RVE for the transverse thermal conductivity is a square with a size L = 300 μm. The authors also highlight that a dispersion of the thermal conductivities in the composite material exists when L is larger than RVE. The distribution functions of the transverse thermal conductivity through the composite thickness ($k_{0°}$) and in the plane ($k_{90°}$) are given in Figure 2.13. The plots highlight that a variation of the thermal conductivities within the range 0.49–0.56 $W.m^{-1}.K^{-1}$and their maxima are shifted from one to the other because of the slight anisotropy of the spatial arrangement of fibers.

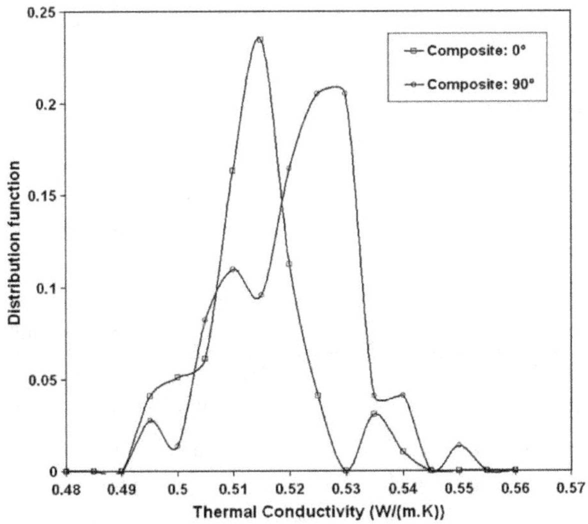

Figure 2.13. *Distribution functions of the transverse thermal conductivities of a UD composite for image sizes higher than RVE [THO 08]*

2.4. Conclusions

The thermophysical properties of polymer and composite materials can be determined from standard characterization devices. However, some of these properties are not easy to measure, depending on the physical state of the material, the temperature to reach as well as the experimental protocol associated with the characterization method. As a first example, we underlined the current difficulty in obtaining experimental data on the specific volume for both thermosets and thermoplastics associated with the right protocol. Secondly, a large part of this chapter described the existing experimental apparatuses to determine the component of the thermal conductivity tensor as well as the lacks for composite characterization, despite recent efforts. Additional developments have to be made in the research laboratories as well as in collaboration with industries to be close to their needs. Different numerical approaches were also presented to predict this important thermal property, starting from analytical models to mathematical homogenization concepts based on idealized or real microstructures.

This chapter is only a snapshot of the experimental and numerical capabilities, being far from exhaustive when we know the huge amount of data available in the literature. However, it can help readers have a better view of what it is possible to do in a standard way as well as of the required future developments.

2.5. Bibliography

[ABO 09] ABOU-MSALLEM Y., JACQUEMIN F., BOYARD N. *et al.*, "Material characterization and residual stresses simulation during the manufacturing process of epoxy matrix composites", *Composites Part A: Applied Science and Manufacturing*, vol. 41, no. 1, pp. 108–115, 2009.

[ANA 10] ANANTHARAMAN A., Analyse mathématique de quelques modèles en calcul de structures électroniques et homogénéisation, PhD Thesis, Paris-Est University, 2010.

[ÅNG 63] ÅNGSTRÖM A.J., "A new method of determining the thermal conductibility of bodies", *Philosophical Magazine Series 4*, vol. 25, no. 166, pp. 130–142, 1863.

[AUR 09] AURIAULT J.-L., BOUTIN C., GEINDREAU C., *Homogenization of Coupled Phenomena in Heterogeneous Media*, ISTE, London and John Wiley & Sons, 2009.

[AUT 05] AUTRIQUE L., SERRA J.J., SCHEER E., "Microscale thermal characterization by inverse method in the frequency domain", *World Congress IFAC*, pp. 64–70, 2005.

[BAI 96] BAILLEUL J.-L., GUYONVARCH G., GARNIER B. *et al.*, "Identification of glass fiber/thermosetting resins composites thermal properties. Application to the optimization of molding processes", *Revue Générale de Thermique*, vol. 35, pp. 65–77, 1996.

[BAK 89] BAKHVALOV N., PANASENKO G., *Homogenization: Averaging Processes in Periodic Media*, Kluwer Press, 1989.

[BEN 78] BENSOUSSAN A., LIONS J.L., PAPANICOLAOU G., *Asymptotic Analysis for Periodic Structures*, North-Holland, 1978.

[BOY 03] BOYARD N., VAYER M., SINTUREL C. *et al.*, "Analysis and modelling of PVTX diagram of an unsatured polyester, thermoplastic additive and mineral fillers blend", *Journal of Applied Polymer Science*, vol. 88, pp. 1258–1267, 2003.

[BOY 04] BOYARD N., VAYER M., SINTUREL C. *et al.*, "ModellingPvTx diagrams: application to various blends based on unsaturated polyester-influence of thermoplastic additive, fillers and reinforcements", *Journal of Applied Polymer Science*, vol. 92, pp. 2976–2988, 2004.

[BUD 65] BUDIANSKY B., "On the elastic moduli of some heterogeneous materials", *Journal of the Mechanics and Physics of Solids*, vol. 13, pp. 223–227, 1965.

[CAP 63] CAPE J.A., LEHMAN G.W., "Temperature and finite-pulse-time effects in the flash method for measuring thermal diffusivity", *Journal of Applied Physics*, vol. 34, no. 7, pp. 1909–1913, 1963.

[CHA 81] CHARLES J.A., WILSON D.W., "A model of passive thermal nondestructive evaluation of composite laminates", *Polymer Composites*, vol. 2, no. 3, pp. 105–111, 1981.

[CHA 02] CHAKRAVORTY S., "PVT testing of polymers under industrial processing conditions", *Polymer Testing*, vol. 21, no. 3, pp. 313–317, 2002.

[DEG 86] DEGIOVANNI A., LAURENT M., "Une nouvelle méthode d'identification de la diffusivité thermique par la méthode flash", *Revue de Physique Appliquée*, vol. 21, pp. 229–237, 1986.

[DEG 94] DEGIOVANNI A., "Conductivité et diffusivité thermique des solides", *Techniques de l'ingénieur*, vol. R2850, pp. 1–15, 1994.

[DEG 96] DEGIOVANNI A., BATSALE J.C., MAILLET D., "Mesure de la diffusivité longitudinale de matériaux anisotropes", *Revue Générale de Thermique*, vol. 35, pp. 141–147, 1996.

[DEM 97] DEMANGE D., BEAUCHIKE P., BEJET M. *et al.*, "Mesure simultanée de la diffusivité thermique selon les deux directions principales d'un matériau", *Revue Générale de Thermique*, vol. 36, pp. 755–770, 1997.

[DON 75] DONALDSON A.B., TAYLOR R.E., "Thermal diffusivity measurement by a radial heat flow method", *Journal of Applied Physics*, vol. 46, pp. 4584–4589, 1975.

[DYM 88] DYMOND J.H., MALHOTRA R., "The Tait equation: 100 years on", *International Journal of Thermophysics*, vol. 9, no. 6, pp. 941–951, 1988.

[ELG 06] EL GANAOUI K., Homogénéisation de modèles de transferts thermiques et radiatifs: application au coeur des réacteurs à caloporteur gaz, PhD Thesis, Ecole Polytechnique, 2006.

[ESH 57] ESHELBY J.D., "The determination of the elastic field of an ellipsoidal inclusion", *Proceedings of the Royal Society of London*, vol. A241, pp. 376–396, 1957.

[FOR 09a] FORSTNER R., PETERS G.W.M., MEIJER H.E.H., "A novel dilatometer for PVT measurements of polymers at high cooling – and shear rates", *International Polymer Processing*, vol. 24, no. 2, pp. 114–121, 2009.

[FOR 09b] FORSTNER R., PETERS G.W.M., RENDINA C. *et al.*, "Volumetric rheology of polymers", *Journal of Thermal Analysis and Calorimetry*, vol. 98, no. 3, pp. 683–691, 2009.

[GOB 04] GOBBE C., ISERNA S., LADEVIE B., "Hot strip method: application to thermal characterisation of orthotropic media", *International Journal of Thermal Sciences*, vol. 43, pp. 951–958, 2004.

[GOY 04] GOYANES S., SALGUEIRO W., SOMOZA A. *et al.*, "Direct relationships between volume variations at macro and nanoscaleinepoxy systems. PALS/PVT measurements", *Polymer*, vol. 45, pp. 6691–6697, 2004.

[GRI 99] GRIESINGER A., HEIDEMARM W., HALME E., "Investigation on measurement accuracy of the periodic hot-wire method by means of numerical temperature field calculations", *International Communications in Heat and Mass Transfer*, vol. 26, no. 4, pp. 451–465, 1999.

[GUS 91] GUSTAFSSON S.E., "Transient plane source techniques for thermal conductivity and thermal diffusivity measurements of solid materials", *Review of Scientific Instruments,* vol. 62, no. 3, pp. 797–804. 1991.

[HAL 67] HALPIN J.C., TSAI S.W., Environmental factors in composite design, Air Force Materials Laboratory, Technical Report, pp. 67–423, 1967.

[HAS 87] HASSELMAN J., "Effective thermal conductivity of composites with interfacial thermal barrier resistance", *Journal of Composite Materials*, vol. 21, pp. 508–515, 1987.

[HAT 85] HATTA H., TAYA M., "Effective thermal conductivity of a misoriented short fiber composite", *Journal of Applied Physics*, vol. 58, no. 8, pp. 2478–2486, 1985.

[HIE 02] HIEBER C.A., "Modeling/simulating the injection molding of isotactic polypropylene", *Polymer Engineering and Science*, vol. 42, no. 7, pp. 1387–1409, 2002.

[HIL 65] HILL R., "A self-consistent mechanics of composite material", *Journal of the Mechanics and Physics of Solids*, vol. 13, pp. 213–222, 1965.

[JAR 01] JARNY Y., GUILLEMET P., "Estimation simultanée de la conductivité thermique et de la chaleur spécifique de matériaux orthotropes", *French National Congress On Thermal Science (SFT Congress)*, 2001.

[JAR 03] JARNY Y., "The adjoint method to compute the numerical solutions of inverse problems", in WOODBURY K.A. (ed.), *Inverse Engineering Handbook*, CRC Press LLC, pp. 103–218, 2003.

[KRA 04] KRAPEZ J.C., SPAGNOLO L., FRIE M. *et al.*, "Measurement of in-plane diffusivity in non-homogeneous slabs by applying flash thermography", *International Journal of Thermal Sciences,* vol. 43, pp. 967–977, 2004.

[KUL 97] KULKARNI M.R., BRADY R.P., "A model of global thermal conductivity in laminatedcarbon/carbon composites", *Composites Science and Technology,* vol. 57, pp. 277–285, 1997.

[LAC 91] LACHI M., DEGIOVANNI A., "Détermination des diffusivités thermiques des matériaux anisotropes par méthode flash bidirectionnelle", *Journal de Physique III*, vol. 1, pp. 2027–2046, 1991.

[LEB 08] LE BOT C., DELAUNAY D., "Rapid solidification of indium: modelling subcooling", *Materials Characterization*, vol. 59, pp. 519–527, 2008.

[LEG 11] LE GOFF R., DELAUNAY D., BOYARD N. *et al.*, "Thermal conductivity of an injected polymer and short glass fibers composite part: measurement and model", *World Congress of the Polymer Processing Society*, 2011.

[LOR 07] LORETTE C., Outils de caractérisation thermophysique et modèles numériques pour les composites thermostructuraux, PhD Thesis, University of Bordeaux, 2007.

[LUY 00] LUYE J.-F., REGNIER G., LE BOT P. *et al.*, "PVT measurement methodology for semicrystalline polymers to simulate injection-molding process", *Journal of Applied Polymer Science*, vol. 79, pp. 302–311, 2000.

[MAT 13] MATINE A., BOYARD N., CK P. *et al.*, "Modeling of thermophysical properties in heterogeneous periodic media according to a multi-scale approach: effective conductivity tensor and edge effects", *International Journal of Heat and Mass Transfer*, vol. 62, pp. 586–603. 2013.

[MAX 73] MAXWELL J.C., *Treatise on Electricity and Magnetism*, Clarendon Press, 1873.

[MEN 09] MENCZEL J.D., JUDOVITS L., PRIME R.B. *et al.*, "Differential scanning calorimetry (DSC)", in MENCZEL J.D., PRIME R.B. (eds), *Thermal Analysis of Polymers, Fundamentals and Applications*, Wiley and Sons, pp. 7–239, 2009.

[MIL 01] MILLISCHER A., DELAUNAY D., JARNY Y., "Thermomechanical coupling in BMC injection process : characterization, modelization and experimental validation", *3rd Canadian International Conference on Composites*, 2001.

[MOR 73] MORI T., TANAKA K., "Average stress in matrix and average elastic energy of material with misfitting inclusions", *Acta Metallurgica*, vol. 21, pp. 571–574, 1973.

[MUL 07] MULLER Y., HAUSSLER L., PIONTECK J., "ABS-Modified epoxy resins-curing kinetics, polymerization induced phase separation, and resulting morphologies", *Macromolecular Symposium*, vol. 254, pp. 267–273, 2007.

[NAW 12a] NAWAB Y., Characterization and modelling of cure dependent properties of thermoset composites – application to the simulation of residual stresses, PhD Thesis, University of Nantes, 2012.

[NAW 12b] NAWAB Y., TARDIF X., BOYARD N. *et al.*, "Determination and modelling of the cure shrinkage of epoxy vinylester resin and associated composites by considering thermal gradients", *Composites Science and Technology*, vol. 73, pp. 81–87, 2012.

[NAW 14] NAWAB Y., BOYARD N., JACQUEMIN F., "Effect of pressure and reinforcement type on thevolume chemical shrinkage in thermoset resin and composite", *Journal of Composite Materials*, vol. 48, pp. 3191–3199, 2014.

[PAR 61] PARKER W.J., JENKINS R.J., BUTLER C.P. *et al.*, "Flash method of determining thermal diffusivity, heatcapacity and thermal conductivity", *Journal of Applied Physics*, vol. 32, no. 9, pp. 1679–1684, 1961.

[PHE 86] PHEDRAN R.C., "Transport properties of cylinder pairs and of the square array of cylinders", *Proceedings of the Royal Society of London*, vol. A 408, pp. 31–43, 1986.

[PIG 15] PIGNON B., TARDIF X., LEFEVRE N. *et al.*, "A new PvT device for high performance thermoplastics: heat transfer analysis and crystallization kinetics identification", *Polymer Testing*, vol. 45, pp. 152–160, 2015.

[PLA 04] PLANA V., Caractérisation par méthode inverse et modélisation des propriétés thermophysiques orthotropes des matériaux composites, PhD Thesis, SUPAERO – National Higher School of Aeronautics and Space, 2004.

[PRA 95] PRADERE C., GOYHENECHE J.M., BATSALE J.C. *et al.*, "Thermal diffusivity measurements on a single fiber with microscale diameter at very high temperature", *International Journal of Thermal Sciences*, vol. 45, pp. 443–451, 1995.

[PRO 76] PROGELHOF R.C., THRONE J.L., RUETSCH R.R., "Methods for predicting the thermal conductivity of composite systems: a review", *Polymer Engineering and Science*, vol. 16, no. 9, pp. 615–625, 1976.

[RAM 05] RAMOS J.A., PAGANI N., RICCARDI C.C. *et al.*, "Cure kinetics and shrinkage model for epoxy-amine systems", *Polymer*, vol. 46, pp. 3323–3328, 2005.

[RID 09] RIDES M., MORIKAWA J., HALLDAHL L. *et al.*, "Intercomparison of thermal conductivity and thermal diffusivity methods for plastics", *Polymer Testing*, vol. 28, pp. 480–489, 2009.

[ROC 05] ROCHAIS D., HOUEDEC H., ENGUEHARD F. *et al.*, "Microscale thermal characterization at temperatures up to 1000°C by photoreflectance microscopy. Application to the characterization of carbon fibres", *Journal of Physics D: Applied Physics*, vol. 38, pp. 1498–1503, 2005.

[SAN 92] SANCHEZ-HUBERT J., SANCHEZ-PALENCIA E.J., *Introduction aux méthodes asymptotiques et à l'homognénéisation*, Masson, 1992.

[SER 05] SERRA J.J., AUTRIQUE L., "Microscale thermal characterization of reinforced composites by photothermal microscopy data inversion", *5th International Conference on Inverse Problems in Engineering: Theory and Practice*, 2005.

[SPE 50] SPENCER R.S., GILMORE G.D., "Equation of state for high polymers", *Journal of Applied Physics*, vol. 21, pp. 523–526, 1950.

[STA 91] STANDARD ISO 8302:1991, Isolation thermique – détermination de la résistance thermique et des propriétés connexes en régime stationnaire – méthode de la plaque chaude gardée, 1991.

[TAI 88] TAIT P.G., "Physics and chemistry of the voyage of H.M.S challenger", *HMSO*, vol. 2, no. 4, 1888.

[TAM 95] TAMMANN G., "Über die Abhängigkeit der Volumina von Lösungen vom Druck", *Zeitschrift für Physikalische Chemie*, vol. 17, pp. 620–636, 1895.

[TAR 12a] TARDIF X., Cristallisation et transferts thermiques dans un polymère thermoplastique semi-cristallin en refroidissement rapide sous pression, PhD Thesis, University of Nantes, 2012.

[TAR 12b] TARDIF X., AGAZZI A., SOBOTKA V. *et al.*, "A multifunctional device to determine specific volume, thermal conductivity and crystallization kinetics of semi-crystalline polymers", *Polymer Testing*, vol. 31, no. 6, pp. 819–827, 2012.

[THO 08a] THOMAS M., Propriétés thermiques de matériaux composites: caractérisation expérimentale et approche microstructurale, PhD Thesis, University of Nantes, 2008.

[THO 08b] THOMAS M., BOYARD N., PEREZ L. *et al.*, "Representative volume element of anisotropic unidirectional carbon – epoxy composite with high fibre volume fraction", *Composites Science and Technology*, vol. 68, nos. 15–16, pp. 3184–3192, 2008.

[THO 10] THOMAS M., BOYARD N., LEFEVRE N. *et al.*, "An experimental device for the simultaneous estimation of the thermal conductivity 3-D tensor and the specific heat of orthotropic composite materials", *International Journal of Heat and Mass Transfer*, vol. 53, no. 23–24, pp. 5487–5498, 2010.

[TOR 98] TORQUATO S., "Morphology and effective properties of disordered heterogeneous media", *International Journal of Solids and Structures*, vol. 35, pp. 2385–2406, 1998.

[VAN 05] VAN DER BEEK M.H.E., PETERS G.W.M., MEIJER H.E.H., "The influence of cooling rate on the specific volume of isotactic poly(propylene) at elevated pressures", *Macromolecular Materials and Engineering*, vol. 290, no. 5, pp. 443–455, 2005.

[VAN 06] VAN DER BEEK M.H.E., PETERS G.W.M., MEIJER H.E.H., "Influence of shear flow on the specific volume and the crystalline morphology of isotactic polypropylene", *Macromolecules.*, vol. 39, pp. 1805–1814, 2006.

[VIL 12] VILLIERE M., LECOINTE D., SOBOTKA V. *et al.*, "Experimental determination and modeling of thermal conductivity tensor of carbon/epoxy composite", *Composites Part A: Applied Science and Manufacturing*, vol. 46, pp. 60–68, 2012.

[WAT 03] WATANABE K., SUZUKI T., MASUBUCHI Y. *et al.*, "Crystallization kinetics of polypropylene under high pressure and steady shear flow", *Polymer*, vol. 44, no. 19, pp. 5843–5849, 2003.

[WON 01] WON J.C., FULCHIRON R., DOUILLARD A. *et al.*, "Effect of the pressure on the crystallization behavior of polyamide", *Journal of Applied Polymer Science*, vol. 80, pp. 1021–1029, 2001.

[ZEM 03] ZEMAN J., Analysis of composite materials with random microstructure, PhD Thesis, Czech Technical University, 2003.

[ZOL 76] ZOLLER P., BOLLI P., PAHUD V. *et al.*, "Apparatus for measuring pressure– volume–temperature relationships of polymers to 350°C and 2200 kg/cm^2", *Review of Scientific Instruments*, vol. 47, no. 8, pp. 948–952, 1976.

[ZOL 89] ZOLLER P., KEHL T., STARKWEATHER H.W. *et al.*, "The equation of state and heat of fusion of Poly(ether ether ketone)", *Journal of Polymer Science Part B: Polymer Physics*, vol. 27, no. 5, pp. 997–1007, 1989.

3

Experimental Determination and Modeling of Transformation Kinetics

This chapter reviews the state-of-the-art of characterization techniques and the main modeling approaches of polymer transformation kinetics in static conditions, i.e. cure kinetics of thermosetting polymers and crystallization kinetics of thermoplastic semi-crystalline polymers.

In section 3.2, we briefly introduce the most suitable and widely used techniques for monitoring the progress of such transformations. For thermoplastics, the limited achievable cooling rates represent a big issue. Nanocalorimetry, a promising technique for measurements at high cooling rates, is also addressed.

Section 3.3 is dedicated to the modeling of cure kinetics of thermosetting resins following either phenomenological or mechanistic approaches. The most popular models to describe autocatalytic and mixing-activated cure behaviors of several classes of thermoset resins (epoxy, polyester and polyurethanes) are reviewed and discussed, including the possible effect of glass transition.

Crystallization kinetics of thermoplastic polymers is presented in section 3.4. After a brief reminder about the basics of polymer crystallization, the models which arise from classical Avrami and Evans theories, completed by the isokinetic Nakamura model for non-isothermal crystallization analysis, are presented. Approaches using differential equation system, such as Schneider rate equations, are also discussed since they represent an alternative approach to describe the evolution of the relative crystallinity. Specific models, able to take into account specific phenomena (crystallization in confined space, transcrystallinity and secondary crystallization) are also listed.

Chapter written by Nicolas BOYARD, Jean-Luc BAILLEUL and M'hamed BOUTAOUS.

3.1. Introduction

The solidification of polymer-based composite materials is an important step during their manufacturing and depends on the chemical nature of the resin, which involves distinct mechanisms and thus specific kinetic models.

Thermosetting resins (unsaturated polyesters, epoxies, vinylesters, polyurethanes, etc.) are composed of monomers or short polymer chains with reactive groups, which crosslink by heating in the presence of initiator and/or catalyzer. The chemical reaction is irreversible and leads to the formation of a three-dimensional (3D) amorphous network formed by covalent bonds. Depending on the resin, reaction mechanisms (free radical polymerization, stepwise polymerization, etc.) are different and complex with many elementary reactions and possible secondary or parasite reactions. Composite forming processes use generally curing cycles constituted with heating ramps and isothermal steps.

Thermoplastics are constituted with long polymer chains and can reach a fluid state by heating. These materials can be either amorphous (no order at large scale, flow will occur during heating, beyond the glass transition temperature, no melting) or semi-crystalline, for which melting point has to be reached to be in the fluid state. In the latter case, the solidification (reversible, out of equilibrium) occurs during the cooling, according to a mechanism of nucleation of growth of semi-crystalline microstructures. Classical processes (injection, extrusion and thermoplastic resin transfer molding (RTM)) are mostly characterized by high cooling rates.

Design of composite parts and molds, optimization of thermal cycle, modeling to obtain target properties (homogeneous degree of cure, mechanical properties, etc.) are among the main concerns of the composite material manufacturing. All these topics require a complete characterization of composite thermal–physical properties and the mathematical description of the phase change kinetics of polymer matrix. As a result, we can also expect to predict residual stresses, thermal history and the field of conversion degree in the molded part during the process cycle.

The modeling of a reaction rate introduces the question of the most appropriate kinetic model compared to what we expect. For a thermosetting resin, a chemist emphasizes a mechanistic approach for an accurate description of the kinetics, whereas an engineer will prefer to use a phenomenological model to save time computations, but still needs reliable results. For semi-crystalline thermoplastics, global models of crystallization are used and they are based on a microscopic approach (nucleation and growth mechanism).

This chapter aims to make an overview of the experimental approaches to quantify the reaction rate and the most popular equations to model the phase change kinetics, considering only a temperature dependence. Their limits of use and specific problems are also addressed.

3.2. What are the most suitable devices to analyze a reaction rate?

Many devices exist to determine the reaction rate for both thermoset and thermoplastic resins. All of these apparatuses record a physical or chemical phenomenon related to the transformation. Thus, they can measure:

– a thermal effect with calorimeter;

– mechanical properties evolution in fluid and solid state (viscosity and viscoelasticity);

– radiative properties evolution during transformation by spectroscopy or X-ray visualization;

– electrical properties.

For the crystallization of semi-crystalline thermoplastics, it is possible to follow the growth of semi-crystalline entities.

The abovementioned technologies have been developed for many years and commercial devices exist for each one. However, these apparatuses are mainly developed to study the neat resin only (i.e. without reinforcement) and the characterization level is often far enough from the industrial processing conditions. As such, the question is raised about the influence of fibers and process conditions on the modeled kinetic obtained from these devices.

In this section, we will first discuss the classical methods to determine transformation kinetics and their limitations. In a second part, we will discuss original methods developed to obtain required properties as close as possible to industrial conditions. To conclude, a discussion will present the reinforcement influences on kinetics models.

3.2.1. *Conventional methods*

In this section, we will first present two general methods for thermosets and thermoplastics resins. Then, we will focus on peculiar methods depending on the nature of the transformation (chemical for thermosetting polymers and physical for thermoplastic polymers).

3.2.1.1. *General methods*

3.2.1.1.1. Devices based on thermal measurement

Crystallization of semi-crystalline polymers and the crosslinking of thermosetting resins are characterized by an exothermal enthalpy released during these transformations. Recorded temperature variations and/or heat flux versus time (and temperature) are used to determine transformation kinetics if we assume that the released enthalpy is directly proportional to the chemical or physical transformation rate.

Differential thermal analysis (DTA) and differential scanning calorimetry (DSC) are the two main approaches involving heat transfer measurements. For both devices, a sample (about 10 mg) is placed inside a furnace with an imposed temperature cycle composed of isothermal steps, heating and cooling ramps. With the DTA, the difference of temperature between the sample and a reference is plotted versus of the programmed temperature. The general principle of DSC consists of heating a sample with a linear temperature ramp and in measuring the heat transfer in this sample relative to a reference. The dimension of the sample must be small enough to assume a uniform temperature at each instant, but large enough so that its properties are representative of the homogeneous material (see Chapter 2, section 2.2).

Such approaches have existed for a long time. For example, Kissinger [KIS 56] observed the displacement of the temperature peak with the cooling rate. A differential thermal analysis (DTA) device is used [KOT 75, GIL 74, CRI 14] to study thermoplastic crystallization and thermoset polymerization. Melting and crystallization thermograms were obtained in a high pressure apparatus by DTA which enabled us to obtain thermograms at pressures of up to 8,000 kg/cm^2 over the range of heating and cooling rates of 1–30°C/min. Figure 3.1 presents an example of data, which are analyzed to determine the pressure influence on the crystallization of polyethylene films from 2,000 to 7,000 kg/cm^2 [KOT 75]. Cycles 1 and 3 involve melting under pressure while heating the sample to a temperature 20–30°C higher than the melting point. Cycle 2 represents crystallization after cycle 1. Cycle 4 involves melting at atmospheric pressure. Increasing the pressure, two peaks appear on the thermograms and the melting temperature increases from 180 to 270°C.

Another way to obtain information by thermal analysis is to carry out experiments with a differential scanning calorimeter (DSC). Typical DSC curves are presented in Figure 3.2(a) for non-isothermal measurement [JEZ 78] and in Figure 3.2(b) for isothermal measurement [HAY 76]. We observe an exothermal signal due to the crystallization on these two figures. Time integration of the surface between the baseline and the curves represents the transformation enthalpy. The conversion degree at a given time t is determined by dividing the phase change enthalpy at this time by the total amount of energy released during the transformation.

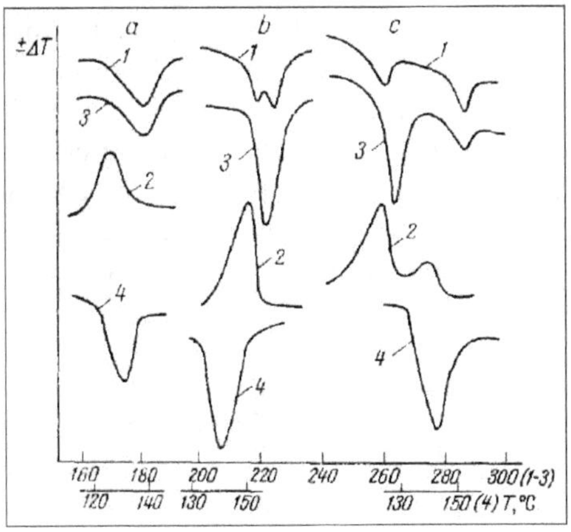

Figure 3.1. *Thermograms of melting of isotropic PE films at pressure of 2,000 (a), 4,000 (b) and 7,000 kg/cm² (c) [KOT 75]*

a)

b)

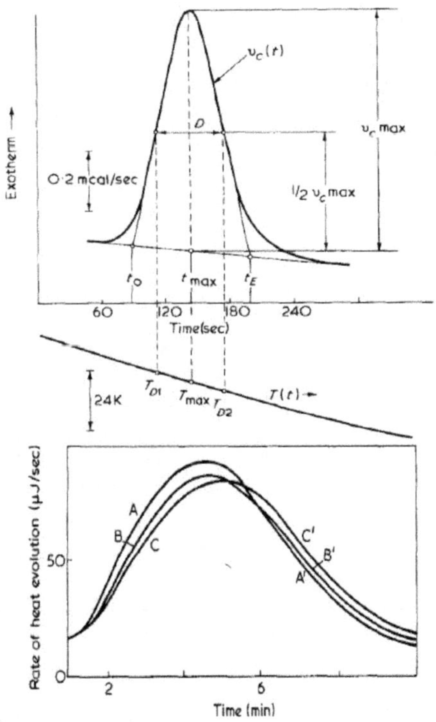

Figure 3.2. *a) Typical DSC thermograms of non-isothermal crystallization of polyethylene terephthalate [JEZ 78]; b) Temperature dependence of crystallization isotherms AA': 396.1K; BB': 396.0K; CC': 395.9K [HAY 76]*

An advantage of these methods is the use of small samples, allowing for fast heating and cooling. However, the phase change kinetics can only be measured in a limited range of temperature in the case of crystallization. To extend this range, a new device is developed and is called a flash scanning calorimeter (FSC) due to its high cooling and heating rates (up to 10,000 K/s for cooling and 20,000 K/s for heating) allowed by the sample size (from few tens to hundreds of nanograms). With such an apparatus, it becomes possible to identify the kinetics close enough industrial conditions [TAR 14].

Finally, it is important to note that it is not possible to measure the transformation kinetics in the presence of fibers, due to the small sample size. This point is important since fibers can influence the kinetics, especially in polymer crystallization. Moreover, surface effects can appear in tiny samples, as it is already known in thin polymer films, modifying the kinetics compared to bulk specimens.

3.2.1.1.2. Mechanical methods

Dynamic mechanical analysis (DMA) and dynamic mechanical thermal analysis (DMTA) are also conventional methods to measure the mechanical properties evolution versus time or temperature during the matrix transformation on rather small samples (few mm). This measurement of the viscoelasticity is thus well suited for polymer mechanical characterization. As mechanical properties evolve significantly during transformation, it could be used to determine transformation kinetics. Hence, the measurements of storage modulus and tanδ can be used (Figure 3.3).

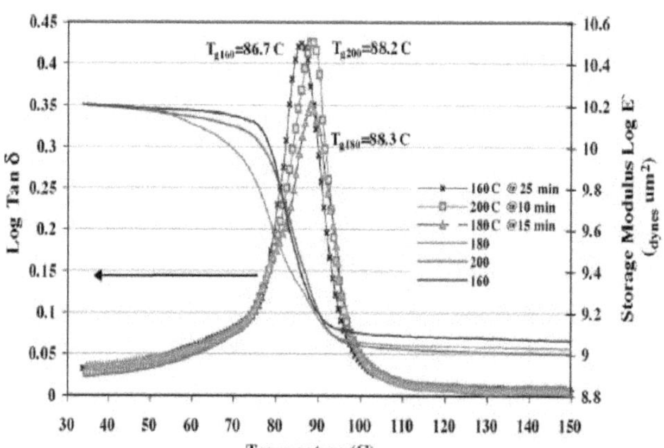

Figure 3.3. *DMTA graphs of one polyester resin [MAF 05]. For a color version of the figure, see www.iste.co.uk/boyard/heat.zip*

For example, Martin *et al.* [MAR 00] present some results obtained by rheometry (thermal scanning rheometry (TSR)) and mechanical modulus measurement (storage and loss modulus and tanδ) for a vinylester resin. These measurements can be held for different temperature cycles and for different frequency solicitations to identify the conversion degree (see Figure 3.3 for a polyester resin [MAF 05]). The rapid evolution of the storage modulus is due to the transformation, like tanδ evolution. Therefore, their analyses lead to the rate of reaction.

For thermoplastic resins, the readers can refer to the results of Hsu [HSU 86] and Long [LON 95] for examples. Like for thermal devices, samples are small and we measure the mechanical properties evolution to determine the transformation kinetics. Thus, for composites, we must use a sample with a representative size to obtain reliable results.

3.2.1.1.3. Dielectric method

During the transformation of the thermoset resins, the chemical structure is modified due to the increase in the molecular weights. During the crystallization of the thermoplastics, its structure is modified due to the growth of semi-crystalline entities. As a result, their dielectric properties are modified [SCH 89].

Kortaberia [KOR 06] and more recently Pissis [PIS 15] determined the reaction kinetic of an epoxy resin (isothermal curing between 150 and 250°C) by measurement of the conductance and capacitance during the transformation. The same method is used by Pathmanathan [PAT 93] for a thermoplastic resin in isothermal condition and within the temperature range 100–177°C.

3.2.1.1.4. Infrared Fourier transform and Raman spectroscopy (FTIR and FT-Raman)

Optical properties of a polymer depend on their chemical and physical structures, which can be analyzed by FTIR and Raman spectroscopies. It is possible to link the conversion degree associated with a phase change to band shifts of the FTIR or Raman spectra recorded during a temperature cycle.

For example, Morgan [MOR 87] and Wang [WAN 12] used the FTIR method for thermosets resin. Rocks [ROC 04] used the FT-Raman method for thermoset resin too.

Figure 3.4 provides an example from Zhang [ZHA 04] of spectra evolution during the crystallization of isotactic polystyrene.

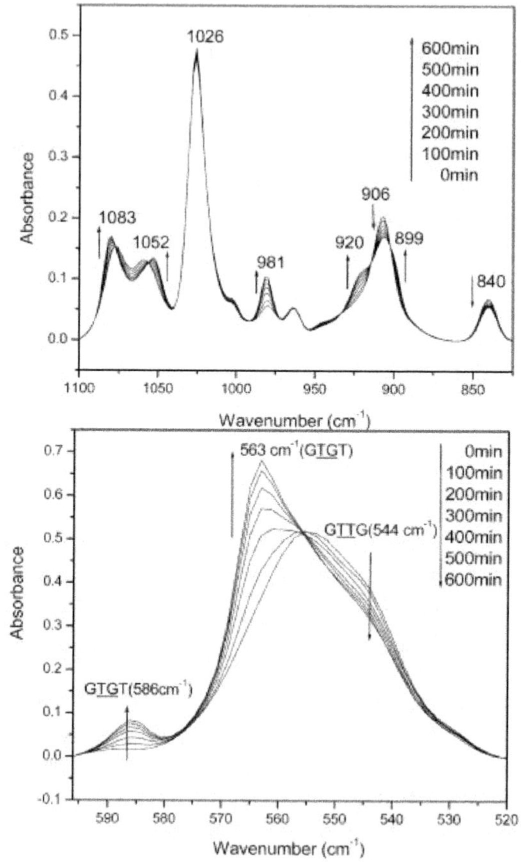

Figure 3.4. *Temporal changes of IR spectrum in the wavenumber ranges of 800–1,100 cm^{-1} and 600–520 cm^{-1} during the crystallization of iPS at 130°C [ZHA 04]*

3.2.1.2. *Particular methods*

3.2.1.2.1. High pressure liquid chromatography (HPLC): gas chromatography mass spectroscopy (GCMS) for thermosets resin

This method is based on a molecular concentration measurement and is mainly dedicated to thermoset resins. For relatively simple thermoset resin formulation, it is possible to predict a pathway of transformation and therefore to determine reaction kinetics [MIJ 92, VER 90]. However, this method needs to know the initial chemical formulation of the resin.

3.2.1.2.2. Polarized optical microscopy (thermoplastic resin)

Historically, polarized optical microscopy was used to observe the phenomenon of crystal germination and growth. Due to the difference of optical properties between crystalline and amorphous phases and the birefringence of the crystalline lamellae, it is possible to follow, with polarized light, the nucleation and growth of semi-crystalline entities [MAG 62, BAN 63, KEI 64]. This method requires a hot stage with a cooling system for very thin samples. We present in Figure 3.5 the spherulite growth of polyethylene.

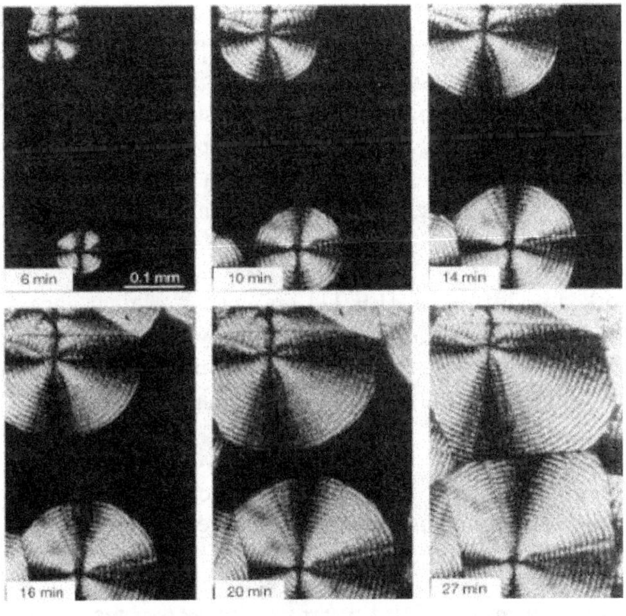

Figure 3.5. *Polyethylene spherulite growth at different times using polarized light optical microscopy [BAR 82]*

Nakamura [NAK 73] developed a sample holder (equipped with heating system) to measure the evolution of the sample crystallinity from WAXS method which was also used in many other works [HSU 86, GUP 92, JON 97] for thermoplastics and blends. SAXS experiments were also carried out (see, for example, [REN 04]) to study the evolution of the microstructure. Like previous methods, these characterization techniques necessitate small resin sample without fibers.

3.2.1.2.3. Ultrasound method

In 1971, Grechishkin [GRE 71] demonstrated the influence of the crystalline structure on the velocity of ultrasound. Thus, a few authors used this way to measure the kinetics of transformation [ADA 81, ALI 98, MAL 11].

3.2.2. *Original methods*

All previous methods are easily accessible using commercial devices. However, one issue is that experimental conditions such as temperature are often far to be representative from actual ones. To answer this important need, specific devices are developed in research laboratories. Brucato *et al.* [BRU 91] built an apparatus for fast cooling to analyze the associated morphologies and phase structures. They observed γ-crystalline form, whereas γ-crystalline structure was found dominant for lower cooling rate. Following the same approach, Lamberti *et al.* [LAM 04] proposed their own device to measure the evolution of the relative crystallinity from the interactions between the light and a crystallizing polymer during fast cooling, closer to industrial conditions.

In their work, Le Goff *et al.* [LE 11] present an original apparatus for a short fiber reinforced composite, with the aim to identify the kinetic function of a semi-crystalline polymer using an inverse estimation method. Another interest of this study is to work on an industrial injection mold equipped with several sensors.

Other recent works [PIG 15, TAR 12] develop a new pressure, volume, temperature (PVT) device and specific protocols to identify crystallization kinetics for high-performance thermoplastics in a large range of pressure.

3.2.3. *A first assessment of the current characterization methods*

Many polymer properties can be used to measure the reaction kinetics or the evolution of the conversion degree. Another interesting point is that several process parameters (temperature, pressure, mechanical stresses, etc.) also modify the transformation kinetics.

As an example, some authors [ITO 95, CUO 11] present the influence of the pressure on relative crystallinity determined from PvT measurements, as illustrated in Figure 3.6. They concluded that pressure accelerates the crystallization rate. As a result, the kinetics identified by DSC must be modified to take into account the pressure effect. In the same way, Sowinski [SOW 15] highlights large crystalline structure modifications.

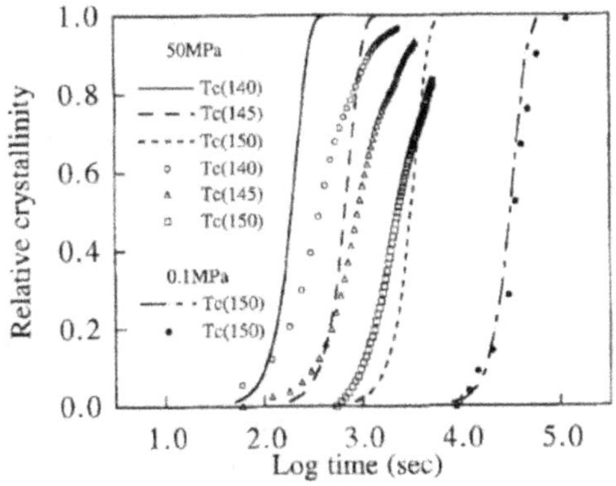

Figure 3.6. *Experimental and calculated crystallinities versus time under 0.1 and 50 MPa [TO 95]*

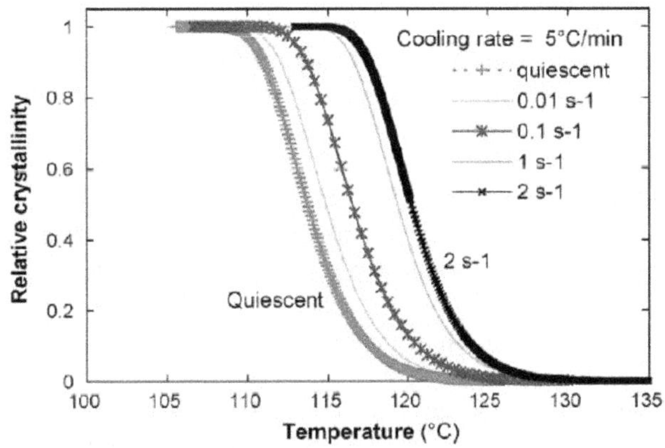

Figure 3.7. *Effect of the shear rate on the crystallization (cooling 5°C/min) [BOU 10]. For a color version of the figure, see www.iste.co.uk/boyard/heat.zip*

Boutaous [BOU 10] also studied the influence of shear on the crystallization rate (Figure 3.7) in the case of a Couette shear flow of a polypropylene between two co-axial cylinders. The results show the enhancement of the crystallization kinetics due to the shearing. Similar phenomenon was observed before by Ishizuka [ISH 77] and Patel [PAT 91] due to polymer chain orientation.

Fibers in thermoplastic composites can influence the transformation kinetics [DUR 15]. Indeed, besides a shearing increase, the interfaces (sizing) between the semi-crystalline polymer and fibers can locally modify the transformation (transcrystallinity, for example). Thus, for complex composite structure, the transformation mechanisms are potentially different inside fibers tows and between them.

3.3. Modeling of the cure kinetics of thermosetting resins

Thermosets are characterized by an exothermal irreversible chemical reaction (crosslinking) during a thermal cycle (typically composed of heating ramps and/or isothermal steps). Chemically speaking, prepolymer chains react with monomer and/or reactive species (hardeners) and other growing chains (according to the type of polymerization mechanism), leading to a continuous increase in the molecular weight and the viscosity. The crosslinking progressively leads to the formation of a network, which finally percolates throughout the system. This critical transition, called gelation, is an important characteristic of thermoset resins since it defines the change from a liquid-like state to a solid-like behavior defined by a gel time t_{gel} and the associated degree of cure α_{gel}. This last parameter is a constant for a thermoset resin.

As the reaction is taking place, the amorphous network of crosslinked resin is densified. The reaction rate increases up to a maximum before decreasing for several reasons: lower concentrations of reactants, diffusion-controlled regime (elementary steps are affected by the size of the reactive species, specifically in chainwise polymerizations) and vitrification of the system.

If gelation does not change the reaction kinetics, vitrification is another physical phenomenon which leads to an overall diffusion control, whatever the polymerization mechanism, since the segmental movements of macromolecules are largely restricted. In this case, the glass transition temperature (Tg) is close or equal to the imposed temperature. Tg defines the limit between the rubbery state (T>Tg) and the glassy state (T<Tg), where the cooperative motions of large polymer segments are not possible anymore and the relaxation times are very long. Under these conditions, polymerization kinetics is strongly reduced and tends to stop. This point will be discussed specifically in section 2.3.3.

The goal of this section is to present the main kinetic models developed to predict the extent of reaction with a focus on the empirical models, which are very useful and easy to use from an engineering point of view.

3.3.1. *Mechanistic models: complexity versus accuracy*

These models enable us to calculate the evolution of the concentrations of reactants and products from a set of equations, which describe the elementary reaction paths involved in the crosslinking mechanism. Such kind of rigorous models is thus quite accurate, provided that all chemical reactions are taken into account. This is often a tricky task since secondary and parasite reactions generally exist. Moreover, all rate constants of each elementary reaction have to be determined.

Mechanistic models for free-radical polymerization such as unsaturated polyester/styrene reaction were studied by Stevenson [STE 86], Ng *et al.* [NG 89] and improved by Ramis and Salla [RAM 99]. Such resin is mainly used in sheet molding compound (SMC) and bulk molding compound (BMC) composite materials and also sometimes in resin transfer molding (RTM) process.

	Reaction	Kinetics
Initiation	$I \rightarrow 2R^{\bullet}$	$\frac{\partial I}{\partial t} = -k_d I$
Inhibition	$R^{\bullet} + Z \rightarrow Z^{\bullet}$	$\frac{\partial Z}{\partial t} = -k_z Z R^{\cdot}$
Propagation	$R_n^{\cdot} + M \rightarrow R_{n+1}^{\cdot}$	$\frac{\partial M}{\partial t} = -k_p M R^{\cdot}$
Termination	$R_n^{\cdot} + R_p^{\cdot} \rightarrow R_{n+p}$	/
Additional equation	$\alpha = \frac{M_0 - M}{M_0}$ where $M = M_0(1-\alpha)$	$\frac{\partial \alpha}{\partial t} = -k_p(1-\alpha)R^{\cdot}$ with $\frac{\partial R^{\cdot}}{\partial t} = f k_d I - q k_z Z R^{\cdot}$

Table 3.1. *A mechanistic description of free-radical polymerization. k is the reaction rate constant (described with an Arrhenian law), f and q are the radical formation and consumption efficiencies, respectively*

To capture the main concepts of a rigorous description and thus keeping a satisfying modeling, several simplifications are made. First, reactive species (carbon double bonds of unsaturated polyester and styrene) are merged and denoted by *M*. The initiator and the inhibitor are denoted by *I* and *Z*, respectively, whereas α represents the conversion degree. The main steps are summarized in Table 3.1. Even if approximations are made, this first approach can provide a good description of the kinetics, as displayed in Figure 3.8.

Some simplifications are often considered:

– radicals are assumed to have the same reactivity;

– termination reaction is often ignored due to the difficulty to monitor the evolution of radical concentration;

– the limitation of the propagation due to diffusion is not taken into account. This phenomenon has a deep impact when the polymer system vitrifies (i.e. the glass transition temperature is equal to the process temperature). However, several elementary reactions can be affected and/or controlled by diffusion in both rubbery and liquid states. Termination is the most affected step but propagation reaction can also be diffusion-sensitive. In this latter case, the rate constant k_p can be described by the following phenomenological equation:

$$k_p = k_{p0}(1 - {}^{\alpha}/_{\alpha_f})^n \qquad [3.1]$$

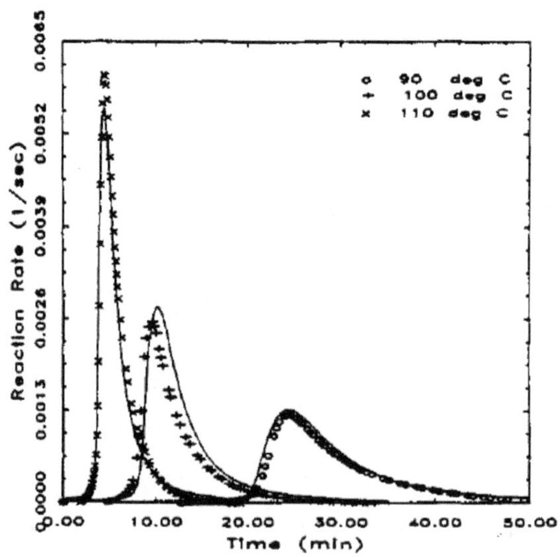

Figure 3.8. *Comparison of kinetic model simulations with isothermal DSC results. Symbols: DSC data; full line: simulation [TWU 93]*

The identification of the equation parameters is often done using a nonlinear optimization method from DSC experiments [TWU 93], leading to rather good fittings (see Figure 3.8). Activation energies and pre-exponential factors were then obtained by plotting the natural logarithm of each rate constant versus (1/T).

Epoxy/amine systems are a classical example of the stepwise polymerization. The study of the mechanism is studied since the 1970s. It is commonly accepted that the main reactions are those associated with primary and secondary amine hydrogen with epoxy groups and an etherification reaction. The set of differential equations used to calculate the concentrations of the species (reactants and products) and the conversion degree for a stoichiometric reaction epoxy/diamine is presented in several papers [HOR 70, DUS 75, RIC 84] and thus will not be discussed in this chapter.

3.3.2. *Description of the kinetics with empirical models: the engineer approach*

From a practical point of view, the mathematical description of the cure kinetics has to be as simple as possible, while containing the global behavior of the complex polymerization mechanism. In the specific case of composite part manufacturing, it is strongly coupled to heat transfer equation and should provide reliable quantitative data (fields of temperature and degree of cure) with enough accuracy.

All phenomenological models lie in the principle of a variable separation assumption:

$$\frac{d\alpha}{dt} = k(T) \times g(\alpha) \qquad [3.2]$$

The appearance of this equation is quite attractive but it is based on the strong assumption that for a given set of conversion and temperature values there is unique value of the reaction rate. However, it is often not true, which limits its rigorous use. As stated by Pascault *et al.* [PAS 02], epoxy/amine systems, which crosslink by stepwise polymerization, are generally good candidates to use the empirical kinetic models. On the contrary, free-radical polymerization of unsaturated polyester with styrene since it involves different initial reactivities of chemical functions and several reaction paths.

Fortunately, experiments show that equation [3.2] provides in most cases a rather good fitting within a restricted range of temperatures and heating rates, even if theory should discourage the use of such models. Extrapolations lead, however, to large deviations from kinetic measurements.

Nowadays, the practical interest of such approach for describing the kinetics of thermoset systems is well recognized and the following models presented in the next sections are widely used.

3.3.2.1. *nth-order model*

For thermoset materials that follow nth-order kinetics, the rate of conversion obeys the general rate equation:

$$\frac{d\alpha}{dt} = k(1-\alpha)^n \qquad [3.3]$$

where n is the reaction order and k is the temperature-dependent kinetic constant according to the Arrhenius law:

$$k(T) = A\,exp\left(\frac{-E_a}{RT}\right) \qquad [3.4]$$

where A is the pre-exponential (s^{-1}), E_a is the activation energy ($J.mol^{-1}$), R is the ideal gas constant (8.314 $J.K^{-1}.mol^{-1}$) and T is the absolute temperature (K).

This model is originally described by Borchardt and Daniels [BOR 56] for solutions and is extended later [PRI 73]. This model assumes that the reaction is only proportional to the concentration of reactants (the products do not modify the kinetics) and also implies that the maximum conversion rate occurs at t = 0 for an isothermal curing. Experimentally, the maximum of the peak reaction is generally close to t = 0. The identification of the model parameters (n, A and E_a) can be done from DSC experiments in isothermal or dynamic conditions. Practical details are given in [BOY 14].

3.3.2.2. *Autocatalytic model*

Contrary to the nth-order model, the autocatalyzed reaction assumes that at least one of the reaction products participates in the crosslinking. Assuming a zero initial cure state, the kinetics of such reactions are described according to the following equation:

$$\frac{d\alpha}{dt} = k\alpha^m(1-\alpha)^n \qquad [3.5]$$

where m and n are the reaction orders and k is the specific rate constant.

From this model, the reaction rate starts from zero (or very small value) and typically reaches its maximum for a conversion degree ranging between 20 and 40%.

A set of four parameters has now to be determined (n, m, E_a, A) following the protocols given in [BOY 14, RYA 79]. To simplify the analysis, it is often assumed that the total reaction order (m + n) is equal to two (restraining the range of applications).

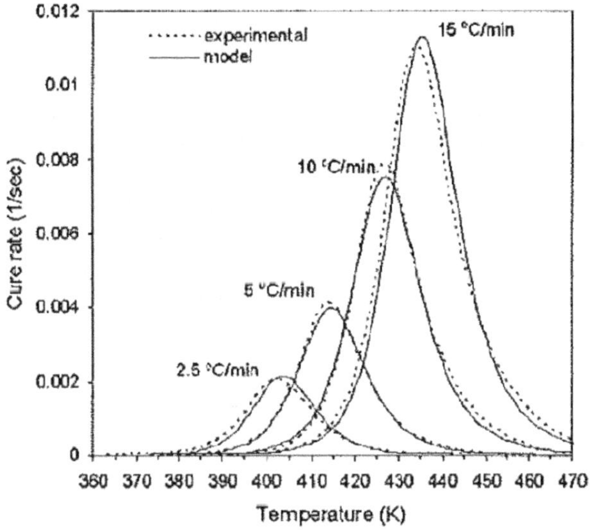

Figure 3.9. *Comparison of experimental dα/dt versus temperature curves and those calculated with the method based on the isoconversion of the peak temperature [HAY 11]*

The results are presented in Figure 3.9. A very good agreement is observed between the experimental data and the computed ones.

3.3.2.3. *Kamal and Sourour's model*

The initial rate of autocatalytic reactions may not be necessarily equal to zero, as induced by the previous model. Reactants can be converted into products via alternative paths. Kamal and Sourour [KAM 74, SOU 76] have shown that these characteristics can be taken into account using the following generalized semi-empirical expression:

$$\frac{d\alpha}{dt} = \left(k_1 + k_2\alpha^m\right)(1 - \alpha)^n \qquad [3.6]$$

where k_1 and k_2 are the catalyzed and autocatalyzed reaction rate constants, respectively, which follow an Arrhenius relationship, m and n are the kinetic exponents and their sum is the overall reaction order. In this case, the influence of the catalytic products on the conversion rate is given by the term $k_2\alpha^m$. This equation is an alternative when the Horie's equation [HOR 70] becomes inadequate.

This phenomenological model is widely and successfully applied to numerous autocatalytic epoxy polymerization reactions [HAR 07, FRA 03] as well as for some polyester resin systems [MIL 01]. The parameters to determine (m, n, k_1, k_2) are estimated from nonlinear regression analysis, even if k_1 can be determined as the extrapolated reaction rate at $\alpha = 0$. Some authors also propose to fix the reaction orders or the total reaction order (m + n) to simplify the analysis.

An example of the kinetic identification for aeronautic RTM6 resin is presented in Figure 3.10 (isothermal cures) from the PhD work of Nakousi [NAK 12]. Modeled and experimental data fit well only up to a given conversion degree using equation [3.7]. Above this point, the model overestimates the evolution of the conversion degree.

a)
b)

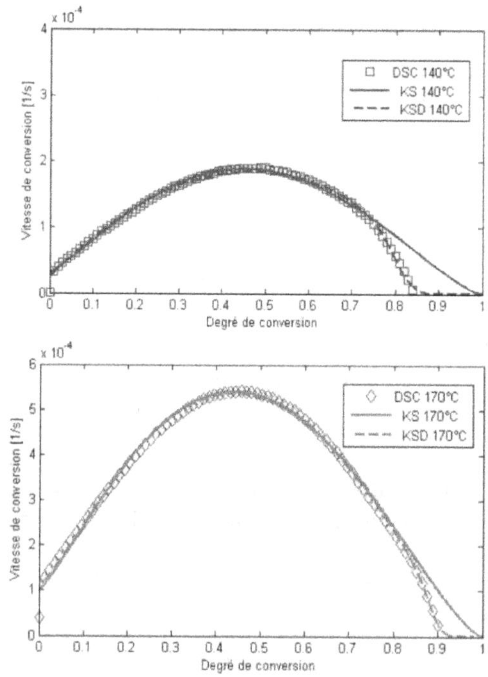

Figure 3.10. *Comparison of DSC data recorded during isothermal cure of RTM6 with Kamal and Sourour's model, neglecting (full line) or not (dashed line) the effect of diffusion. a) T_{iso} = 140°C, b) T_{iso} = 170°C. For a color version of the figure, see www.iste.co.uk/boyard/heat.zip*

One drawback of the previous models is that the resin is considered as fully cured ($\alpha = 1$) at the end of the experiment. They did not take into account the effect of diffusion induced by the glass transition, especially when isothermal crosslinking temperature does not exceed Tg∞. In Figure 3.10, the author adds the vitrification effect using the Rabinovitch's equation for k_1 and k_2 (see section 3.3.2.6), leading to a good agreement between measured and computed reaction rates.

Some authors [GON 89] also propose to modify equation [3.6]:

$$\frac{d\alpha}{dt} = \left(k_1 + k_2\alpha^m\right)\left(\alpha_{max} - \alpha\right)^n \quad [3.7]$$

$$\alpha_{max} = \alpha_0 + B \times T_{curing} \quad [3.8]$$

where α_0 and B are the parameters to determine after data treatment of DSC experiments. Another answer is to use a specific function to consider diffusion effect, as discussed in section 3.3.2.4.

Some complex crosslinkings can also be modeled by combining an nth-order-type reaction and an autocatalytic-type reaction, as proposed by Lee *et al.* [LEE 92]:

$$\frac{d\alpha}{dt} = k_a(1-\alpha)^p + k_b\alpha^m(1-\alpha)^n \quad [3.9]$$

where k_a and k_b are the non-catalytic and autocatalytic rate constants, respectively, and p, n and m are the respective reaction orders.

3.3.2.4. *Bailleul's model*

Although the model of Kamal and Sourour contains several interesting features, it is limited to two simple rate constants. The description of more complex crosslinking reactions requires other models, which can be adapted to this complexity. A new model based on the generic form of the kinetic equation given by equation [3.2] was thus proposed by Bailleul [BAI 96] where K(T) is an Arrhenius function and $G(\alpha)$ has a polynomial form:

$$\frac{d\alpha}{dt} = K(T) \times G(\alpha) \quad [3.10]$$

$$K(T) = K_{ref}exp\left[-B\left(\frac{T_{ref}}{T} - 1\right)\right] \quad [3.11]$$

$$G(\alpha) = \sum_{i=0}^{n} a_i\alpha^i \quad [3.12]$$

where T_{ref} is a temperature fixed by the experimenter, K_{ref} is an Arrhenius constant associated with T_{ref}, n is an integer and B is a parameter to determine.

This model is often used to describe the crosslinking kinetics of epoxy systems dedicated to aeronautics [ABO 10] and for other thermosetting matrices such as vinylesters and polyesters resins [RUI 05, NAW 12] characterized by free radical polymerization. The parameters K_{ref}, B and a_i are estimated either from numerical methods to fit non-isothermal DSC data (several commercial software exist such as Curekinetics®), or by using the isoconversion principle at the peak temperature[1]. Details are given in [BOY 14].

It is sometimes necessary to include the effect of an inhibitor, which avoids the beginning of the reaction during a storage time and/or the filling of a heated mold. An induction time $I_d(t,T)$ is thus defined as a time integral of the thermal history:

$$I_d(t,T) = t_{ref} - \int_0^t exp\left[-C\left(\frac{T_{ref}}{T} - 1\right)\right] dt \qquad [3.13]$$

where t_{ref} is a time associated with T_{ref} and C is an induction constant.

Bailleul's model is thus modified by multiplying [3.12] by a new function $K_2(I_d)$, where $K_2(I_d) = 0$ if $I_d \geq 0$ or $K_2(I_d) = 1$ if $I_d < 0$

Figures 3.11(a–b) illustrate the results which can be obtained to model the crosslinking kinetics of a vinylester resin (DERAKANE MOMENTUM 411-350) used for infusion and an unsaturated polyester resin especially formulated for RTM process (high content of styrene). The kinetic model applied by authors to reproduce the reaction rate of the vinylester resin (Figure 3.11(a)) requires the use of an induction and diffusion functions

$$\frac{d\alpha}{dt} = K_2(I_d) \times K(T) \times G(\alpha) \times f(\alpha) \qquad [3.14]$$

Figure 3.11(b) shows that two reaction peaks are present on the heat flux curve (proportional to the reaction rate): the first peak can be attributed to the copolymerization of styrene and UP double bonds, while the second peak is only due to styrene reaction (homopolymerization)). It is more difficult to model such behavior since two kinetics are overlapped but Dupuy *et al.* [DUP 05] demonstrate that the global reaction rate can be described by the following equation:

$$\frac{d\alpha}{dt} = \sum_{i=1}^{i=2} K_2(I_{d,i}) \times K_i(T) \times G_i(\alpha_i) \qquad [3.15]$$

1 Isoconversional methods exist for the identification of model parameters, even for complex reactions [SBI 07, VYA 11].

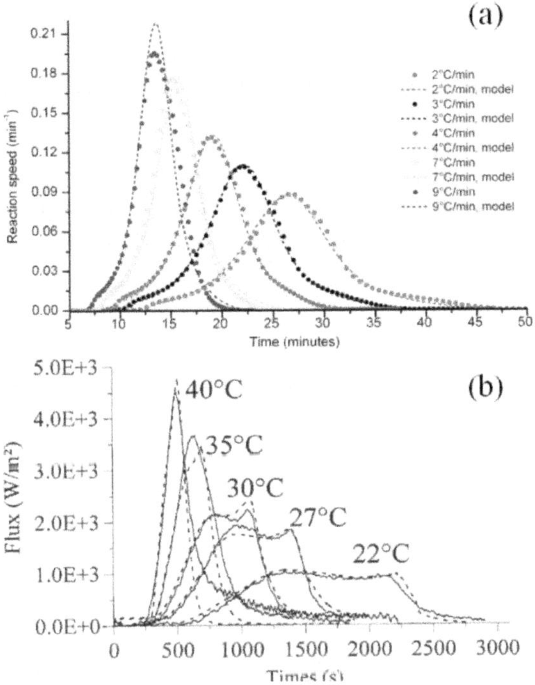

Figure 3.11. *Kinetics modeling from the Bailleul's model for a) a vinylester resin [NAW 12], b) an unsaturated polyester resin with a high content of styrene monomer [DUP 05]. For a color version of the figure, see www.iste.co.uk/boyard/heat.zip*

3.3.3. *Modeling of the diffusion induced by vitrification*

When the resin vitrification occurs ($T_{cure} \approx Tg_{system}$), the crosslinking is strongly slowed down and finally stops before full conversion is achieved (see [PAS 02] for more details). This phenomenon is classical during isothermal curing when $T < Tg_{\infty}$ but can also occur during non-isothermal cure at slow heating rates (usually < 2 K/min). Experimental conversion and reaction rates, at diffusion controlled step, are lower than those predicted by the previous models, which do not consider the vitrification effect.

A particularity of thermoset polymers is that the glass transition temperature (Tg) is not constant during the crosslinking: it rises as the cure progresses according to the Pascault–Williams equation:

$$T_g = \frac{(1-\alpha)T_{g0}+\lambda\alpha T_{g\infty}}{(1-\alpha)+\lambda\alpha} \qquad [3.16]$$

where Tg_0 and Tg are the glass transition temperature when $\alpha = 0$ and $\alpha = 1$, respectively, and λ is an adjustable structure-dependent parameter between 0 and 1. This equation shows that a unique relationship exists between Tg and the degree of cure.

The effect of T_g is thus included in equation [3.2] by stating that the reaction of diffusion acts in series with the chemical one. Thus, the overall polymerization rate constant K(T) is expressed by the Rabinowitch model [RAB 37]:

$$\frac{1}{K} = \frac{1}{k_c} + \frac{1}{k_d} \qquad [3.17]$$

where k_c is the chemical rate constant (identical to the one in equation [3.2]) and k_d is the diffusion rate constant.

A difficulty is to model for k_d which reproduces the experimental evolution of the reaction close to or in the glassy state. A simple semi-empirical expression is proposed by Chern and Poehlein [CHE 87] based on free volume concept:

$$k_d = k_c\, exp[-C(\alpha - \alpha_c)] \qquad [3.18]$$

where C is an adjustable parameter and α_c is a critical conversion value at which the diffusion is the prominent factor. Both parameters are temperature-dependent [COL 91]. However, it is known that this equation involves a rather abrupt onset of diffusion control whereas it is more gradual in reality. An example of results is given in Figure 3.10.

$$f(\alpha) = \frac{1}{1+exp[-C(\alpha-\alpha_c)]} \qquad [3.19]$$

The combination of equations [3.17] and [3.18] leads to the expression of the diffusion factor (equation [3.19]). The effective rate constant is thus computed by multiplying k_c by $f(\alpha)$.

3.4. Overall crystallization kinetics of semi-crystalline thermoplastics

Polymer crystallization can be described by overall crystallization kinetic theories. These models are based on a general description of the transformation of the polymers from the liquid to the solid state. From the molten state, many polymers develop a semi-crystalline spherulitic morphology. In this kind of microstructure, the crystallization starts at particular points of the volume: the nuclei. These are the starting points of mono-crystalline lamellae growing and multiplying in all directions of space, including between them some amorphous polymers. The combination of the crystalline lamellae and the amorphous polymer defines, in three dimensions, a spherical semi-crystalline entity: the spherulite. These growing entities tend to occupy the whole volume without covering each other (Figure 3.12). This description is the basic principle of the overall kinetic theories, neglecting the possibility of a secondary crystallization of the amorphous part. The studied quantity is generally the transformed fraction α, defined as the ratio of the volume occupied by spherulites to the total volume. This quantity is 0 at the beginning and reaches 1 at the end of the transformation. Thus, it has to be differentiated from the crystalline fraction, as this one will never be equal to 1.

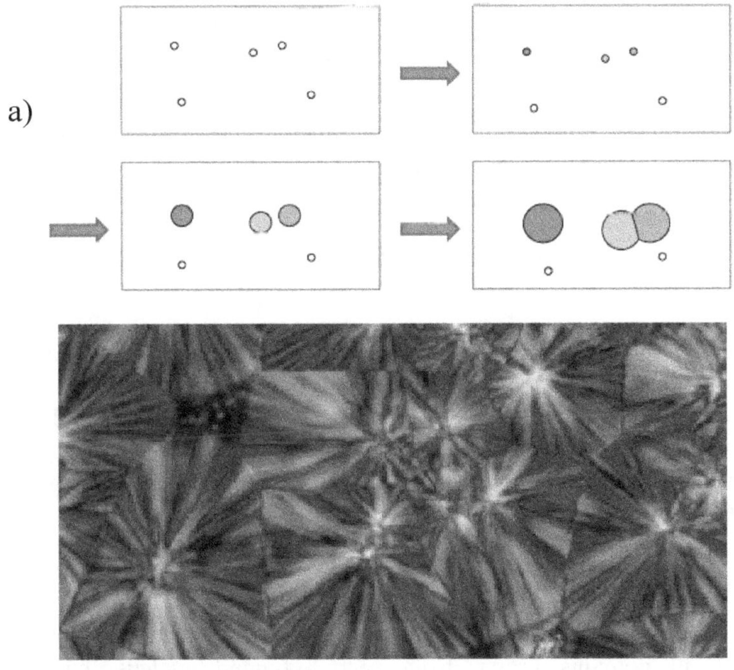

Figure 3.12. *Polymer crystallization. a) Principle of nucleation and growth mechanism. b) Example of iPP (isotactic polypropylene) spherulites observed by polarized-light microscopy. For a color version of the figure, see www.iste.co.uk/boyard/heat.zip*

All overall kinetics theories are based on the same assumptions concerning nucleation and growth in an infinite volume. There is only one kind of nuclei, uniformly distributed. All the entities are growing at the same rate, and are spheres in 3D, disks in 2D and rods in 1D. There are pre-existing sites in the molten liquid, called potential nuclei, with an initial density N_0 and function of temperature. Each potential nucleus in a non-transformed area can be activated, i.e. generate an entity (e.g. a spherulite). On the other hand, a potential nucleus that has been absorbed by an entity before getting activated can no longer be activated. The activation frequency q is defined as the probability per unit time for any nucleus to be activated. The new entity generated by this activation immediately starts to grow. The spherulite growth rate G is the lamella growth rate. The impingement of two growing spherulites stops their growth at the contact point, restricts the growth to a complex microstructure with sharp and/or bent interspherulitic boundaries. G and q are the only functions of temperature, as the crystalline growth is governed by secondary nucleation phenomena, and not by diffusion phenomena. In the limiting case of instantaneous nucleation, all the potential nuclei are instantaneously activated and the number of spherulites per unit volume remains constant and equal to N_0. In the other limiting case of isothermal sporadic in time nucleation, the number of activated nuclei per unit time and unit volume is constant: $\frac{\mathrm{d}N_a}{\mathrm{d}t} = qN = constant$ (with N the number of remaining potential nuclei per unit volume). This implies that q is small and the initial number of nuclei N_0 is high.

3.4.1. *Most popular crystallization kinetics models for process simulations*

The best-known approaches are those developed by Avrami and Evans. Avrami's model [AVR 39, AVR 40, AVR 41] is based on the concept of the so-called extended volume, whereas Evans considers the probability theory [EVA 45]. However, both approaches are strictly equivalent [BIL 90]. Avrami's theory applies to isothermal and non-isothermal conditions using the well-known general equation:

$$\alpha(t) = 1 - \exp(-\tilde{\alpha}) \qquad [3.20]$$

where $\tilde{\alpha}$ is the extended volume fraction (the extended volume is the sum of volumes of all domains grown from each activated nucleus, neglecting the impingement). Extended volume also includes the "phantom" entities created from nucleation which occurred (virtually) in the solidified areas.

In the specific case of isothermal crystallisation, G and q are constants and equation [3.20] can be simplified for instantaneous or sporadic nucleation, under the classical form:

$$\alpha(t) = 1 - \exp(-K_{Av}t^n) \qquad [3.21]$$

where K_{Av} and n are the Avrami kinetic function and the Avrami exponent, respectively. They can be written as a function of N_0 and/or q, depending on the dimensionality of crystallization (from 1D to 3D) and the nucleation mode (see Table 3.2).

Growth geometry	Sporadic nucleation	Instantaneous nucleation
Spheres	$K_{Av} = \frac{\pi q N_0 G^3}{3}$, n = 4	$K_{Av} = \frac{4\pi N_0 G^3}{3}$, n = 3
Disks	$K_{Av} = \frac{\pi q N_0 G^2}{3}$, n = 3	$K_{Av} = \pi N_0 G^2$, n = 2
Rods	$K_{Av} = \frac{q N_0 G}{2}$, n = 2	$K_{Av} = N_0 G$, n = 1

Table 3.2. *Expression of the Avrami kinetic function K_{Av} for instantaneous and sporadic nucleation, in 3D and 2D. N_0 is the initial number of potential nuclei*

Isothermal crystallization being far from the process temperature cycle, other models have extended equation [3.20]. Among them, Ozawa [OZA 71] proposes to describe non-isothermal crystallization at a constant cooling rate V, based on Evans' theory:

$$\alpha(t) = 1 - \exp\left(-\frac{K_{Oz}(T)}{V^n}\right) \quad [3.22]$$

This model was generalized by Billon *et al.* [BIL 91] to non-constant cooling rates. Note that it is formally equivalent to the Nakamura's equation [NAK 73, NAK 72] for non-isothermal conditions, which is deduced from the Avrami approach with isokinetic assumption (G/q=constant):

$$\alpha(t) = 1 - \exp\left(-\left[\int_0^t K_{Nak}(T(t))dt\right]^n\right) \quad [3.23]$$

The three kinetic functions are related to each other following the expression:

$$K_{Nak} = K_{Av}^{1/n} = -\frac{d\left(K_{Oz}^{1/n}\right)}{dT} \quad [3.24]$$

Patel [PAT 91] thus proposes the differential form of the Nakamura equation, which is more adapted for numerical solving:

$$\frac{d\alpha}{dt} = nK_{Nak}(T)(1-\alpha)\left[\ln\left(\frac{1}{1-\alpha}\right)\right]^{(n-1)/n} \quad [3.25]$$

Generally speaking, such simplified models should replace the three parameters (N_0, q, G) by two "more easily" measurable ones. From a practical point of view, Nakamura's approach is widely used in the modeling of the crystallization kinetics coupled with heat transfer in polymer and/or composite processes [LAF 86, TIT 97, GUO 99]. The main reasons are (1) only two parameters have to be experimentally determined (i.e. K_{Nak} and n) and (2) the relation between K_{Nak} and K_{Av} allows a direct application of isothermal crystallization analysis to the modeling of crystallization for any temperature cycle.

Other models/theories also exist in the literature and are all based on statistical approaches allowing the calculation of the average transformed volume fraction α. However, they generally remain in restrictive cases such as crystallization in isothermal or constant cooling rate conditions. Note also that the Tobin model [TOB 74, TOB 76, TOB 77] should not be used since it is erroneous due to an incorrect assumption [PIO 06]. A critical review of overall crystallization kinetics theories has been done by Piorkowska *et al.* [PIO 13] and Di Lorenzo *et al.* [DIL 99].

3.4.2. *Systems of differential equations*

Other models were derived from the general form of the Avrami theory (equation [3.20]) to avoid the simplifications discussed above. Schneider *et al.* [SCH 88] propose to differentiate the expression with respect to time, leading to a system of first-order differential equations, where each of them can be written as $\dot{\varphi}_{i-1} = G(t)\varphi_i(t)$.

$$\frac{\partial \varphi_3}{\partial t} = 8\pi \frac{\partial \tilde{N}}{\partial t} \text{ with } \varphi_3 = 8\pi \widetilde{N_a} \quad [3.26]$$

$$\frac{\partial \varphi_2}{\partial t} = G\varphi_3 \text{ with } \varphi_2 = 8\pi \tilde{R} \quad [3.27]$$

$$\frac{\partial \varphi_1}{\partial t} = G\varphi_2 \text{with } \varphi_1 = 4\pi \widetilde{R^2} \quad [3.28]$$

$$\frac{\partial \varphi_0}{\partial t} = G\varphi_1 \text{with } \varphi_0 = \tilde{\alpha} \quad [3.29]$$

$\widetilde{N_a}$ is the virtual number of activated nuclei in the extended volume, φ_0 and φ_1 are the extended volume fraction and the total surface of spherulites per unit of volume, respectively, neglecting the impingement but taking into account phantom (i.e. virtual) spherulites. $\tilde{R}$ is the sum of all radii in unit volume. The calculations of the transformed fraction and/or the associated kinetics are given by the relations:

$$\alpha = 1 - \exp(-\varphi 0) \qquad [3.30]$$

$$\frac{\partial \alpha}{\partial t} = G\varphi_1 \exp(-\varphi_0) = G\varphi_1(1-\alpha) \qquad [3.31]$$

The question is thus to know $\widetilde{N_a}$ for solving this set of equations. The answer is easy only in the case of instantaneous nucleation since $\widetilde{N_a}$ is equal to the measured initial number of nuclei for different temperatures. The growth rate G(T) has also to be determined.

Haudin and Chenot [HAU 04] proposed another approach to calculate α the number of activated nuclei N_a, generalizing Avrami's model for any temperature cycle, any nucleation condition (i.e. not only instantaneous or sporadic nucleation) and writing it in a differential form. The basic assumptions and the general framework of the theory are kept, but it includes the possibility to generate new nuclei in the liquid fraction due to a temperature decrease (per unit volume). The authors finally obtained a nonlinear set of equations:

$$\begin{aligned}
\frac{dN}{dt} &= -N\left(q + \frac{1}{1-\alpha}\frac{d\alpha}{dt}\right) + (1-\alpha)\frac{dN_0(T)}{dT}\frac{dT}{dt} \\
\frac{d\alpha}{dt} &= 4\pi(1-\alpha)G(F^2\tilde{N}_a - 2FP + Q) \\
\frac{dN_a}{dt} &= q(t)N(t) \\
\frac{dP}{dt} &= F\frac{d\tilde{N}_a}{dt} \\
\frac{d\tilde{N}_a}{dt} &= \frac{q(t)N(t)}{1-\alpha} \\
\frac{dQ}{dt} &= F^2\frac{d\tilde{N}_a}{dt} \\
\frac{dF}{dt} &= G(t)
\end{aligned} \qquad [3.32]$$

With initial conditions: $N(0) = N_0(T_0), \alpha(0) = N_a(0) = \tilde{N}_a(0) = F(0) = P(0) = Q(0) = 0$.

N, N_a are the number of potential and activated nuclei per unit volume at a given time. F, P and Q are the auxiliary functions to get a first-order ordinary differential system, which can be solved numerically with a Runge–Kutta algorithm, for example. Note that only three parameters are required to feed this model: the initial potential number density N_0, the growth rate G(T) and the activation frequency q(T).

These two models developed with a differential form have also the interest to include flow effects, which enhance the nucleation rate and introduce two types of morphologies. Despite their attractive nature due to their mathematical formalisms and their capacity to operate under any process conditions (cooling, flow, etc.), such models are not among the most used for simulation [ZUI 01, HAU 08]. The difficulty to experimentally measure the parameters and their dependence on the temperature (and flow, when required) remains a brake.

3.4.3. *Measurements of crystallization kinetics and associated parameters*

One big experimental challenge in quiescent crystallization is to succeed in getting the temperature dependence of parameters used as input data in the previous models: activated nucleus density, growth rate, activation frequency and kinetic function. Despite academic efforts, this remains a tricky task and data are still lacking for "low" temperatures and high cooling rates, which could validate (or not) theory predictions, as well as at high pressures. In this context, several specific devices exist to directly observe the development of the microstructures (optical microscopy coupled to hot stage, X-ray diffraction and scattering) or to follow macroscopic classical apparatus.

Many works deal with some devices combining optical microscopy and hot stage prototype. A detailed bibliography can be found in [LE 11, BOY 10]. Broadly speaking, the cooling rate of thin polymer samples is often controlled with a gas flow used as heat exchanger. For example, it is possible to reach a cooling rate of 2,500°C/min with a 150 μm-thick sample and up to 5,000°C/min when the polymer size is three times smaller [SUP 02]. Other researchers recently developed optical cells, in which the heating and cooling are managed with a blend of hot and cold air flows [BOY 12], or where high pressure can be imposed [BOY 14]. Figure 3.13 depicts the experimental growth rates down to 60°C and nucleation density for a polypropylene, from DSC and microscopy experiments at high cooling rates. The shape of the growth rate curve G(T) seems to exhibit a plateau at the lowest temperatures accessible with the devices. However, we can expect a decrease at a lower temperature level due a decrease in the polymer chain mobility.

The nucleation density is particularly difficult to measure. One classical way is to count the number of spherulites (see circles in Figure 3.13(b)). In the case of instantaneous nucleation considered here, the number of activated nuclei per unit volume is the initial density of potential nuclei N_0. A different strategy to determine N_0 is to use the relations existing between K_{Av} or K_{Oz} and N_0 and G (see squares in Figure 3.13(b)).

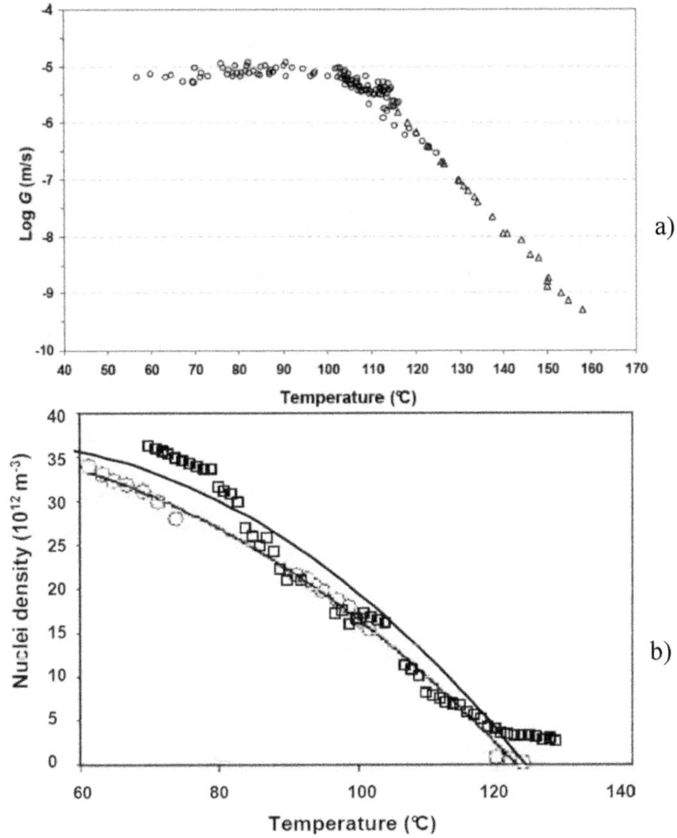

Figure 3.13. *a) Decimal logarithm of the growth rate G of iPP spherulites against the crystallization temperature; experimental results in isothermal (triangles) and nonisothermal conditions (circles). b) Estimates of the initial number of potential nuclei in case of instantaneous nucleation [BOY 12]*

The crystallization kinetics is also widely analyzed by differential scanning calorimetry (DSC) to measure the evolution of the relative crystallinity (i.e. transformed fraction) from the exothermal heat flux induced by the solidification, as well as to determine the kinetic functions (K_{Av} or K_{Oz}) defined in simplified models (see section 3.4.1). However, classical DSC is limited in cooling rates, which should not exceed several tens degrees per minute to avoid temperature gradient in the sample. As a result, it also restricts the temperature range in which the kinetic function can be estimated.

To overcome this problem, one way is to increase the cooling rate while keeping the DSC principles. It induces some thermal issues, including the reduction of the thermal inertia of the system, the temperature homogeneity within the sample and the consideration of the thermal contact resistance between the sample and the pan during the crystallization. Pijpers *et al.* [PIJ 02] thus modify standard power compensation DSC and reduce the sample mass (between 0.1 and 1 mg) to create a high-performance apparatus (HPer DSC) with higher cooling rates (up to 500 K/min). It allows the study of rapid transformations in real time. To reach higher cooling rates, an ultrafast thin film calorimeter, also called nanocalorimeter of fast scanning calorimeter (FSC), was proposed by Adamovsky [ADA 03] to cool a very small sample (several tens of nanograms) at a rate of 10,000 K/s. It was then technologically improved by a team at Rostock University [ZHU 10] by developing power compensation FSC, which is now commercialized by Mettler Toledo with the name Flash DSC1. Such a device has opened a new gateway on a deeper analysis of polymerphysics as well as on the crystallization kinetics at low temperature and/or high cooling rates. A recent work analyzes the crystallization rate of PEEK over a large temperature range (170–320°C) within the framework of the Avrami–Evans theory [TAR 14X]. The kinetic function K_{Av} is deduced from the measured crystallization half-time (Figure 3.14), without considering secondary crystallization. The K_{Av} curve versus temperature has a bell-shape in agreement with theory, and mirrors the crystallization half-time evolution. It has one maximum located at about 230°C. According to the Avrami's theory, values of n (close to 3) tend to demonstrate that crystallization occurs following instantaneous nucleation and a three dimensional growth of spherical entities.

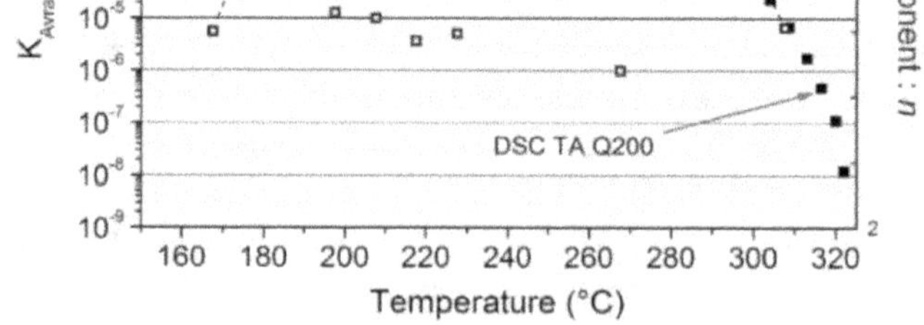

Figure 3.14. *Evolution of the Avrami kinetic function and the Avrami exponent versus temperature for PEEK [TAR 14]. Black unfilled squares: Flash DSC data, black filled squares: standard DSC data. For a color version of the figure, see www.iste.co.uk/boyard/heat.zip*

It is interesting to note that some Flash DSC and standard DSC data points are overlapped, suggesting the consistency of Flash DSC results. However, the low sample mass and thus the small size of the sample implies a high surface/volume ratio, raising some questions on the possible impact on the crystallization rate (case similar to thin film crystallization) and on the microstructure.

Specific devices [LE 11, PIG 15, TAR 12] are also developed specifically for the identification of the kinetic function defined in simplified overall crystallization models. Most of them are built to mimic the process temperature conditions. The recorded data (temperature, specific volume, etc.) are then analyzed using inverse methods coupled to heat transfer models to identify the values of the kinetic function with respect to temperature.

3.4.4. *Models for specific crystallization phenomena and geometries*

Even if the Avrami–Evans' theory is suitable to describe the crystallization kinetics of many polymers (polypropylene is a typical example), its application for other thermoplastics such as polyethylene, polyamide, polyethylene terephtalate as well as polyetheretherketone is more delicate. After a given value of the transformed fraction, simulated curve diverges from the experimental one since the theory predicts crystallization rate. It is demonstrated that a secondary crystallization is responsible for this divergence and generally occurs once the spherulites have impinged on one another. Secondary crystallization represents an insertion of new, lamellae between the already grown ones and/or in further development of the already formed crystalline structure (e.g. lamella thickening). Depending on the polymer nature, primary and secondary crystallization can occur simultaneously or successively. Several authors try to describe this crystallization behavior using overall kinetics models. Among them, Hillier [HIL 65] proposes a model for an isothermal condition and based on the Avrami's theory and considers that both crystallization mechanisms occur simultaneously:

$$\alpha_{tot} = \alpha_1(t) + \alpha_{2m} K_{Av2} \int_0^t (1 - \exp(-K_{Av1}\theta^n))\left(\exp(-K_{Av2}(t-\theta))\right) d\theta \qquad [3.33]$$

where $\alpha_1(t) = \alpha_{1m}(1 - \exp(-\mathrm{K_{Av1}t^n}))$.

Velisaris and Seferis [VEL 86] model primary and secondary crystallizations as the sum of two Avrami equations and is used to model the PEEK crystallization. However, this approach can be criticized knowing the assumptions of the Avrami theory about the secondary crystallization. Dietz [DIE 81] proposes to include the secondary crystallization through an additional term in the differential form of Avrami equation to slow the phase change rate:

$$\frac{d\alpha}{dt} = nK_{Av}(1-\alpha)t^{n-1}\exp\left(\frac{-\gamma\alpha}{1-\alpha}\right) \qquad [3.34]$$

where γ is an adjustable parameter ranging between 0 and 1.

For high density polyethylene (HDPE), the secondary crystallization is associated with lamella thickening. Based on an experimental observation relating to the lamellar thickness to the logarithm of time, Marand *et al.* [MAR 04] propose a new model where the global transformed fraction is a combination of both primary and secondary crystallizations occurring simultaneously:

$$\alpha = \alpha_{prim} + \alpha_{second} = \alpha_{prim} - C\int_0^t \alpha_{prim}(t')\left(-\frac{d\log(1+\frac{t-t'}{\tau})}{dt'}\right)dt' \qquad [3.35]$$

where C is a parameter depending on the thickening rate, $\alpha_{prim}(t')$ is the primary crystallinity fraction at time t' < t and τ is a time constant. Giboz [GIB 09] has the same approach for non-isothermal cases but with a primary crystallization described by the Tobin's equation (known to be incorrect).

Sometimes the presence of surfaces or inclusions will generate a fast and massive local nucleation. In this case, the numerous spherulites limit the growth of their neighbors and then oblige them to grow along one direction perpendicular to the surface on which they were generated. Then, they form a crystallization front propagating at the spherulite growth rate *G*. This phenomenon is called transcrystallinity and can lead to very different crystallization kinetics from those obtained in bulk infinite volume. It can play an important role in DSC measurement results, and thus has to be taken into account in their interpretations [ESC 84]. In these models, transcrystallinity is introduced through a second set of parameters (number of potential nuclei per surface unit N_s (m^{-2}) and activation frequency *s*, equivalent to N_0 (m^{-3}) and *q* in the volume, *G* remaining the same), describing the behavior of nuclei generated on surfaces. These parameters are chosen so that the nucleation on the surfaces is faster than in the volume, in agreement with experimental observations. Some experimental methods, based on these simulations, have been proposed to find the overall kinetics parameters [MON 04, BIL 93].

Sample geometry, especially volume restriction, as in thin films, can also lead to changes in crystallization kinetics, and models have to be adapted to the sample geometry to perform accurate predictions. This was done first by Escleine *et al.* [ESC 84], followed by other extensions, which offer the possibility to consider transcrystallinity on the film surfaces and non-isothermal cases [BIL 90, BIL 94, BIL 89]. An extension for thin films of the differential model of Haudin (section 3.3.4.2) was proposed by Chenot *et al.* [DUR 15, CHE 05] for any temperature and nucleation conditions (see Figure 3.15). Furthermore, it has been shown that transcrystallinity and volume restriction effects are tightly connected [BIL 94].

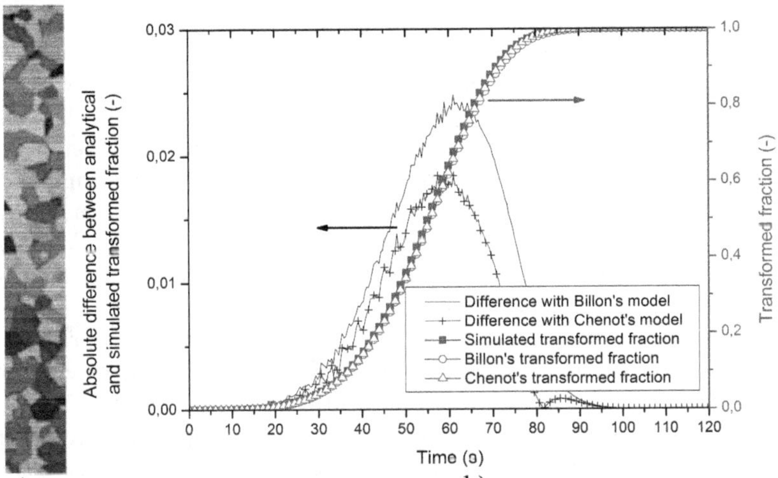

a) b)

Figure 3.15. *a) 2D slice of a crystallized thin film virtual PP sample obtained with pixel coloring method; each color represents a spherulite. b) Comparison between transformed fraction obtained by the numerical method and by analytical ones (Billon et al. [BIL 90], Chenot et al. [CHE 05]) in a PP thin film: (in a non-isothermal case (initial temperature: 110°C, cooling rate: 5°C/min). From [DUR 15]. For a color version of the figure, see www.iste.co.uk/boyard/heat.zip*

3.5. Concluding remarks

Simulation of composite and polymer part processing is one of the main concerns for industries. In a context of competitiveness and productivity, the prediction of the final properties of the part and the optimization of the manufacturing conditions can be viewed as a new Graal. Important questions include: how to switch from the physical test to the virtual test, allowing a validation of the products and how to master the quality of the parts.

Accurate predictions will save time and money, whereas experimental tests will remain cumbersome, expensive and overall unnecessary. However, process simulation is complex since it requires a multiphysic approach with couplings. The quality of the numerical results is based on the accuracy of both the physical models and the input data (thermo-physical, mechanical, rheological properties, boundary conditions, etc.). Among all these parameters, the transformation kinetics of polymer matrices (thermosets and thermoplastics) is very important since the whole material properties dramatically change with the evolution of the transformed fraction and it has an impact on the temperature fields. Reliable modeling of the reaction rate is thus necessary. This chapter highlights some basics in the main approaches proposed in the literature to describe crosslinking and crystallization.

For simulation purposes, empirical models are indubitably of real interest for engineers since they are easy to handle and simplified approaches often give good results. In the case of thermosetting resins, these models aim to globally describe a set of reaction complex mechanisms. Depending on the polymerization nature, they do not fulfill all theoretical requirements. Moreover, the description of the vitrification effect is also a tricky task. Therefore, kinetic models are more or less representative of the actual mechanism (e.g. for polyester resins), leading to validity only over a temperature range scanned for their determinations. Extrapolations to high or low temperature levels often provide poor agreement with experimental reaction rates. We can consider it as the main drawback for thermoset resins. On the contrary, a positive point is that several tools exist to facilitate the determination of the model parameters from a set of several experiments performed in isothermal or dynamic conditions. Finally, we can concede that these empirical approaches often give rather good results and are reliable in the process temperature range.

Considering thermoplastics crystallization, the general Avrami theory is a good framework to describe the solidification rate, as demonstrated in section 3.4.2. Such models are developed within the form of a differential system, which facilitates numerical integration. A big issue remains the estimation of the model parameters, which are quite difficult to obtain experimentally. A good alternative is the simplified model derived from the Avrami approach. Literature shows that Nakamura's equation is widely used to model crystallization in process simulations. Some other papers based their studies on Ozawa's equation extended by Billon *et al.*. Despite several assumptions, numerical results are often in good agreement with experimental ones. An undeniable advantage of these models compared to the general form of Avrami's theory is to use more "global" parameters. Their determination is, however, not so straightforward, especially at low temperatures. Even if specific devices are developed in the last 10–20 years to have information of crystallization at high cooling rates, data is still lacking and further efforts are required. All crystallization models are mainly used in simulations to calculate the evolution of the transformed fraction. They can also help to calculate the thickness of the solid layer during injection molding and the mean spherulite size. More recently, models were enriched to predict the microstructure, including spherulites and shish-kebabs [ZUI 01, HAU 08].

New experimental set-ups have to be imagined to obtain more information on kinetic parameters and microstructures with respect to process conditions. The presence of fibers should be now more systematically included in future works in both experimental and modeling approaches.

3.6. Bibliography

[ABO 10] ABOU MSALLEM Y., JACQUEMIN F., BOYARD N. *et al.*, "Material characterization and residual stresses simulation during the manufacturing process of epoxy matrix composites", *Composites Part A: Applied Science*, vol. 41, pp. 108–115, 2010.

[ADA 81] ADACHI K., HARRISON G., LAMB J. *et al.*, "Ultrasonic investigations of morphology and stress relaxation in drawn polypropylene", *Polymer*, vol. 22, pp. 1026–1031, 1981.

[ADA 03] ADAMOVSKY S.A., MINAKOV A.A., SCHICK C., "Scanning microcalorimetry at high cooling rate", *Thermochimica Acta*, vol. 403, pp. 55–67, 2003.

[ALI 98] ALIG I., TADJBAKHSCH S., "Film formation and crystallization kinetics of polychloroprene studied by an ultrasonic shear wave reflection method", *Journal of Polymer Science Polymer Physics*, vol. 36, pp. 2949–2959, 1998.

[AVR 39] AVRAMI M., "Kinetics of phase change I, General theory", *Journal of Chemical Physics*, vol. 7, pp. 1103–1112, 1939.

[AVR 40] AVRAMI M., "Kinetics of phase change II, transformation-time relations for random distribution of nuclei", *Journal of Chemical Physics*, vol. 8, pp. 212–224, 1940.

[AVR 41] AVRAMI M., "Kinetics of phase change III: granulation, phase change, and microstructure", *Journal of Chemical Physics*, vol. 9, pp. 117–184, 1941.

[BAI 96] BAILLEUL J.-L., DELAUNAY D., JARNY Y., "Determination of temperature variable properties of composite materials: methodology and experimental results", *Journal of Reinforced Plastics and Composites*, vol. 15, no. 5, pp. 479–496, 1996.

[BAN 63] BANKS W., HAY J.N., SHARPLES A. *et al.*, "The crystallization of PE II", *Polymer*, vol. 5, pp. 163–175, 1963.

[BAR 82] BARHAM P.J., JARVIS D.A., KELLER A., "A new look at the crystallization of polyethylene. III. Crystallization from the melt at high supercoolings", *Journal of Polymer Science Polymer Physics*, vol. 20, pp. 1733–1748, 1982.

[BIL 89] BILLON N., HAUDIN J.-M., "Overall crystallization kinetics of thin polymer films. General theoretical approach. I. Volume nucleation", *Colloïd Polymer Science*, vol. 267, no. 12, pp. 1064–1076, 1989.

[BIL 90] BILLON N., HAUDIN J.-M., "Simulation de la cristallisation des polymères sur ordinateur", *Annales de Chimie-Science des Matériaux*, vol. 15, pp. 245–274, 1990.

[BIL 91] BILLON N., BARQ P., HAUDIN J.-M., "Modeling of the cooling of semi-crystalline polymers during their processing", *International Polymer Processing*, vol. 6, pp. 348–355, 1991.

[BIL 93] BILLON N., HAUDIN J.M., "Determination of nucleation rate in polymers using isothermal crystallization experiments and computer simulation", *Colloïd Polymer Science*, vol. 271, no. 4, pp. 343–346, 1993.

[BIL 94] BILLON N., MAGNET C., HAUDIN J.-M. *et al.*, "Transcrystallinity effects in thin polymer films. Experimental and theoretical approach", *Colloïd Polymer Science*, vol. 272, no. 6, pp. 633–654, 1994.

[BOR 56] BORCHARDT H.J., DANIELS F., "The application of differential thermal analysis to the study of reaction kinetics 1", *Am. Chem. Soc.*, vol. 79, no. 1, pp. 41–46, 1956.

[BOU 10] BOUTAOUS M., BOURGIN P., ZINET M., "Thermally and flow induced crystallization of polymers at low shear rate", *Journal of Non-Newtonian Fluid Mechanics,* vol. 165, pp. 227–237, 2010.

[BOY 10] BOYER S.A.E., HAUDIN J.-M., "Crystallization of polymers at constant and high cooling rates: a new hot-stage microscopy set-up", *Polymer Testing*, vol. 29, pp. 445–452, 2010.

[BOY 12] BOYER S.A.E, RK P., GANET P. *et al.*, "Crystallization of polypropylene at high cooling rates: Microscopic and calorimetric studies", *Journal of Applied Polymer Science*, vol. 125, no. 6, pp. 4219–4232, 2012.

[BOY 14a] BOYER S.A.E., FOURNIER F.E.J., GANDIN C.A. *et al.*, "CRISTAPRESS: an optical cell for structure development in high pressure crystallization", *Review of Scientific Instruments*, vol. 85, p. 013906, 2014.

[BOY 14b] BOYARD N., SOBOTKA V., DELAUNAY D., "Theoretical modeling of curing process", in THOMAS S., SINTUREL C., THOMAS R. (eds), *Micro- and Nanostructured Epoxy/Rubber Blends*, Wiley-VCH, 2014.

[BRU 91] BRUCATO V., CRIPPA G., PICCAROLO S. *et al.*, "Crystallization of polymer melts under fast cooling. I: Nucleated polyamide 6", *Polymer Engineering & Science*, vol. 31, pp. 1411–1416, 1991.

[CHE 87] CHERN C.S., POEHLEIN G.W., "A kinetic-model for curing reactions of epoxides with amines", *Polymer Engineering & Science*, vol. 27, pp. 788–795, 1987.

[CHE 05] CHENOT J.-L., SMIRNOVA J., HAUDIN J.-M. *et al.*, "Contribution to numerical modeling of polymer crystallization", *8th International ESAFORM Conference on Material Forming*, pp. 813–816, Romania, 2005.

[COL 91] COLE K.C., HECHLER J.J., NOEL D., "A new approach to modeling the cure kinetics of epoxy/amine thermosetting resins. 2. Application to a typical system based on bis (4-diglycidylamino) phenyl]methane and bis(4-aminophenyl) sulfone", *Macromolecules*, vol. 24, pp. 3098–3110, 1991.

[CRI 14] CRIADO M., SOBRADOS I., SANZ J., "Polymerization of hybrid organic-inorganic materials from several silicon compounds followed by TGA/DTA, FTIR and NR techniques", *Progress in Organic Coatings,* vol. 77, pp. 880–891, 2014.

[CUO 11] CUONG LE M., BELHABIB S., NICOLAZO C. *et al.*, "Pressure influence on crystallization kinetics during injection molding", *Journal Materials Processing Technology*, vol. 211, pp. 1757–1763, 2011.

[DIE 81] DIETZ W., "Sphärolithwachstum in polymeren", *Colloïd Polymer Science*, vol. 429, pp. 413–429, 1981.

[DIL 99] DI LORENZO W., SILVESTRE C., "Non-isothermal crystallization of polymers", *Progress Polymer Science*, vol. 24, pp. 917–950, 1999.

[DUP 05] DUPUY J., ADAMI J., MAAZOUZ A. *et al.*, "Kinetic modeling of an unsaturated polyester resin using two complementary techniques: near infrared spectroscopy and heat flux sensors", *Polymer Engineering & Science,* vol. 45, no. 6, pp. 846–856, 2005.

[DUR 15] DURIN A., CHENOT J.-L, HAUDIN J.-M. *et al.*, "Simulating polymer crystallization in thin films: numerical and analytical methods", *European Polymer Journal,* vol. 73, pp. 1–16, 2015.

[DUS 75] DUSEK K., ILAVSKY M., "Statistics of curing of diepoxides with diamines", *Journal of Polymer Science Polymer Physics*, vol. 53, pp. 29–44, 1975.

[ESC 84] ESCLEINE J.-M., MONASSE B., WEY E. *et al.*, "Influence of specimen thickness on isothermal crystallization kinetics. A theoretical analysis", *Colloïd Polymer Science*, vol. 262, no. 5, pp. 366–373, 1984.

[EVA 45] EVANS U.R., "The laws of expanding circles and spheres in relation to the lateral growth of surface films and the grain-size of metals", *Transactions of the Faraday Society,* vol. 41, pp. 365–374, 1945.

[FRA 03]. FRANCIS B., VANDEN POEL G., POSADA F. *et al.*, "Cure kinetics and morphology of blends of epoxy resin with poly (ether ether ketone) containing pendant tertiary butyl groups", *Polymer*, vol. 44, pp. 3687–3699, 2003.

[GIB 09] GIBOZ J., De l'injection traditionnelle à la micro-injection de pièces en polymères thermoplastiques: Divergences et similitudes, PhD Thesis, University of Savoie, France, 2009.

[GIL 74] GILBERT M., HYBART F.J., "Effect of chemical structure on crystallization rates and melting of polymers: 2 aliphatic polyesters", *Polymer*, vol. 15, pp. 407–412, 1974.

[GON 89] GONZALEZ-ROMERO V.M., CASILLAS N., "Isothermal and temperature programmed kinetic studies of thermosets", *Polymer Engineering & Science*, vol. 29, no. 5, pp. 295–301, 1989.

[GRE 71] GRECHISHKIN V.A., KAZARYAN L.G., PEREPECHKO I.I., "Influence of orientation and crystallization on the velocity of ultrasound in polyethylene terephthalate", *Ultrasonics*, vol. 9, p. 123, 1971.

[GUP 92] GUPTA A.K., RANA S.K., DEOPURA B.L., "Crystallization behavior of high-density polyethylene/linear low-density polyethylene blend", *Journal of Applied Polymer Science*, vol. 44, pp. 719–726, 1992.

[GUO 99] GUO X., ISAYEV A.I., GUO L., "Crystallinity and microstructure in injection moldings of isotactic polypropylenes. Part 1: A new approach to modeling and model parameters", *Polymer Engineering & Science*, vol. 39, pp. 1096–2114, 1999.

[HAR 07] HARSCH M., KARGER-KOCSIS J., HOLST M., "Influence of fillers and additives on the cure kinetics of an epoxy/anhydride resin", *European Polymer Journal*, vol. 43, pp. 1168–1178, 2007.

[HAU 04] HAUDIN J.-M., CHENOT J.-L., "Numerical and physical modeling of polymer crystallization: part I: theoretical and numerical analysis", *International Polymer Processing*, vol. 20, no. 3, pp. 267–274, 2004.

[HAU 08] HAUDIN J.-M., SMIRNOVA J., SILVA L. *et al.*, "Modeling of structure development during polymer processing", *Polymer Science A*, vol. 50, no. 5, pp. 538–549, 2008.

[HAY 76] HAY J.N., FITZGERALD P.A., WILES M., "Use of differential scanning calorimetry to study polymer crystallization kinetics", *Polymer*, vol. 17, pp. 1015–1118, 1976.

[HAY 11] HAYATY M., BEHESHTY M.H., ESFANDE H., "Cure kinetics of a glass/epoxy prepreg by dynamic differential scanning calorimetry", *Journal of Applied Polymer Science*, vol. 120, pp. 62–69, 2011.

[HIL 65] HILLIER I.H., "Modified Avrami equation for the bulk crystallization kinetics of spherolitic polymers", *Journal of Polymer Science Part A: Polymer Chemistry*, vol. 3, pp. 3067–3078, 1965.

[HOR 70] HORIE K., HIURA H., SAWADA M. *et al.*, "Calorimetric investigation of polymerization reactions. III. Curing reaction of epoxides with amines", *Journal Polymer Science Al: Polymer Chemical*, vol. 8, no. 6, pp. 1357–1372, 1970.

[HSU 86] HSU C.C., GEIL P.H., MIYAJI H. *et al.*, "Structure and properties of polypropylene crystallized from the glassy state", *Journal of Polymer Science: Part B: Polymer Physics*, vol. 24, pp. 2379–2401, 1986.

[ISH 77] ISHIZUKA O., KOYAMA K., "Crystallization of running filament in melt spinning of polypropylene", *Polymer*, vol. 18, pp. 913–918, 1977.

[ITO 95] ITO H., TSUTSUMI Y., MINAGAWA K. *et al.*, "Simulations of polymer crystallization under high pressure", *Colloïd Polymer Science*, vol. 273, pp. 811–815, 1995.

[JEZ 78] JEZIORNY A., "Parameters characterizing the kinetics of the non-isothermal crystallization of poly(ethylene terephthalate) determined by D.S.C.", *Polymer*, vol. 19, pp. 1142–1144, 1978.

[JON 97] JONES N.A., ATKINS E.D.T., HILL M.J. *et al.*, "Chain-folded lamellar crystals of aliphatic polyamides. Investigation of nylons 4 8, 4 10, 4 12, 6 10, 6 12, 6 18 and 8 12", *Polymer*, vol. 38, pp. 2689–2699, 1997.

[KAM 74] KAMAL M.R., "Thermoset characterization for moldability analysis", *Polymer Engineering & Science*, vol. 14, pp. 231–239, 1974.

[KEI 64] KEITH H.D., PADDEN F.J., "Spherulitic crystallization from the melt. II. Influence of fractionation and impurity segregation on the kinetics of crystallization", *Journal Applied Physics*, vol. 35, pp. 1286–1296, 1964.

[KIS 56] KISSINGER H.E., "Variation of peak temperature with heating rate in differential thermal analysis", *J. of Research of the Nat. Bureau of Standards*, vol. 57, no. 4, pp. 217–221, 1956.

[KOR 06] KORTABERIA G., SOLAR S., JIMENO A. *et al.*, "Curing of an epoxy resin modified with nanoclay monitored by dielectric spectroscopy and rheological measurements", *Journal of Applied Polymer Science*, vol. 102, pp. 5927–5933, 2006.

[KOT 75] KOTOV N.M., BAKEYEV N.F., BELOV G.P. *et al.*, "Study of melting and crystallization of" polyethylene at high pressure", *Polymer Science USSR*, vol. 17, pp. 1312–1317, 1975.

[LAF 86] LAFLEUR P.G., KAMAL M.R., "A structure-oriented computer simulation of the injection molding of viscoelastic crystalline polymers part I: Model with fountain flow, packing, solidification", *Polymer Engineering & Science,* vol. 26, pp. 92–102, 1986.

[LAM 04] LAMBERTI G., DE SANTIS F., BRUCATO V. *et al.*, "Modeling the interactions between light and crystallization polymer during fast cooling", *Applied Physics A*, vol. 78, pp. 895–901, 2004.

[LE 11] LE GOFF R., BOYARD N., SOBOTKA V. *et al.*, "Inverse estimation of the crystallization kinetic function of semi-crystalline polymers and short fibre reinforced composites in moderate cooling conditions", *Polymer Testing*, vol. 30, pp. 678–687, 2011.

[LEE 92] LEE S.N., CHIU M.T., LIN H.S., "Kinetic model for the curing reaction of a tetraglycidyl diamino diphenyl methane / diamino diphenyl sulphone (TGDDM/DDS) epoxy resin system", *Polymer Engineering & Science*, vol. 32, no. 15, pp. 1037–1046, 1992.

[LON 95] LONG Y., SHANKS R.A., STACHURSKI Z.H., "Kinetics of polymer crystallization", *Prog. Polym. Sci.*, vol. 20, pp. 651–701, 1995.

[MAF 05] MAFI R., MIRABEDINI S.M., ATTAR M.M. *et al.*, "Cure characterization of epoxy and polyester clear powder coating using differential scanning calorimetry (DSC) and dynamic mechanical thermal analysis (DMTA)", *Progress in Organic Coatings,* vol. 54, pp. 164–169, 2005.

[MAG 62] MAGILL J.H., "Crystallization of polyamides II-Nylon 6 and Nylon 66", *Polymer*, vol. 35, no. 3, pp. 367–371, 1962.

[MAL 11] MALEKY F., MARANGONO A., "Ultrasonic technique for determination of the shear elastic modulus of polycrystalline soft materials", *Crystal Growth & Design*, vol. 11, pp. 941–944, 2011.

[MAR 00] MARTIN J.S., LAZA J.M., MORRAS M.L. *et al.*, "Study of the curing process of a vinyl ester resin by means of TSR and DMTA", *Polymer*, vol. 41, pp. 4203–4211, 2000.

[MAR 04] MARAND H., HUANG Z., "Isothermal lamellar thickening in linear polyethylene: correlation between the evolution of the degree of crystallinity and the melting temperature", *Macromolecules*, vol. 37, 6492–6497, 2004.

[MIJ 92] MIJOVIC J., FISHBAIN A., WIJAYA J., "Mechanistic modeling of epoxy-amine kinetics. 1. Model compound study", *Macromolecules*, vol. 25, pp. 979–985, 1992.

[MIL 01] MILLISCHER A., DELAUNAY D., "Experimental and numerical analysis of heat transfer in bulk molding compound injection process", *J. Reinf. Plast. Comp.*, vol. 20, no. 6, pp. 495–512, 2001.

[MON 04] MONASSE B., SMIRNOVA J., HAUDIN J.-M. *et al.*, "Numerical and physical modeling of polymer crystallization. Part II: isothermal and non-isothermal crystallization kinetics of a polypropylene in 2 dimensions – experiments and simulation", *International Polymer Processing*, vol. 20, no. 3, pp. 275–286, 2004.

[MOR 87] MORGAN R.J., MONES E.T., "The cure reactions, network structure, and mechanical response of diaminophenyl sulfone-cured tetraglycidyl4,4' diaminodiphenyl methane epoxies", *Journal of Applied Polymer Science*, vol. 33, pp. 999–1020, 1987.

[NAK 72] NAKAMURA K., WATANABE T., KATAYAMA K. *et al.*, "Some aspects of nonisothermal crystallization of polymers. I. Relationship between crystallization temperature, crystallinity, and cooling conditions", *Journal of Applied Polymer Science*, vol. 16, no. 5, pp. 1077–1091, 1972.

[NAK 73] NAKAMURA K., KATAYAMA K., AMANO T., "Some aspects of nonisothermal crystallization of polymers. II. Consideration of the isokinetic condition", *Journal of Applied Polymer Science*, vol. 17, pp. 1031–1041, 1973.

[NAK 12] NAKOUSI S., Modélisation du procédé de cuisson de composites infusés par chauffage infra-rouge, PhD Thesis, University of Toulouse 3, France, 2012.

[NAW 12] NAWAB Y., TARDIF X., BOYARD N. *et al.*, "Determination and Modeling of the cure shrinkage of epoxy vinylester resin and associated composites by considering thermal gradients", *Composites Science and Technology*, vol. 73, pp. 81–87, 2012.

[NG 89] NG H., MANAS-ZLOCZOWER I., "A nonisothermal differential scanning calorimetry study of the curing kinetics of an unsaturated polyester system", *Polymer Engineering & Science,* vol. 29, pp. 1097–1102, 1989.

[OZA 71] OZAWA T., "Kinetics of non-isothermal crystallization", *Polymer*, vol. 12, no. 3, pp. 150–158, 1971.

[PAS 02] PASCAULT J.-P., SAUTEREAU H., VERDU J. *et al.*, *Thermosetting Polymers*, Marcel Dekker Inc., New York, 2002.

[PAT 91] PATTEL R.M., SPRUIELL J.E., "Crystallization kinetics during polymer processing – Analysis of available approaches for process modeling", *Polymer Engineering & Science*, vol. 31, pp. 730–738, 1991.

[PAT 93] PATHMANATHAN K., JOHARI G.P., "The effect of increased crystallization on the electrical properties of nylon-12", *Journal of Polymer Science Polymer Physics*, vol. 31, pp. 265–271, 1993.

[PIG 15] PIGNON B., TARDIF X., LEFEVRE N. *et al.*, "A new PvT device for high performance thermoplastics: heat transfer analysis and crystallization kinetics identification", *Polymer Testing*, vol. 45, pp. 152–160, 2015.

[PIJ 02] PIJPERS M.F.J., MATHOT V.B.F., GODERIS B. *et al.*, "High-speed calorimetry for the analysis of kinetics of vitrification, crystallization and melting of macromolecule", *Macromolecules*, vol. 35, no. 9, pp. 3601–3613, 2002.

[PIO 06] PIORKOWSKA E., GALESKI A., HAUDIN J.-M., "Critical assessment of overall crystallization kinetics theories and predictions", *Progress Polymer Science*, vol. 31, no. 6, pp. 549–575, 2006.

[PIO 13] PIORKOWSKA E., GALESKI A., "Overall crystallization kinetics", in PIORKOWSKA E., RUTLEDGE G.C. (eds), *Handbook of Polymer Crystallization*, John Wiley and Sons Inc., 2013.

[PIS 15] PISSIS P., PANDIS C., MAROULAS P. *et al.*, "Electrical/dielectric measurements for monitoring polymerization, morphology and mechanical integrity in polymer nanocomposites", *Proceeding Engineering,* vol. 114, pp. 598–605, 2015.

[PRI 73] PRIME R.B., "Differential scanning calorimetry of the epoxy cure reaction", *Polymer Engineering & Science*, vol. 13, pp. 365–371, 1973.

[RAB 37] RABINOWITC H E., "Collision, co-ordination, diffusion and reaction velocity in condensed systems", *Trans. Faraday Soc.*, vol. 33, pp. 1225–1233, 1937.

[RAM 99] RAMIS X., SALLA J.M., "Effect of the initiator content and temperature on the curing of an unsaturated polyester resin", *Journal of Polymer Science B: Polymer Physics*, vol. 37, pp. 751–768, 1999.

[REN 04] REN M., MO Z., CHEN Q. *et al.*, "Crystallization kinetics and morphology of nylon 1212", *Polymer*, vol. 45, pp. 3511–3518, 2004.

[RIC 84] RICCARDIH C.C., ADABBO E., WILLIAMS R.J.J., "Curing reaction of epoxy resins with diamines", *Journal of Applied Polymer Science*, vol. 29, no. 8, pp. 2481–2492, 1984.

[ROC 04] ROCKS J., RINTOUL L., VOHWINKEL F. *et al.*, "The kinetics and mechanism of cure of an amino-glycidyl epoxy resin by a co-anhydride as studied by FT-Raman spectroscopy", *Polymer*, vol. 45, pp. 6799–6811, 2004.

[RUI 05] RUIZ E., TROCHU F., "Thermal and mechanical properties during cure of ... elastic andviscoelastic modeling", *Journal of Reinforced Plastics and Composites*, vol. 39, no. 10, pp. 881–916, 2005.

[RYA 79] RYAN M.E., DUTTA A., "Kinetics of epoxy cure: a rapid technique for kinetic parameter estimation", *Polymer*, vol. 20, pp. 203–206, 1979.

[SBI 07] SBIRRAZZUOLI N., VINCENT L., MIJA A. *et al.*, "Integral, differential and advanced isoconversional methods: complex mechanisms and isothermal predicted conversion–time curves" *Chemometrics and Intelligent Laboratory Systems*, vol. 96, no. 2, pp. 219–226, 2007.

[SCH 88] SCHNEIDER W., KOPPL A., BERGER J., "Non-isothermal crystallization: crystallization of polymers: system of rate equations", *International Polymer Processing*, vol. 3–4, pp. 151–154, 1988.

[SCH 89] SCHULTE K., BARON C., "Load and failure analyses of CFRP laminates by means of electrical resistivity measurement", *Composites Science and Technology*, vol. 36, pp. 63–76, 1989.

[SOU 76] SOUROUR S., KAMAL M.R., "Differential scanning calorimetry of epoxy cure: isothermal cure kinetics", *Thermochimica Acta*, vol. 14, pp. 41–59, 1976.

[SOW 15] SOWINSKI P., PIORKOWSKA E., BOYER S.E.A. *et al.*, "The role of nucleating agents in high-pressure-induced gamma crystallization in isotactic polypropylene", *Colloïd Polymer Science*, vol. 293, no. 3, pp. 665–675, 2015.

[STE 86] STEVENSO J.F., "Free radical polymerization models for simulating reactive processing", *Polymer Engineering & Science,* vol. 26, no. 11, pp. 746–759, 1986.

[SUP 02] SUPAPHOL P., SPRUEILL J.E., "Nonisothermal bulk crystallization of high-density polyethylene via modified depolarized light microscopy technique: further analysis", *Journal of Applied Polymer Science*, vol. 86, pp. 1009–1022, 2002.

[TAR 12] TARDIF X., AGAZZI A., SOBOTKA V. *et al.*, "A multifunctional device to determine specific volume, thermal conductivity and crystallization kinetics of semi-crystalline polymers", *Polymer Testing*, vol. 31, pp. 819–827, 2012.

[TAR 14] TARDIF X., PIGNON B., BOYARD N. *et al.*, "Experimental study of crystalllization of polyetheretherketone (PEEK) over a large temperature range using nano-calorimeter", *Polymer Testing*, vol. 36, pp. 10–19, 2014.

[TIT 97] TITOMANLIO G., SPERANZA V., BRUCATO V., "On the simulation of thermoplastic injection moulding process", *International Polymer Processing*, vol. 12, pp. 45–53, 1997.

[TOB 74] TOBIN M.C., "Theory of phase transition kinetics with growth site impingement. I. Homogeneous nucleation", *Journal of Polymer Science Polymer Physics*, vol. 12, pp. 399–406, 1974.

[TOB 76] TOBIN M.C., "Theory of phase transition kinetics with growth site impingement. II. Heterogeneous nucleation", *Journal of Polymer Science Polymer Physics*, vol. 17, pp. 2253–2257, 1976.

[TOB 77] TOBIN M.C., "Theory of phase transition kinetics with growth site impingement. III. mixed heterogeneous-homogeneous nucleation and noninegral exponents of the time", *Journal of Polymer Science part B: Polymer Physics*, vol. 15, pp. 2269–2270, 1977.

[TWU 93] TWU J.-T., HILL R.R., WANG T.J. *et al.*, "Numerical simulation of non-isothermal SMC (sheet molding compound) molding", *Polymer Composites*, vol. 14, no. 6, pp. 503–514, 1993.

[VEL 86] VELISARIS C., SEFERIS J.C., "Crystallization kinetics of polyetheretherketone (PEEK) matrices", *Polymer Engineering & Science*, vol. 26, pp. 1574–1587, 1986.

[VER 90] VERCHERE D., SAUTEREAU H., PASCAULT J.P. *et al.*, "Build-up of epoxycycloaliphatic amine networks. Kinetics, vitrification and gelation", *Macromolecules*, vol. 23, pp. 725–731, 1990.

[VYA 11] VYAZOVKIN S., BURNHAM A.K., CRIADO J.M. *et al.*, "ICTAC Kinetics Committee recommendations for performing kinetic computations on thermal analysis data", *Thermochimica Acta*, vol. 520, nos. 1–2, pp. 1–19, 2011.

[WAN 12] WANG J.Y., YANG S.Y., HUANG Y.L. *et al.*, "Synthesis and properties of trifluoromethyl groups containing epoxy resins cured with amine for low D_k materials", *Journal of Applied Polymer Science*, vol. 124, pp. 2615–2624, 2012.

[ZHA 04] ZHANG J., DUAN Y., SHEN D. *et al.*, "Structure changes during the induction period of cold crystallization of isotactic polystyrene investigated by infrared and two-dimensional infrared correlation spectroscopy", *Macromolecules*, vol. 37, pp. 3292–3298, 2004.

[ZHU 10] ZHURAVLEV E., SCHICK C., "Fast scanning power compensated differential scanning nanocalorimeter: 1. The device", *Thermochimica Acta*, vol. 505, pp. 1–13, 2010.

4

Phase Change Kinetics within Process Conditions and Coupling with Heat Transfer

This chapter carries on from the previous one by presenting how we can take into account the effect of material forming conditions (heating/cooling rates, mechanical stresses (shear flow, elongational flow and pressure) and the addition of other materials (nucleating agents, catalysts, reinforcement: particles and fibers) in transformation kinetics. In order to control the final product properties, a sound understanding of these coupled effects is essential, especially for thermoplastic semi-crystalline materials. Both thermal and flow-induced crystallization modeling are presented. They act either in synergy or in competition (i.e. enhancing or cancelling each other), which can result in either promoting or hindering the transformation kinetics. A focus is also made on the effect of crystallization on the rheological behavior of a semi-crystalline polymer, on the induced microstructures observed in the solid state and on the final mechanical, thermophysical and optical properties.

4.1. Introduction

The study of the relationship between polymer materials or composites, their structure and behavior on the one hand and their processing conditions on the other hand, is a multidisciplinary science. Its modeling requires multiphysic coupling and considerations at various scales: microscopic for the chemical structures of the macromolecules constituting the material and macroscopic for the global behavior and the process–material interactions.

Chapter written by M'hamed BOUTAOUS, Matthieu ZINET, Nicolas BOYARD and Jean-Luc BAILLEUL.

Such studies often require important experimental developments to understand the basic aspects of these interactions, as well as mathematical modeling often inspired by experimental observations. Laboratory experimental conditions are unfortunately not always representative of those material experiences during processing. Mathematical modeling in this case is, therefore, an essential tool to explore varied or extreme conditions and offers detailed parametric studies for the optimization of the process–structure–properties triangle.

The heat transfer during processing of polymer and/or composites materials plays an important role. The molds are truly heat exchangers that are determinant for the quality of the finished products (Figure 4.1). It is necessary to model the molding cavity thermal behavior, taking into account several physical phenomena which occur, due to the interactions between the mold and the material.

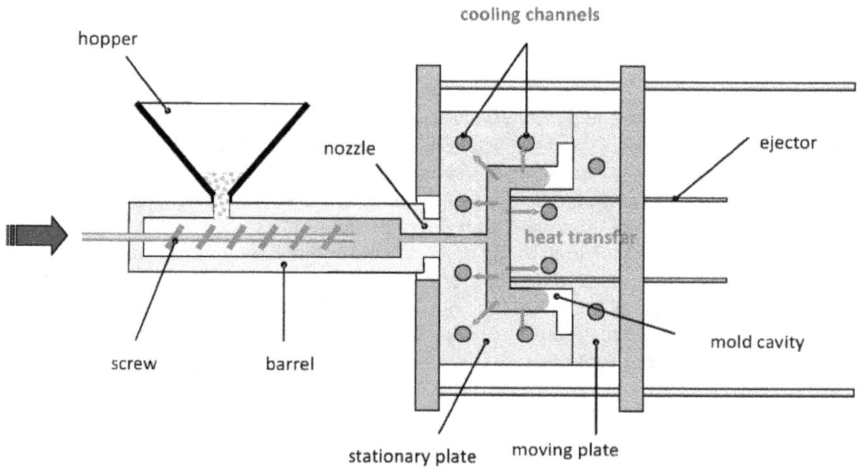

Figure 4.1. *Principle of injection molding process. For a color version of the figure, see www.iste.co.uk/boyard/heat.zip*

Numerical simulation of thermoplastic polymer behavior during processing requires to account for a large amount of physical data and phenomena. In semi-crystalline polymers, the properties of the final product are strongly dependent on the thermomechanical history experienced by the material during processing. Structural heterogeneities such as rigidity gradients and shrinkage anisotropy are directly related to the crystalline microstructure. Therefore, to model and simulate heat transfer in polymer materials during their processing, it is necessary to take into account all the thermodynamical and physical transformations. One of them is the crystallization phenomenon, which specifically concerns semi-crystalline polymers commonly used in numerous applications.

The thermomechanical history of semi-crystalline polymers strongly affects their crystallization kinetics. The mechanical and thermal stresses that are experienced modify both the crystallization morphology and kinetics. They align macromolecular chains, improve nucleation, accelerate crystallization and generate oriented morphologies.

A variety of approaches have been proposed to model flow effects on crystallization kinetics [ZIN 10a]. The most interesting ones explicitly take into account the nucleation and growth processes. Indeed, experimental observations have clearly shown that the number of nuclei per unit volume (nucleation density) is considerably increased under shear conditions [NAK 72, MON 95]. This additional nucleation rate can be linked to various parameters: free energy-related quantities [TAN 03], shear rate [ITO 96], combination of shear strain and shear rate [BUS 96] and recoverable strain tensor [VER 96].

A more detailed survey on crystallization kinetics modeling has been presented by Zinet *et al.* [ZIN 10a, ZIN 10b], where the proposed numerical model is able to simulate the crystallization of polymers under non-isothermal flow condition. Based on the earlier work of Marrucci [MAR 72], and later Mazinani [MAZ 09] on the molecules orientation under flow, the authors make the assumption that the polymer melt elasticity, quantified by the trace of the extra-stress tensor, is the driving force of flow-induced extra nucleation. Thus, the sensitivity of the crystallization kinetics to the flow is not only a consequence of the flow kinematics, but is also determined by the viscoelastic rheological behavior of the melt.

In addition, polymers usually serve as a matrix in composite materials, where filler and/or fibers with different type and shapes are added to form a polymer-based composite. The interactions of the matrix with these additional interfaces lead to other phenomena, which have to be taken into account in the modeling or the characterization of such materials: transcrystallinity, additional nucleation, interfacial molecular diffusion, etc.

The objectives of this chapter are to provide tools or data on polymer solidification during processing conditions. Understanding and mastering the polymer or composite transformation kinetics are key factors in controlling the final properties of these materials. First, key experimental observations concerning crystallization in processing conditions, especially flow, are presented. Then, various modeling approaches based on these observations are reviewed and discussed. Extensive theoretical developments on the crystallization of polymers subjected to non-isothermal flow are reported. An illustration of the process–structure relationship is proposed through recent theories and the presented results and/or observations are dealing with the effect on crystallization kinetics of thermoplastic polymers, in order to quantify and discriminate, respectively, the thermal and flow effects on the material behavior.

Note that more works are still needed for a better understanding of the matrix fiber interactions, and/or to extend more accurately the kinetics theories to composites materials. Illustrations will be presented and commented.

4.2. Flow-induced crystallization: experimental observations

In most plastics processing methods, the molten polymer is subjected to flows of varying intensity, duration and geometrical complexity. In the case of semi-crystalline polymers, final material properties are strongly affected by the thermomechanical history experienced by the material during forming. This explains why flow-induced crystallization has prompted a large number of experimental studies so far.

Basically, the flows encountered in industrial processes fall into two categories: elongational flows and shear flows. For elongational flows, all the non-diagonal elements of the strain rate tensor D and the stress tensor τ are zero, whereas for shear flows, the non-diagonal elements can be non-zero [BIR 87]. Elongational flows are found in fiber spinning, film blowing or injection molding in the melt front, while shear flows are encountered in extrusion and injection molding (excluding melt front). We will focus more specifically here on the next sections.

4.2.1. *Relevant experimental techniques*

Shear-induced crystallization can be studied by various means: rheometric flows or configurations closer to actual processes coupled with optical methods (microscopy, transmitted light intensity measurement and birefringence) [LIE 93], X-ray scattering (WAXD[1] and SAXS[2]) [KUM 99, SOM 00] or light scattering (SALS[3]) techniques [POG 01].

In most studies, a short preshear treatment is applied at the beginning of the test and substantially all of the crystallization develops in the quiescent state [LIE 93,

1 Wide angle X-ray diffusion.
2 Small angle X-ray scattering.
3 Small angle light diffusion.

KUM 99, KOS 02]. Nevertheless, some experimenters choose to shear the sample continuously for the whole duration of crystallization [WAS 00, ACI 03]. Both techniques are of interest, the choice being dictated by the targeted industrial application. For instance, in injection molding, the polymer is subjected to a high shear rate for a short time, while in extrusion, shearing is applied continuously at lower rates.

The rotational rheometer is by far the most common shearing method, mainly cone-plate geometry [KOS 02, HAA 69] and plane-plane geometry [WAS 00, ACI 03]. The main drawback of this technique is the limitation of the shear rate to a few tens of s^{-1}. In the cone-plate geometry, the shear rate is independent of the radial location, however, the gap (linked to the sample thickness) must be perfectly controlled, which can be difficult if the experimental protocol includes large temperature (hence specific volume) variations.

4.2.2. *Effect of flow on crystallization kinetics*

4.2.2.1. *Seminal work*

The study of flow-induced crystallization of molten polymers began in the late 1960s. The first notable works on the subject were those of Haas and Maxwell [HAA 69], which demonstrate an accelerated crystallization kinetics in a polyethylene (PE) and a linear poly-1-butene (PB-1) subjected to constant stress creep flow. These authors also reported a significant increase in the number of nuclei and the formation of columnar-type structures oriented in the flow direction.

These observations are found in most of the later works despite the diversity of materials and devices used in the studies. In particular, several investigations of the overall crystallization kinetics by the Avrami model showed an increase in the rate constant *k* under shear stress [HAA 69, FRI 74, ULR 76, SHE 77, WOL 78].

From isothermal crystallization tests of a PE after shearing in a parallel plate rheometer (shear rate of 0.03–30 s^{-1}), Lagasse and Maxwell [LAG 76] demonstrated the existence of a critical shear rate, independent of the crystallization temperature, at which the induction time[4] decreases. The curves in Figure 4.2 are typical of flow-induced crystallization kinetics. Note that in this study, the definition of induction

---

4 In crystallization studies, the induction time is generally defined as the elapsed time between the start of the test and the time when crystallization becomes observable. Therefore, it is clear that the induction time is highly dependent on the observation method and cannot be considered as an objective measure of crystallization kinetics. It is rather a benchmark for comparison of the various test runs within a given study.

time is rather inaccurate, so the conclusions are essentially qualitative. The same trends were obtained by Kobayashi and Nagasawa using a coaxial cylinder rheometer [KOB 70].

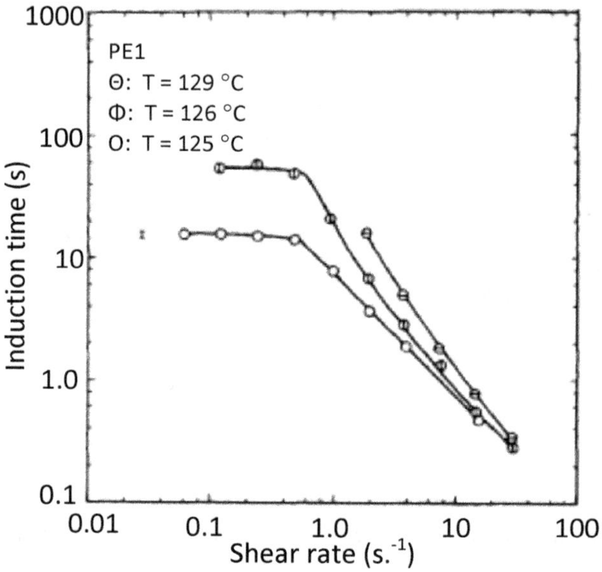

Figure 4.2. *Decreased induction time of crystallization beyond a critical shear rate, for a PE [LAG 76]*

Finally, using a rotational plane-plane rheometer, Wolkowicz [WOL 78] showed that the main effect of shear on the crystallization process of a poly(1-butene) was an increase in the nucleation rate while the crystalline growth rate remained the same as in quiescent conditions.

4.2.2.2. *Recent developments*

With the increasing availability of commercial rheometers and apparatus allowing optical observation of the sample during the crystallization process, such as the multipass rheometer [MAC 95] and the Linkam cell [MAC 99], the study of crystallization under shear was developed significantly from the 1990s.

In the work of Vleeshouwers and Meijer [VLE 96], an isotactic polypropylene was subjected to a low amplitude oscillatory shearing (shear rate up to 10 s^{-1}) during a time t_s ($0 < t_s < 200$ s) at a high temperature, then rapidly cooled to the crystallization temperature. The acceleration of crystallization is as notable as if shearing was applied directly to the cooled sample. In addition, the respective roles of the shear rate and shearing time are not the same: a short shearing applied at a high rate seems to have more effect than a long shearing at a lower rate.

Similar experimental methods have been used by Bove and Nobile [BOV 02] to study the influence of mild shearing (0.01 – 1 s^{-1}) on poly(1-butene) of several molecular masses. A flow sensitivity factor has been defined as the ratio of the Avrami constants obtained, respectively, after shearing and under quiescent conditions. It appears that the highest molecular weight PB-1s have a higher sensitivity to flow, while in quiescent conditions, the molecular weight does not influence the crystallization kinetics.

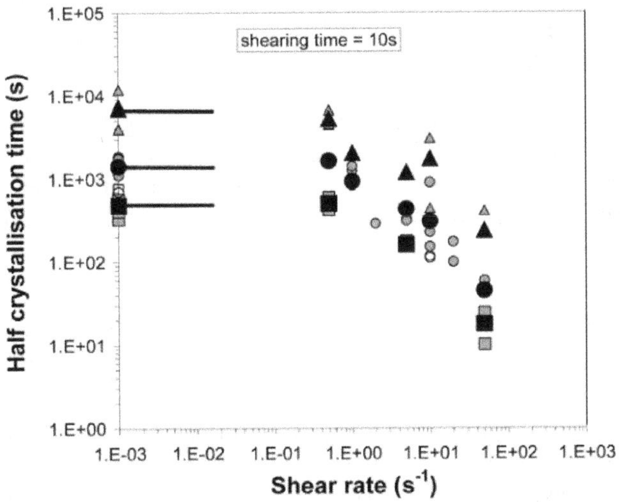

Experimental half crystallization times versus shear rate. ▲: Shearing hot stage, $T_c = 140$ °C; ●: shearing hot stage, $T_c = 134$ °C; ■: shearing hot stage, $T_c = 130$ °C; ○: rheometer, $T_c = 134$ °C; □: rheometer, $T_c = 130$ °C; ▲: average values, $T_c = 140$ °C; ●: average values, $T_c = 134$ °C; ■: average values $T_c = 130$ °C. Horizontal lines show the average values at zero shear ($x = -\infty$ on a logarithmic scale) calculated from DSC, rheometry, and microscopy experiments (DSC results are not shown).

Figure 4.3. *Effect of preshear treatment on crystallization kinetics of polypropylene [KOS 02]*

The effects of a preshear treatment on the crystallization kinetics of isotactic polypropylenes of various molecular weights were characterized by Koscher and

Fulchiron [KOS 02] using a Linkam cell and a standard rheometer (plane-plane geometry). After melting, the samples were brought to the desired crystallization temperature (140, 134 or 130°C), then subjected to preshear (10^{-3} – 50 s^{-1}) for a time t_s. Crystallization was then monitored by several methods (Figure 4.3). The plot of the average crystallization half-time versus shear rate reveals the acceleration of the kinetics beyond a certain critical shear rate. The microscope observation of the sample during crystallization showed an increase in the nucleation density with the shear rate. In addition, if the shearing time is sufficient, an alignment of nuclei appears, leading to fibrillar-shaped crystallites.

This study also showed that high average molecular weight iPPs crystallize more slowly than those with lower average molecular weight in quiescent conditions, but they are much more sensitive to shearing and have their crystallization kinetics increased in larger proportions. However, the polymer polydispersity index does not appear as a major factor.

Acierno *et al.* [ACI 03] applied continuous shear (range: 10^{-3}–10 s^{-1}) to isotactic poly (1-butene) (iPB) of various molecular weights (with similar polydispersity index) by means of a Linkam cell with plane-plane geometry. Measurement of the transmitted light intensity (turbidity measurement) is used to define a dimensionless crystallization time:

$$\Theta = \frac{t_{0.5,\dot{\gamma}}}{t_{0.5,q}} \qquad [4.1]$$

where $t_{0.5,\dot{\gamma}}$ is the turbidity half-time (i.e. the time for which the transmitted intensity decreases by 50%) measured under shear and $t_{0.5,q}$ is the turbidity half-time measured under quiescent conditions. The Weissenberg number, which represents the ability of shearing (rate: $\dot{\gamma}$) to orient the chain segments of a given polymer, is defined herein as a function of the longest relaxation time of the polymer τ_{max}:

$$We = \dot{\gamma}\tau_{max} \qquad [4.2]$$

Note that τ_{max} is strongly dependent on the molecular weight. Figure 4.4 clearly shows that all the data points (Θ, *We*) can be plotted on a single master curve exhibiting a plateau followed by a decrease beyond a first critical value of *We* (≈20), imputed to a simple chain orientation. In addition, microscopic observations highlight a second critical *We* (≈150) associated with chain stretching, beyond which the formed crystalline morphologies are no longer spherulites but alignments of nuclei (row nucleation).

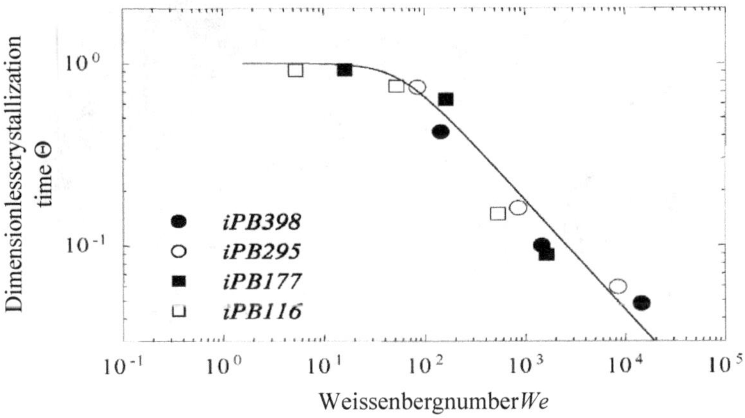

Figure 4.4. *Dimensionless crystallization time of isotactic poly(1-butene) of various molecular weight versus Weissenberg number, at 103°C [ACI 03]*

During the 1990s, the group of G. Eder and H. Janeschitz–Kriegl focused on the study of isothermal crystallization under short time but high-intensity shearing. They developed a special apparatus for injecting a polymer into a thermoregulated pipe of rectangular section, with a controlled constant pressure and a controlled injection time [LIE 93]. The device is mounted to the output of a standard extruder. The shear rate at the walls is ranging 60–140 s^{-1}. The configuration of the apparatus eliminates the elongation effects due to the fountain flow pattern usually met in the presence of a flow front [TAD 74], as well as the influence of the thermomechanical history experienced by the polymer in the extruder. The pipe is equipped with windows for optical monitoring of crystallization (polarized light-transmitted intensity). After completion of crystallization, a section of the solidified sample is cut and examined by microscopy (Figure 4.5).

From the surface to the core, three zones can be identified:

– close to the wall, a highly oriented layer;

– an isotropic layer with a very fine granular structure;

– in the center, an isotropic zone formed of larger radius spherulites.

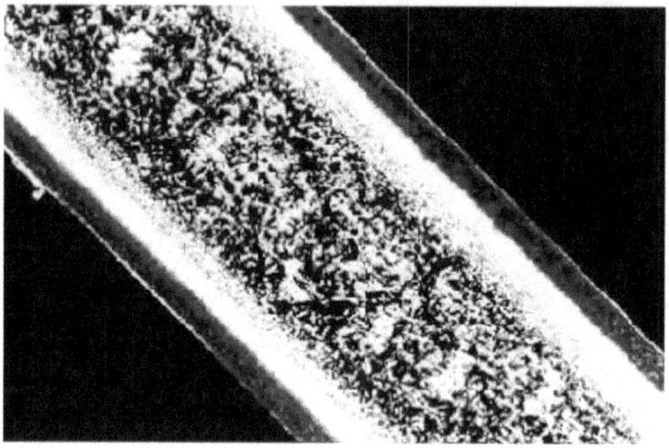

Figure 4.5. *Polarized light micrograph of a solidified isotactic polypropylene sample. The section is parallel to the flow direction [EDE 97]*

The thickness of each layer depends on both the shear rate $\dot{\gamma}$ and the duration t_s of the applied shearing. In particular, the shear rate seems to be more influent than the shearing time in an extent that depends on the considered layer. Thus, at the interface between the granular layer and the spherulitic core, the critical shear rate $\dot{\gamma}_c$ is such that the product $\dot{\gamma}_c^A t_s^2 = cte$, while at the interface between the highly oriented layer and the granular layer, the shear rate has more weight: $\dot{\gamma}_c^A t_s = cte$. A two-step nucleation mechanism was assumed by Liedauer *et al.* [LIE 93] to explain these observations:

– first step: formation of point-like primary nuclei, with a frequency proportional to $\dot{\gamma}^2$. This process does not depend on the shear direction, but only on its intensity;

– second step: if shearing is sufficiently intense and/or long, part of the primary nuclei become fibrillary precursors ("thread-like precursors") oriented in the flow direction, responsible for the formation of the highly oriented layer.

Both nucleation processes have been described by the following differential equations:

$$\dot{N} = \left(\frac{\dot{\gamma}}{\dot{\gamma}_n}\right)^2 g_n - \frac{N}{\tau_n} \qquad [4.3]$$

and

$$\dot{L}=\left(\frac{\dot{\gamma}}{\dot{\gamma}_l}\right)^2 g_l-\frac{L}{\tau_l} \qquad [4.4]$$

where N is the primary nucleation density, L is the length of the oriented precursors, $\dot{\gamma}_n$ and $\dot{\gamma}_l$ are the characteristic shear rate of each process, g_n and g_l are the factors of respective dimension $m^{-3}s^{-1}$ and $m^{-1}s^{-1}$, τ_n and τ_l are the relaxation times associated with each process. The solution of these equations, under assumptions on the relaxation times, is consistent with experimental results: the total length L_{tot} of the precursors is proportional to $\dot{\gamma}^4 t_s^{\,2}$ or $\dot{\gamma}^4 t_s$, depending on the particular case. In continuation of this work, Jerschow and Janeschitz-Kriegl [JER 97] showed that iPP with high molecular weight or containing nucleating agents tends to form a thick highly oriented layer, even at low shear rates.

The principle of the device developed by Liedauer *et al.* was taken up and miniaturized by Kumaraswamy *et al.* [KUM 99, KUM 00, KUM 02] in order to enable monitoring of the crystallization by WAXD. In particular, a saturation effect of the crystallization kinetics beyond a critical shearing time has been found, which was explained by the reaching of a maximum orientation and stretching state of the macromolecular chains.

4.2.3. *Flow effect on crystalline morphology*

Polymers crystallized from a melt subjected to flow generally exhibit unique crystalline structures, very different from the spherulitic morphology obtained in the quiescent state. A type of supramolecular cylindrical structure called "shish-kebab", first observed in polymers subjected to elongational flows [KEL 67], is in the form of a bundle of chains extended and aligned in the flow direction ("shish") on which the crystalline lamellae nucleate and grow epitaxially ("kebab"), apart from a few nanometers. It was subsequently shown [ULR 76, HIS 83] that, due to chain entanglements, this type of morphology could also appear during the crystallization of molten polymers subjected to shear flow.

Later, the use of transmission electron microscopy on PE samples, sheared between two parallel disks [HOS 95], and on iPP samples, sheared by glass fiber traction [MON 96, WHI 98], has enabled a more detailed observation of the crystalline structures at the lamellar scale. In particular, the existence of lamellae oriented parallel to the central portion (shish) was shown, all the more numerous as the radial distance to the shish axis is greater. The effect of this particular growth mode would be to increase the final crystallinity by occupying the volumes left free by the lamellae oriented perpendicular to the shish.

The small angle X-ray scattering (SAXS) technique was used by Somani *et al.* [SOM 00] in order to study the crystallization of isotactic polypropylenes subjected to short-term shearing (rate: 10–100 s^{-1}) in a modified Linkam cell at 140°C. Deconvolution of the scattering patterns enabled us to monitor in real time the evolution of the isotropic crystalline fraction (spherulites) and oriented crystalline fraction. The total strain was set to be constant for each run: $\dot{\gamma} \cdot t_s = 1428\%$. These experiments showed, for $\dot{\gamma} = 100$ s^{-1}, an enhancement of the overall crystallization kinetics of about 2 decades compared to the quiescent state (Figure 4.6(a)). An increase in the oriented crystalline fraction was also observed, but only below a certain value of the shear rate. The authors link this to the concept of "orientation critical molecular mass" M^*: for a given shear rate, only the chains with a molecular mass higher than M^* can form oriented structures (Figure 4.6(b)), while smaller chains return to the random coil state when flow stops. A nucleation and growth mechanism is proposed: shearing promotes the alignment of chain segments and therefore the appearance of primary nuclei on which crystalline lamellae grow perpendicular to the flow (secondary nucleation). Attention is drawn to the fact that structures in the form of long segments of extended chains (shish) were not observed, probably due to insufficient total strain. However, this does not prevent the formation of oriented crystalline structures (kebabs) from primary nuclei of very short length.

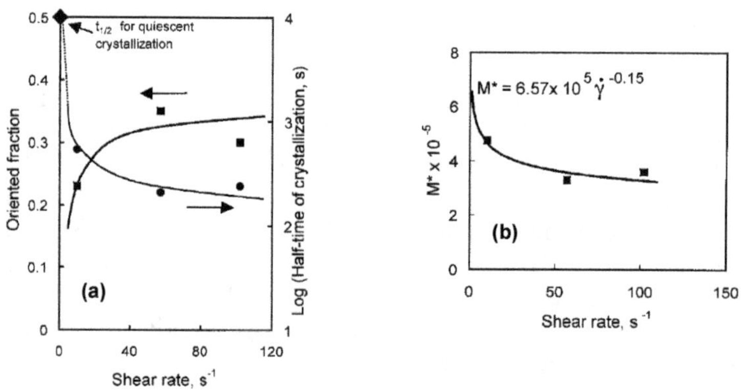

Figure 4.6. *a) Effect of shear rate on the oriented fraction and half-time of crystallization of iPP at 140°C, for a constant 1,428% strain; b) orientation critical molecular mass versus shear rate, at 140°C, for a constant 1,428% strain [SOM 00]*

Using an *in situ* atomic force microscopy (AFM) technique, Hobbs and Miles [HOB 01a, HOB 01b] visualized the formation of shish-kebab structures during the crystallization of polyethylene film subjected to flow. In particular, they showed a variation of the lamellar growth rate from one crystal to the other and an overall slowing growth over time. These observations have not been fully explained. In addition, a deflection phenomenon of the lamellar growth direction which causes the "interlacing" of kebab belonging to adjacent crystals was reported.

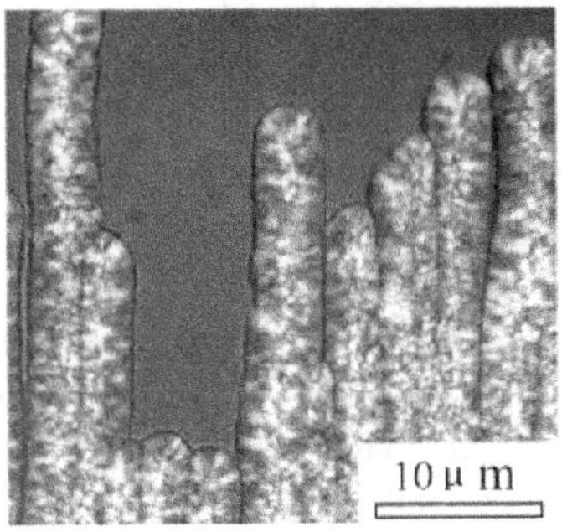

Figure 4.7. *Polarized optical microscopy visualization of shish-kebab morphology in iPP, obtained after 5 s shearing at 1 s^{-1} and crystallization at 140°C [ZHA 05]*

Recently, Zhang *et al.* [ZHA 05] observed by optical microscopy the formation of shish-kebabs in iPP after relatively low rate shearing (of the order of 1 s^{-1}) and isothermal crystallization (Figure 4.7). Like Somani *et al.* [SOM 00], they proposed a mechanism of formation of these structures (Figure 4.8): shearing causes a stretching and alignment of the entangled chain network, creating secondary nucleation sites for the free segments of adjacent chains. Subsequently, growth continues laterally, perpendicular to the shish axis.

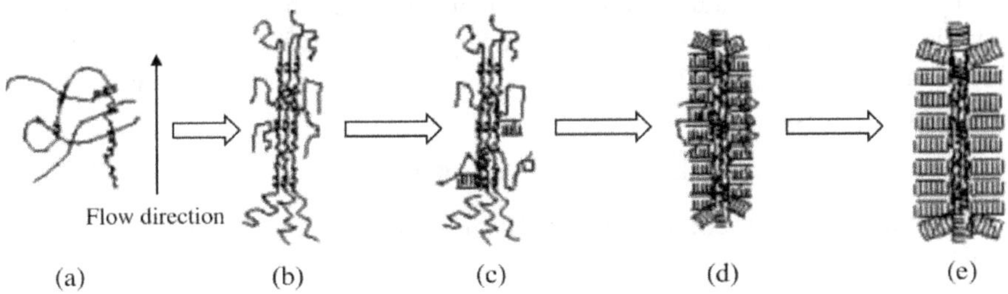

Figure 4.8. *Schematic illustration of the morphological development during shear-induced crystallization of iPP: a) entangled random coils before shear (amorphous melt, no orientation) with possible helical segments; b) stretching and alignment of chains in the flow direction, formation of shear-induced nuclei ("shish"); c) growth from row ordered nuclei; d) epitaxial growth of "shish-kebab"; e) final morphology after shear-induced crystallization [ZHA 05]*

4.2.4. *Flow effect on crystalline growth rate*

While the enhancement of nucleation is generally observed when the polymer is subjected to shear, the effect of flow on crystalline growth rate is still a matter of disagreement.

The fiber pulling method developed by Monasse [MON 95, MON 92] consists of pulling a glass fiber through a sample of polymer contained in a thermos-regulated cell to generate high rate shearing of up to 400 s^{-1}. The device is placed under the microscope, thereby monitoring the crystallization by optical methods. Due to this set-up, Monasse showed, for PE, that the growth rate is significantly increased compared to that of quiescent crystallization. It is multiplied by 4 between crystallization in quiescent conditions and crystallization under a 4 s^{-1} shearing. He reported that the effect on the growth rate is very weak for shear rates ranging from 0 to 0.5 s^{-1}. Later, Tribout *et al.* [TRI 96], Jay *et al.* [JAY 99] and Duplay *et al.* [DUP 00] came to the same conclusion for iPP (sixfold increase in the growth rate for 10 s^{-1} shearing). However, it should be noted that in these works, crystallization takes place under continuous shearing.

However, several studies also conclude that there is no significant variation in the crystalline growth rate due to shearing, as reported by Lee and Kamal [LEE 99]. We can also mention the works of Wolkowicz [WOL 78] on poly(1-butene), Sherwood *et al.* [SHE 77] on poly(ethylene oxide) and poly(ε-caprolactone), and Koscher and Fulchiron [KOS 02] on isotactic poly(propylene).

4.2.5. *Effect of flow on rheological properties*

Only a limited number of studies have investigated the crystallinity effect on rheology. Indeed, simultaneous measurements of the changes in the degree of crystallinity and rheological functions during crystallization are very difficult to achieve. In most cases, the measurements are performed in two separate isothermal crystallization experiments:

– evolution of the crystallinity over time using DSC;

– evolution of the rheological function of interest (shear viscosity and complex viscosity) over time using rheometry.

This experimental methodology is not without its own problems, including: reproducibility of the crystallization kinetics due to uncertainties on the temperature control in both devices, and reproducibility of the thermal history of the polymer melt prior to the test runs (melting and cooling for crystallization).

According to the results of several authors, Figure 4.9 shows the evolution of the normalized rheological function defined as the ratio between the values of any rheological function during crystallization $f(\alpha,T)$ and in the fully amorphous state before the beginning of the crystallization $f_0(T)$:

$$\Gamma = \frac{f(\alpha,T)}{f_0(T)} \qquad [4.5]$$

Here, f is the shear viscosity or the complex viscosity, depending on the type of measurement performed: respectively, steady shearing (rate $\dot{\gamma}$) and oscillatory shearing (frequency ω). All of this work clearly shows an abrupt increase in viscosity when the crystallization occurs. However, there is no real quantitative agreement between the different results: the critical degree of space filling, which corresponds to a one-decade increase in Γ, varies from a few percent [TIT 97, POG 98, PAN 01] to over 40% [BOU 98].

Moreover, it is difficult to draw clear conclusions about the effect of temperature: while [PAN 01] practically finds no temperature dependence of the solidification curve, the results of [TIT 97] show that the viscosity increase is steeper if the crystallization temperature is lower. Similarly, the measurements of [POG 98] and [ACI 03] show that solidification occurs at a lower crystallinity degree when the oscillatory shearing frequency increases.

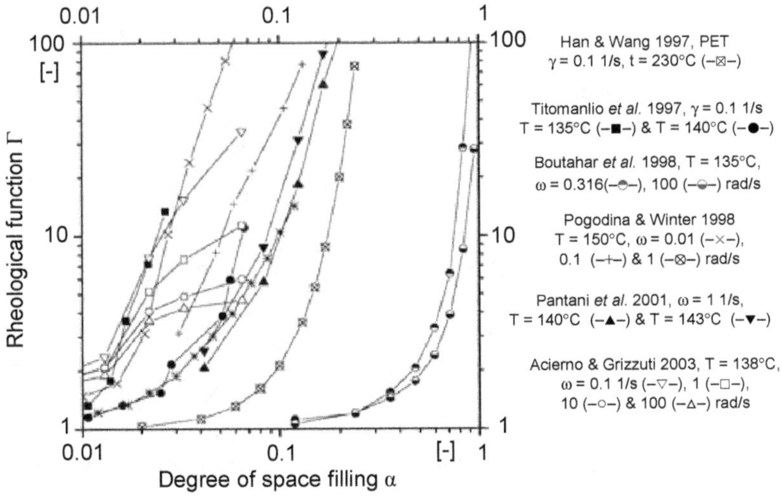

Figure 4.9. *Measured evolution of the normalized rheological function Γ versus crystallinity degree α, for several temperatures, in iPP (except [HAN 97], PET) [LAM 07]*

The nature of the solidification observed during iPP crystallization was studied by Pogodina *et al.* [POG 01, POG 98]. Using several experimental techniques (monitoring of spherulitic growth by optical microscopy, morphological monitoring by small angle light scattering SALS and rheometric measurement of the changes in complex modulus), they showed that crystallization can be assimilated to a physical gelation process. At the critical gelation point (marked as the time for which the module reaches 10% of its final value), crystallinity is still low and the spherulites are already visible but not yet in contact with each other. They explain this observation by the formation of an interconnection network between spherulites within the amorphous phase. The rigidity of the critical gel depends on the nucleation density: when it is low, the interconnection network is lost and the gel rigidity is lower. By contrast, for a dense nucleation (which is the case for large degrees of supercooling), the meshes of the network are tighter and the gel is more rigid. Unfortunately, in these kind of studies, no rheological characterizations of samples crystallized under high shear rates are available, mainly for experimental feasibility reasons.

4.2.6. *Summary of experimental observations and guidelines for modeling*

To summarize, the effects of a processing flow (and more specifically of shear flow) on the crystallization of thermoplastics are the following:

– *Effect on the crystallization kinetics*: Almost all studies show a decrease in crystallization time when shear stress increases, as well as the existence of a critical shear rate below which the effect is negligible. The shearing time is also a parameter to be considered, but it seems that its importance is lower: for a given total strain, the acceleration of the kinetics is more sensitive if the shear rate is higher. Similarly, some authors find a kinetics saturation effect for high strain, which could be explained by a state of maximum stretching of the molecular chains [KUM 99].

– *Effect on the nucleation*: All real-time microscopic observations highlight an increase in the nucleation density when flow is applied. This is the main cause invoked to explain the acceleration of the kinetics. Therefore, the main challenge in modeling is to connect this increase in nucleation to characteristic quantities of the flow itself (strain, strain rate, duration, etc.) and/or its effect on the polymer (elastic stress, free energy, etc.). The spatial organization of nuclei is also affected: if the shear rate is sufficiently large, the nuclei tend to appear in columns oriented in the flow direction.

– *Effect on the crystalline growth rate*: This point is still under discussion. Some authors report an increase in the growth rate under shear, while others observed no change. Anyway, this effect seems to have much lower impact on the kinetics than the nucleation enhancement.

– *Effect on crystalline morphologies*: This aspect of flow-induced crystallization is the subject of many studies implementing various observation techniques, from X-ray diffraction to atomic force microscopy. For low shear rates, the spherulitic morphology persists with decreasing crystallite size, due to a larger nucleation density. When shearing increases, nuclei are distributed into alignments parallel to the flow direction, leading to fibrillar crystallites. The shearing time seems to play an important role in the length reached by these structures. Secondary nucleation can occur and lead to the growth of radial lamellae perpendicular to the flow. This is the so-called shish-kebab morphology. Some authors observed that at the end of this process, a fraction of the inter-lamellar volume is filled by the growth of a second type of lamellae oriented parallel to the shish [WHI 98]. From the perspective of modeling, assimilation of these complex structures to cylinders can, therefore, constitute a good approximation. Likewise, as suggested by Eder and Janeschitz-Kriegl [EDE 97], their formation mechanism can be simplified into three sequential steps: primary nucleation, shish extension (axial growth) and increase in its diameter (radial growth).

– *Influence of the polymer molecular parameters*: While crystallization in quiescent conditions is virtually unaffected by the molecular parameters of a given polymer, the effect of flow strongly depends on the chain molecular weight distribution. Long chains with higher molecular weight have a greater ability to conform to the orientation imposed by the shear treatment (longer relaxation time), which leads to increased nucleation density and therefore more pronounced enhancement of crystallization kinetics. From a morphological point of view, the presence of long chains, even in low amounts, appears to promote the formation of elongated structures of the shish-kebab type. This confirms the need to take into account, in the crystallization models, the nature of the polymer and particularly the chain dynamics through its rheological properties.

– *Effect of crystallinity on the rheological behavior*: It is recognized that the rheological properties of the polymer are deeply affected by crystallization. Qualitatively, a steep increase in viscosity (shear or complex) is always observed when the crystallinity degree increases. Quantitatively, the practical difficulties associated with such experiments lead to a large variability in the accuracy of data. In the current state of knowledge, there is no universal model linking the crystallinity degree and the rheological properties of polymers, but various approaches (based on theory of concentrated suspensions, theory of liquid crystal polymers, etc.) exist, whose aim is essentially to model a particular experimental data set.

4.3. Flow-induced crystallization: modeling

The first models describing the effect of flow on polymer crystallization kinetics involved elastomers subjected to deformation (strain-induced crystallization, SIC). They were developed by Flory (theory of increased melting point) [FLO 47] and later by Gaylord and Lohse [GAY 76]. According to these models, strain has the effect of increasing the distance between the crosslinks of the rubber network, thus reducing the entropy of the molecular chains and activating crystallization. In the case of polymer melts, there is no crosslinking but geometric entanglements limiting chain mobility. To illustrate the variety of modeling options, an interesting review presenting a quasi-exhaustive summary of the main existing approaches is detailed in Zinet's PhD Thesis [ZIN 10].

4.3.1. *Overall kinetics modeling*

A variety of approaches have been proposed to model the flow effect on crystallization kinetics. The simplest ones consist of modifying the standard approach of Nakamura [NAK 72]: the Avrami parameter or the crystallization kinetic function is increased by a multiplying factor, which is linked to flow kinematical variables, such as shear intensity [MON 95], shear rate [TAN 03], macroscopic strain [ITO 96] and molecular orientation [BUS 96]. Alternatively, this factor can be defined as a function of stress-type parameters, for instance the Cauchy stress [VER 96], the trace of the extra-stress tensor [DOU 00a, DOU 00b] or an orientation factor [ZIA 88].

Another approach is based on the fact that flow causes the orientation and the extension of the macromolecular chains, leading to a decrease in polymer melt entropy and consequently an increase in melting temperature. According to the Hoffman–Lauritzen expression of the crystalline growth rate [HOF 97], it results in the enhancement of crystallization kinetics. As in the previously stated models, the melting temperature increase must be linked to the stress-strain state of the polymer, using either stress [TIT 97, GUO 01] or strain-related parameters [TIT 04, KIM 05].

The main shortcoming of these Nakamura-based approaches is that they only provide the overall degree of crystallinity, with no additional information on the final structure, like the number and size distributions of the crystallites, or the part of isotropic and oriented entities.

4.3.2. *Explicit nucleation and growth modeling*

More complex models explicitly take into account the nucleation and growth process. Experimental observations have clearly shown that the number of nuclei is considerably increased under shear conditions [POG 01, KOS 02], whereas the effect of flow on the crystalline growth rate is still under investigation, as already mentioned in section 4.2.4.

4.3.2.1. *Growth rate*

In the literature, there are several studies that lead to contradictory results. The reason for this discrepancy can be related to several factors, e.g. the influence of material tacticity [TRI 96], the existence of a deformation rate threshold below which the flow effect is negligible or the wide range of shearing times used in the different studies. Nevertheless, most of the existing crystallization models consider that the growth rate remains unaffected by the flow and ascribe the enhancement of crystallization kinetics to the nucleation density.

4.3.2.2. *Nucleation rate*

The additional flow-induced nucleation rate can be linked to free energy-related parameters. Zheng and Kennedy [ZHE 04] express the molecular chain conformation and thus the flow-induced change of free energy in the amorphous and semi-crystalline phases by means of a finite extensible nonlinear elastic model with Peterlin closure approximation (FENE-P) and a rigid dumbbell model, respectively. The driving force of the additional nucleation is then related to the excess free energy through a Hoffman–Lauritzen type expression. Moreover, the Avrami index is taken as a function of the orientation distribution, thus allowing the crystallization kinetics to depend on the crystalline geometry (spherical growth to rod-like growth).

Another option is to connect the extra nucleation to stress or strain-related parameters. In the milestone work of Eder and Janeschitz-Kriegl [EDE 97], the flow-induced nucleation rate is written as a function of the shear rate. As these additional nuclei are expected to grow into shish-kebab crystallites, they are geometrically represented as cylinders. The Schneider rate equations [SCH 88] are then used to calculate the total degree of crystallinity and also provide some information about the resulting microstructure: mean number, diameter, length and volume of crystallites per material unit volume.

Zuidema *et al.* [ZUI 01] improved the physical basis of Eder's model by replacing the shear rate by the second invariant of the Finger tensor, i.e. the recoverable strain tensor, as the driving force of flow-induced nucleation. In their work, a viscoelastic model was introduced to calculate the strain tensor.

Using experimental rheometric data [WAS 00], Tanner [TAN 05] proposed to link the shear-induced nucleation rate to a combination of the stress and strain rate. Boutaous *et al.* [BOU 09] extended the concentrated suspension theory-based work of Tanner to non-isothermal flows. On the basis of material parameters measurable by rather common techniques (namely DSC and rheometer), they used two distinct sets of modified Schneider equations to differentiate the crystallized volume fraction induced by the flow from that induced by the thermal effects (Table 4.1). They showed that at relatively high shear rate and low cooling rate, shearing dominates the nucleation phenomenon. This shear effect is less effective when the cooling rate is higher and when the initial temperature of the crystallizing polymer is lower. Indeed, when the initial temperature is already low, time is not sufficient for the flow to have an effect before reaching the "gelation" temperature.

		50 °C/min		5°C/min	
	T0 (°C)	Thermally induced nuclei (%)	Shear induced nuclei (%)	Thermally induced nuclei (%)	Shear induced nuclei (%)
0,1 (s^{-1})	200	99,94	0,06	72,31	27,69
	260	99,84	0,16	43,59	56,41
2 (s^{-1})	200	99,6	0,4	25,08	74,92
	260	98,96	1,04	9,82	90,18

Table 4.1. *Simulated percentage of thermally induced and shear-induced nuclei for non-isothermal crystallization of iPP, with various cooling rates, shear rates and initial temperatures [BOU 09]*

4.3.3. *Role of viscoelasticity*

The link between the flow characteristics and the enhancement of nucleation has been described by various phenomenological models, using more or less easily experimentally accessible variables, such as the shear rate, the stress tensor or an energy-based function. It is not yet possible to affirm which option is the best, mainly because the physical basis of the interaction between flow and crystallization is not completely understood.

Nevertheless, the most widespread explanation is that flow tends to orientate and stretch the macromolecular chains constituting the polymer melt, thus decreasing the entropy, increasing the local molecular ordering and the predisposition to give new homogeneous nuclei, whereas in quiescent conditions the heterogeneous nuclei are the precursors of the crystallization process. Based on these considerations, it clearly appears that a realistic flow-induced crystallization model should take into account information about the macromolecular chain dynamics. This is particularly important in the case of short-term flows: after cessation of flow, molecular orientation is not relaxed instantaneously but progressively: an ordered state remains for a more or less long time, depending on the molecular weight distribution (high molecular weight molecules are more sensitive to flow due to their larger relaxation times). As a result, the flow-induced nucleation process can continue despite the melt not flowing. In a macroscopic scale modeling approach, the chain dynamics is reflected by the viscoelastic rheological behavior of the polymer melt.

In other words, the flow causes the macromolecular chains to stretch and store additional energy, resulting in an increase in the melt free energy. The quantity of stored energy is commonly called elastic free energy (ΔG_e) and is related to the flow-induced change in the conformational tensor. Extending the Marrucci theory [MAR 72], several studies link ΔG_e to the trace of the extra-stress tensor (see the review by Mazinani *et al.* [MAZ 09]).

In this framework, Zinet *et al.* [ZIN 10] developed a numerical model able to simulate the crystallization of polymers under non-isothermal flows. They assumed that the polymer melt elasticity, quantified by the trace of the extra-stress tensor $tr(\boldsymbol{\tau}^{VE})$, is the driving force of flow-induced extra nucleation:

$$\dot{N}^f = C\left[tr\left(\boldsymbol{\tau}^{VE}\right)\right]^p \qquad [4.6]$$

where C and p are the two material-related parameters to be determined using experimental crystallization data. Thus, the sensitivity of the crystallization kinetics to the flow is not only a consequence of the flow kinematics, but is also determined by the rheological behavior of the melt. Furthermore, the formulation of the model gives additional information about the final crystalline structure: relative crystalline fraction, distribution and size of thermally and flow-induced crystallites (Figures 4.10 and 4.11).

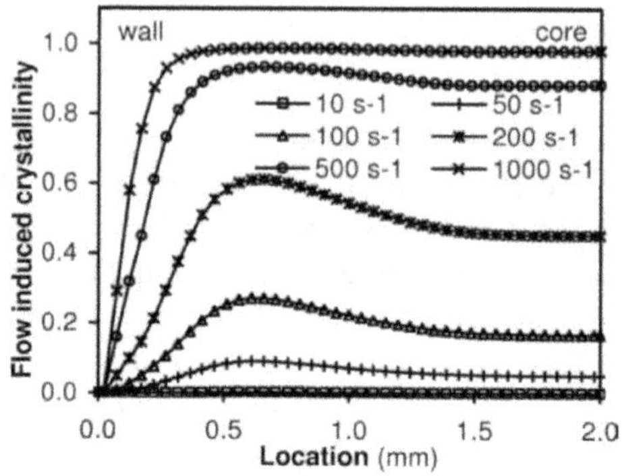

Figure 4.10. *Profile of the flow-induced crystalline fraction obtained at the end of iPP crystallization in a nonisothermal shear flow [ZIN 10]*

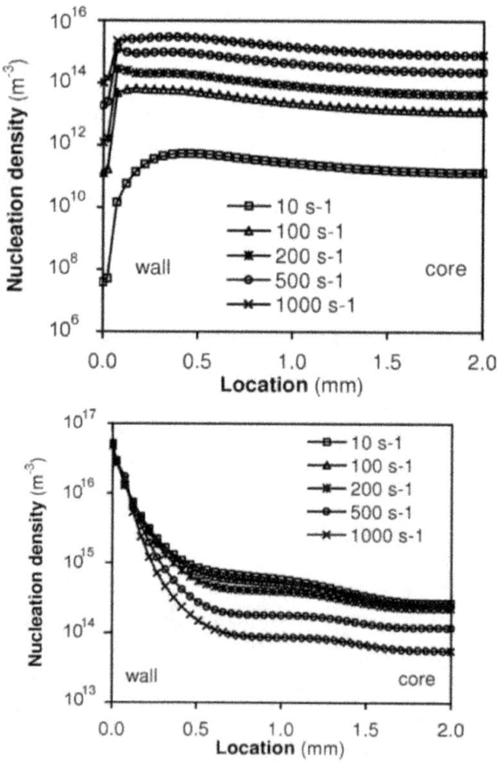

Figure 4.11. *a) Profile of the a) thermally induced nucleation density; b) flow-induced nucleation density obtained at the end of iPP crystallization in a nonisothermal shear flow [ZIN 10]*

4.4. Effect of the composite components

4.4.1. *Effect of nucleating agents*

In polymers and composites industries, nucleating agents are usually added in order to enhance the crystallization kinetics, reduce the thermal kinetics time and to improve some material characteristics as stiffness, transparency, barrier properties.

In thermoplastic materials, nucleating agents generate a higher number of crystallization sites. This enhancement of the nucleation density leads to the reduction of the crystalline structure size. The transformation kinetics are strongly affected with drastic consequences on the final parts characteristics.

In static conditions, it is widely established that nucleating agents, like Talc, can greatly increase the crystallization rate [ZHA 05, GUR 78, REF 14, KAI 05]. The increase in the number of nuclei affects the structure of the crystalline entities and leads to the reduction of their size, which in turn affect the global morphology with, in many situations, special interfacial-oriented crystalline structures: transcrystallinity [KAR 06, JIA 12]. The most popular global theory of crystallization is that formulated by Avrami. As reported in Chapter 3, its principle is based on the nucleation and growth analysis. Of course, these additional phenomena are not considered in the global theories of nucleation and growth, and work is needed to understand the interfacial interaction between the polymer matrix and the additives. Hence, it results in a deviation of the experimental results when the kinetic is analyzed by Avrami theory: a non-integer coefficient "n" is generally obtained [BOV 02] and an increasing of the kinetic factor K is observed.

During processing, the flow acts on the polymer and also on the nucleating particles. Evidently, a synergic effect on the transformation kinetics is expected. In fact, the flow is also an activator of additional nucleation. Many studies are conducted to understand and explain this effect. Early on, Lagasse *et al.* [LAG 76] explored the role of nucleating agents under shear flow. They concluded that the particles and the flow do not act by addition or superposition of their effect, but the flow effect is annealed by the particles in case of low shear. Higher shear rates are necessary in order to observe an enhancement of the crystallization kinetics, compared to the case without nucleating agent. The flow role is easily highlighted in the absence of added particles. In Figure 4.12, representing the induction times for crystallization of nucleated and non-nucleated polymers, we can observe a clear discrimination between the effects of the shear alone, and that of the shear combined with the presence of nucleating agents. It is important to note that for high shear rates, the two curves tend to be confused. The crystallization of the non-nucleated polymer is shear-accelerated over the entire range of shear rates, while the nucleated one shows acceleration under flow after a given shear rate threshold value. This critical shear rate for the onset of shear accelerated crystallization is not governed by rheological factors alone, but also by the nucleating density in quiescent polymer state.

The same observation is also reported for other polymers [NAM 04, REI 94], and we can summarize the observations as follows: if a polymer is nucleated or non-nucleated under quiescent conditions, the critical shear rate value which enhances its crystallization kinetics would be higher for that with the higher nucleation density [LAG 76].

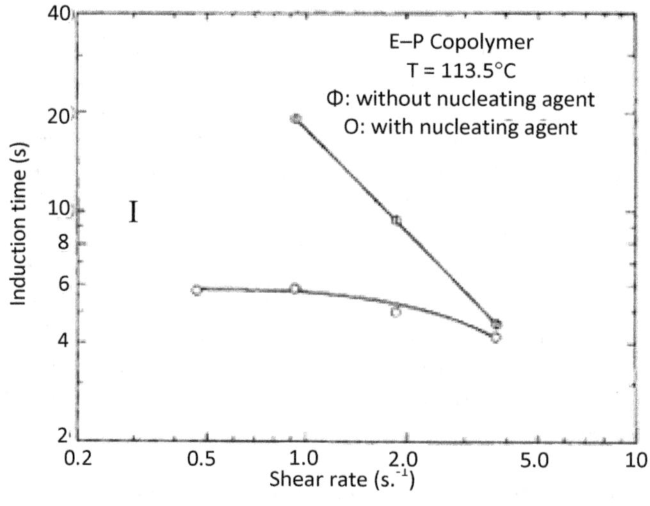

Figure 4.12. *Influence of heterogeneous nucleating agent on crystallization of ethylene-propylene copolymer [LAG 76]*

4.4.2. *Effect of fibers*

In composite materials, fibers are generally added to the polymer matrix in order to obtain better mechanical properties. The fibers could play the role of a nucleating agent and accelerate the nucleation density in the polymer.

Many types of fibers are used, among them the most known are glass fibers, carbon fibers, aramid fibers, etc. Depending on their type and treatment, they are less of high nucleating for the polymer matrix.

In the case of nucleating fibers, a very high number of nuclei appear on the fibers surface, in row organization. This oriented organization forces the activated nuclei to grow only in the transverse direction to the fiber orientation. It forms a special organization of the crystalline entities: columnar unidirectionally oriented at the fibers surface, and spherulitic far from the interface or in the polymer bulk. This structure is named as transcrystalline, as presented in Figures 13(a) and (b).

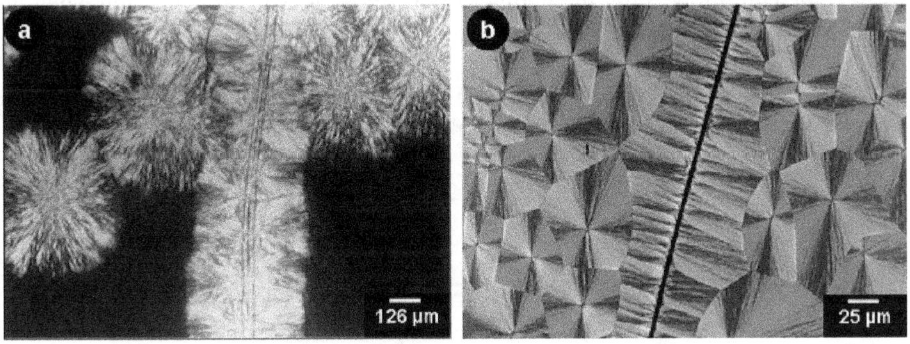

Figure 4.13. *a) Isothermal crystallization of flax fibers/isotactic polypropylene [POM 00] and b) transcrystalline structures in carbon fiber reinforced nylon 66 composite [CLA 94]*

The size of the different zones depends on the fiber–matrix interaction, and many other parameters, related to the matrix, fillers and processing conditions [PIO 13, CHA 91, NAU 07].

Many studies deal with the quantification of the transcrystallinity in fiber-filled polymers [KAR 06, NAU 07, NAU 07, CAI 97, VAR 94] and the effect of the fibers on the crystallization kinetics and forms [MOO 98].

We could globally summarize that the noticeable effects of the fibers on crystallization depend on many factors: the properties of the fibers (roughness, type, size, etc), the fiber–matrix interface, the fiber volume fraction, the fiber diameter, polymer type and properties, and the thermomechanical conditions [BUR 84].

As with nucleating agents, it is established that the incorporation of fibers in a polymer matrix changes the crystallization kinetics. Due to the fundamental assumptions on which the Avrami–Evans models are based, their predictions when polymers are reinforced by fibers frequently deviate from the overall crystallization kinetics. In fact, the strong assumption of random distribution of nuclei is not respected, and the Avrami parameters become adjustable numerical coefficients, without real physical meaning [PIO 13, NAU 07].

Nevertheless, many studies exist in the literature, modeling the crystallization kinetics based on adapted Avrami–Evans theory [CHA 91, PAR 00, DEV 91].

In static conditions, only the number of nuclei N_0 is generally affected when the model is applied to the neat polymer and to its fiber-reinforced composite.

The classical equations coupled with Hauffman–Lauritzen theory are still the same for the analysis of the relative crystallinity:

$$\alpha(t) = 1 - e^{(-K_{Av}t^n)} = 1 - exp(-\frac{4\pi N_0 G^3}{3} t^3) \quad [4.7]$$

$$G = G_0 \exp\left(-\frac{K_g}{T\Delta T}\right) \exp\left(-\frac{U^*}{R(T-T_\infty)}\right) \quad [4.8]$$

where $\alpha(t)$ is the transformed volume fraction (written in the case of a 3D growth and an instantaneous nucleation); $n = 3$ is the Avrami exponent; K_{Av} is the Avrami rate constant; N_0 is the number of nuclei per unit volume; G is the growth rate of crystalline entities.

To take into account the fiber effect, the N_0 values were numerically optimized by parametric identification, from fitting the model prediction to the experimental data, generally obtained by calorimetric curves based on the enthalpy measurement. The dependence of N_0 on the temperature is obtained from these analyses, and usually assumed with a logarithmic evolution.

Moreover, in the case of composites, the flow generates a shear rate distribution between the fibers. The effect of the flow on the transformation kinetics is added with that of the thermal one and acts as additional nucleation rate, expressed with empirical relationships. The associated parameters are all numerically adjustable to fit with the experiment.

In the same spirit, the flow is also perturbed by the presence of the fibers and a principle of local concentrated shear is defined, in contrast with uniform shear distribution in the case of non-filled polymer flow (Figure 4.14). Typically, the flow-added nuclei are linked to the fibers volume as in the following equation [NAU 07]:

$$\dot{N}(t) = |c\,\dot{\gamma}(t)|^d \quad [4.9]$$

where c is a parameter in which the shear concentrated effect, due to the flow between fibers, is implicitly taken into account, chosen with simple expression:

$$c = c_0 + c_1 \frac{\emptyset}{1-\emptyset} \quad [4.10]$$

where $\emptyset$ is the volume fraction of the fiber in the material, ranging between 0 and 1.

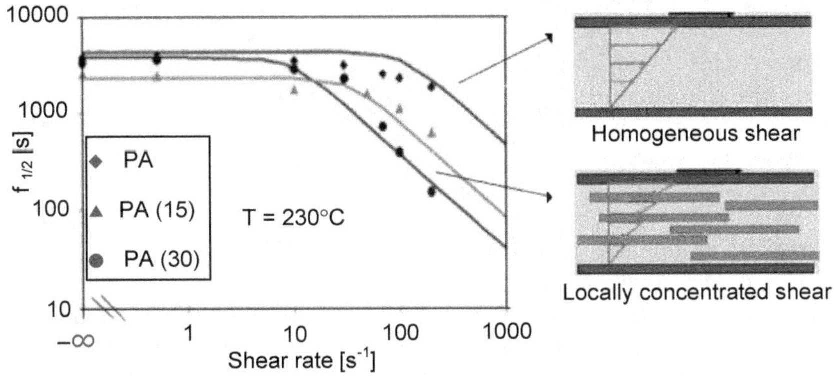

Figure 4.14. *Experimental (symbols) and predicted (lines) half crystallization times for PA, PA(15 fiber vol%) and PA (30 fiber vol%), at 230°C, with 5 s shearing time [NAU 07]. For a color version of the figure, see www.iste.co.uk/boyard/heat.zip*

This type of modeling, even if all the parameters lose their physical meaning, helps in numerical simulations of the polymer-fiber processing, by introducing the kinetics effects in the theoretical prediction and analysis.

4.5. Concluding remarks

Controlling the phase change kinetics during processing of polymers and polymer-based composites is an essential way to improve the performance of any plastic/composites process. In the case of semi-crystalline thermoplastic polymers, the final part characteristics and behavior are closely related to the crystalline microstructure developed during processing. They depend on the thermomechanical history experienced by the material: molten polymer, additives, fibers, etc.

In this chapter, we presented an overview of characterization techniques, experimental observations and the state-of-the-art concerning the transformations experienced by the polymer during processing. The different roles of the flow (shearing and elongation), polymer nature, viscoelasticity, additives, fibers and thermal kinetics, on the microstructure of manufactured parts, are highlighted and discussed. A review of recent theories and models is presented, in order to provide a guide for the readers on the importance of each parameter.

The flow accelerates the crystallization kinetics and affects the size of the crystalline entities. This is explained by an increase in the nucleation density. However, its effect on the crystal growth rate is not yet clear. Quantitatively, the effect on the flow of crystallization depends on the macromolecular dynamic, and appears at the macroscopic scale via the complex viscoelastic rheological behavior.

The literature review has shown a wide variety of possible approaches to model the flow effect on polymer crystallization but not in a complete satisfactory way. Indeed, it is a complex phenomenon and some aspects are still poorly understood.

However, despite the profusion of options proposed by the various authors (including the choice of the coupling variable and the rheological model), a key point should be highlighted: the flow-induced crystallization kinetics enhancement is connected to an increase in nucleation favored by the alignment and the extension of the macromolecular chains. The chain dynamics, itself influenced by the molecular weight distribution, plays a leading role in this mechanism. It is therefore necessary to take it into account through the rheological characterization of the material. In other words, connecting the flow effect to a purely kinematic variable (shear rate, total strain or combination of both) seems insufficient, particularly in the case of flows whose duration is comparable to the relaxation times of the polymer.

Clearly, a detailed modeling taking into account many mechanisms will be very realistic and consistent from a physical point of view, but it will involve a large number of parameters that require heavy characterization of the material. Conversely, an essentially global and phenomenological modeling will be simple to implement, but will not account for some important aspects. In addition, it will have a limited range of validity.

A balanced compromise between realism and simplicity, determined by the intended application and the intended use of the results, should be seeked for. In particular, the models intended for numerical simulation of plastics processing in an industrial setting should feature easily obtainable parameters and integration opportunities in a comprehensive process simulation. From our point of view, these are the decisive criteria for the choice of the modeling assumptions.

Even if the major parts of these process–material interaction parameters are now well understood, and even if many innovative techniques are being developed for experimental analysis and characterization, further work remains if we want to extend models developed in laboratories to real processing conditions of modeling and optimization.

Specifically, in the case of composite materials, considering the rise in their number of applications, the extension and modification of classical global theories of crystallization reveals their limits. We need more developments in order to take into account the high number of phenomena involved during manufacturing: fiber–matrix interaction, transcrystallinity, nature of matrix and fibers of additives, viscoelastic behavior and constitutive equations, fast kinetics, etc.

Obtaining information on the microstructure and its link with the processing kinetics is a very interesting issue. It is an essential step to develop realistic constitutive mechanical behavior laws in connection with the physical structure of the material and the prediction of final thermomechanical properties of semi-crystalline thermoplastic materials.

4.6. Bibliography

[ACI 03a] ACIERNO S., GRIZZUTI N., "Measurements of the rheological behavior of a crystallizing polymer by an 'inverse quenching' technique", *Journal of Rheology*, vol. 47, no. 2, pp. 563–576, 2003.

[ACI 03b] ACIERNO S., PALOMBA B., WINTER H.H. *et al.*, "Effect of molecular weight on the flow-induced crystallization of isotactic poly(1-butene)", *Rheologica Acta*, vol. 42, no. 3, pp. 243–250, 2003.

[BIR 87] BIRD R.B., ARMSTRONG R.C., HASSAGER O., *Dynamics of Polymeric Liquids, Volume 1: Fluid Mechanics*, 2nd ed., Wiley, New York, 1987.

[BOU 98] BOUTAHAR K., CARROT C., GUILLET J., "Crystallization of polyolefins from rheological measurements – relation between the transformed fraction and the dynamic moduli", *Macromolecules*, vol. 31, no. 6, pp. 1921–1929, 1998.

[BOU 09] BOUTAOUS M., BOURGIN P., ZINET M., "Thermally and flow induced crystallization of polymers at low shear rate", *Journal of Non-Newtonian Fluid Mechanics*, vol. 165, no. 5–6, pp. 227–237, 2009.

[BOV 02] BOVE L., NOBILE M.R., "Shear-induced crystallization of isotactic poly(1-butene)", *Macromolecules Symposium*, vol. 185, pp. 135–147, 2002.

[BUR 84] BURTON R.H., DAY T.H., FOLKES M.J., "Localized flow-induced morphological features in fiber reinforced thermoplastics", *Polymer Communications*, vol. 25, pp. 361–362, 1984.

[BUS 96] BUSHMAN A.C., MCHUGH A.J., "A continuum model for the dynamics of flow-induced crystallization", *Journal of Polymer Science part B: Polymer Physics*, vol. 34, no. 14, pp. 2393–2407, 1996.

[CAI 97] CAI Y., PETERMANN J., WITTICH H., "Transcrystallization in fiber-reinforced isotactic polypropylene composites in a temperature gradient", *Journal of Applied Polymer Science*, vol. 65, pp. 67–75, 1997.

[CHA 91] CHABERT B., CHAUCHARD J., "Polymer crystallization in the presence of fibers in fiber-reinforced thermoplastics-matrix composites", *Annales de Chimie-Science des Matériaux*, vol. 16, no. 3, pp. 173–187, 1991.

[CLA 94] CLARK R.L., Master Thesis, Virginia Polytechnic Institute and State University, Blacksburg, VA, 1994.

[DEV 91] DEVAUX E., CHABERT B., "Nature of the transcrystalline interphase of polypropylene glass-fiber reinforced composites after a shear-stress", *Polymer Communications*, vol. 32, pp. 464–468, 1991.

[DOU 00a] DOUFAS A.K., MCHUGH A.J., MILLER C., "Simulation of melt spinning including flow-induced crystallization - Part I. Model development and predictions", *Journal of Non-Newtonian Fluid Mechanics*, vol. 92, no. 1, pp. 27–66, 2000.

[DOU 00b] DOUFAS A.K., MCHUGH A.J., MILLER C. *et al.*, "Simulation of melt spinning including flow-induced crystallization - Part II. Quantitative comparisons with industrial spinline data", *Journal of Non-Newtonian Fluid Mechanics*, vol. 92, no. 1, pp. 81–103, 2000.

[DUP 00] DUPLAY C., MONASSE B., HAUDIN J.M. *et al.*, "Shear-induced crystallization of polypropylene: influence of molecular weight", *Journal of Material Science*, vol. 35, no. 24, pp. 6093–6103, 2000.

[EDE 97] EDER G., JANESCHITZ-KRIEGL H., "Crystallization", in MEIJER H.E.H. (ed.), *Materials Science and Technology, Processing of Polymers*, Wiley-VCH, Weinheim, 1997.

[FLO 47] FLORY P.J., "Thermodynamics of crystallization in high polymers. I. Crystallization induced by stretching", *Journal of Chemical Physics*, vol. 15, pp. 397–408, 1947.

[FRI 74] FRITZSCHE A.K., PRICE F.P., "Crystallization of polyethylene oxide under shear", *Polymer Engineering and Science*, vol. 14, no. 6, pp. 401–412, 1974.

[GAY 76] GAYLORD R.J., LOHSE D.J., "Morphological-changes during oriented polymer crystallization", *Polymer Engineering and Science*, vol. 16, no. 3, pp. 163–167, 1976.

[GOD 07] GODARA A., Polymer crystallization and micromechanical behavior of fiber-reinforced polymer composites, PhD Thesis, University of Aachen, Germany, 2007.

[GUO 01] GUO J.X.,. NARH K.A., "Computer simulation of stress-induced crystallization in injection molded thermoplastics", *Polymer Engineering and Science*, vol. 41, no. 11, pp. 1996–2012, 2001.

[GUR 78] GURATO G., GAIDANO D., ZANNETTI R., "Influence of nucleating agents on the crystallization of 6-Polyamide", *Die Makromolekulare Chemie*, vol. 179, no. 1, pp. 231–245, 1978.

[HAA 69] HAAS T.W., MAXWELL B., "Effects of shear stress on the crystallization of linear polyethylene and polybutene-1", *Polymer Engineering and Science*, vol. 9, no. 4, pp. 225–241, 1969.

[HAN 97] HAN S., WANG K.K., "Shrinkage prediction for slowly-crystallizing thermoplastic polymers in injection molding", *International Polymer Processing*, vol. 12, no. 3, pp. 228–237, 1997.

[HIS 83] HSIUE E.S., ROBERTSON R.E., YEH G.S.Y., "Effects of shearing conditions on crystalline orientation and relaxation in polyethylene", *Polymer Engineering and Science*, vol. 23, no. 2, pp. 74–78, 1983.

[HOB 01a] HOBBS J.K., HUMPHRIS A.D.L., MILES M.J., "In-situ atomic force microscopy of polyethylene crystallization. 1. Crystallization from an oriented backbone", *Macromolecules*, vol. 34, no. 16, pp. 5508–5519, 2001.

[HOB 01b] HOBBS J.K., MILES M.J., "Direct observation of polyethylene shish-kebab crystallization using in-situ atomic force microscopy", *Macromolecules*, vol. 34, no. 3, pp. 353–355, 2001.

[HOF 97] HOFFMAN J.D., MILLER R.L., "Kinetics of crystallization from the melt and chain folding in polyethylene fractions revisited: theory and experiment", *Polymer*, vol. 38, no. 13, pp. 3151–3212, 1997.

[HOS 95] HOSIER I.L., BASSETT D.C., MONEVA I.T., "On the morphology of polyethylene crystallized from a sheared melt", *Polymer*, vol. 36, no. 22, pp. 4197–4202, 1995.

[ITO 96] ITO H., MINAGAWA K., TAKIMOTO J. *et al.*, "Effect of pressure and shear stress on crystallization behaviors in injection molding", *International Polymer Processing*, vol. 11, no. 4, pp. 363–368, 1996.

[JAY 99] JAY F., HAUDIN J.M., MONASSE B., "Shear-induced crystallization of polypropylenes: effect of molecular weight", *Journal of Material Science*, vol. 34, no. 9, pp. 2089–2102, 1999.

[JER 97] ERSCHOWP J., JANESCHITZ-KRIEGL H., "The role of long molecules and nucleating agents in shear induced crystallization of isotactic polypropylenes", *International Polymer Processing*, vol. 12, no. 1, pp. 72–77, 1997.

[JIA 12] JIANG Q., JIA H., WANG J. *et al.*, "Effects of nucleating agents on crystallization behavior and mechanical properties of high-fluid polypropylene", *Iranian Polymer Journal*, vol. 21, no. 3, pp. 201–209, 2012.

[KAI 05] KAI W., HE Y., INOUE Y., "Fast crystallization of poly(3-hydroxybutyrate) and poly(3-hydroxybutyrate-co-3-hydroxy-valerate) with talc and boron nitride as nucleating agents", *Polymer International*, vol. 54, no. 5, pp. 780–789, 2005.

[KAR 06] KARPP-PFORDT S., Cristallisation induite par cisaillement du MXD6 dans différentes formulations: additifs nucléants, fibres de verre, PhD Thesis, University Claude Bernard Lyon1, France, 2006.

[KEL 67] KELLER A., MACHIN M.J., "Oriented crystallization in polymers", *Journal of Macromolecular Science B*, vol. B1, no. 1, pp. 41–91, 1967.

[KIM 05] KIM K.H., ISAYEV A.I., KWON K. *et al.*, "Modeling and experimental study of birefringence in injection molding of semicrystalline polymers", *Polymer*, vol. 46, no. 12, pp. 4183–4203, 2005.

[KOB 70] KOBAYASHI K., NAGASAWA T., "Crystallization of sheared polymer melts", *Journal of Macromolecular Science B*, vol. B42, pp. 331–345, 1970.

[KOS 02] KOSCHER E., FULCHIRON R., "Influence of shear on polypropylene crystallization: morphology development and kinetics", *Polymer*, vol. 43, no. 25, pp. 6931–6942, 2002.

[KUM 99a] KUMARASWAMY G., ISSAIAN A.M., KORNFIELD J.A., "Shear-enhanced crystallization in isotactic polypropylene. 1. Correspondence between in Situ rheo-optics and ex situ structure determination", *Macromolecules*, vol. 32, no. 22, pp. 7537–7547, 1999.

[KUM 99b] KUMARASWAMY G., VERMA R.K., KORNFIELD J.A., "Novel flow apparatus for investigating shear-enhanced crystallization and structure development in semicrystalline polymers", *Review of Scientific Instruments*, vol. 70, no. 4, pp. 2097–2104, 1999.

[KUM 00] KUMARASWAMY G., VERMA R.K., ISSAIAN A.M. *et al.*, "Shear-enhanced crystallization in isotactic polypropylene. Part 2. Analysis of the Formation of the Oriented "Skin", *Polymer*, vol. 41, no. 25, pp. 8931–8940, 2000.

[KUM 02] KUMARASWAMY G., KORNFIELD J.A., YEH F.J. *et al.*, "Shear-enhanced crystallization in isotactic polypropylene. 3. Evidence for a kinetic pathway to nucleation", *Macromolecules*, vol. 35, no. 5, pp. 1762–1769, 2002.

[LAG 76] LAGASSE R.R., MAXWEL B., "Experimental-study of kinetics of polymer crystallization during shear-flow", *Polymer Engineering and Science*, vol. 16, no. 3, pp. 189–199, 1976.

[LAM 07] LAMBERTI G., PETERS G.W.M., TITOMANLIO G., "Crystallinity and linear rheological properties of polymers", *International Polymer Processing*, vol. 22, no. 3, pp. 303–310, 2007.

[LEE 99] LEE O., KAMAL M.R., "Experimental study of post-shear crystallization of polypropylene melts", *Polymer Engineering and Science*, vol. 39, no. 2, pp. 236–248, 1999.

[LIE 93] LIEDAUER S., EDER G., JANESCHITZ-KRIEGL H. *et al.*, "On the kinetics of shear-induced crystallization in polypropylene, *International Polymer Processing*, vol. 8, no. 3, pp. 236–244, 1993.

[MAC 95] MACKLEY M.R., MARSHALL R.T.J., SMEULDERS J.B.A.F., "The multipass rheometer", *Journal of Rheology*, vol. 39, no. 6, pp. 1293–1309, 1995.

[MAC 99] MACKLEY M.R., WANNABORWORN S., GAO P. *et al.*, "The optical microscopy of sheared liquids using a newly developed optical stage", *Microscopy and Analysis*, vol. 69, pp. 25–29, 1999.

[MAR 72] MARRUCCI G., "The free energy constitutive equation for polymer solutions from the dumbbell model", *Transactions of the Society of Rheology*, vol. 16, pp. 321–330, 1972.

[MAZ 09] MAZINANI S., SHARIF F., MOHAMMADI N., "Predicting flow induced change in phase diagram of polymer solutions in simple shear flow", *e-Polymers*, vol. 87, no. 1, pp. 1048–1066, 2009.

[MON 92] MONASSE B., "Polypropylene nucleation on a glass-fiber after melt shearing", *Journal of Material Science*, vol. 27, no. 22, pp. 6047–6052, 1992.

[MON 95] MONASSE B., "Nucleation and anisotropic crystalline growth of polyethylene under shear", *Journal of Material Science*, vol. 30, no. 19, pp. 5002–5012, 1995.

[MON 96] MONKS A.W., WHITE H.M., BASSETT D.C., "On shish-kebab morphologies in crystalline polymers", *Polymer*, vol. 37, no. 26, pp. 5933–5936, 1996.

[MOO 98] MOON C.K., "Effect of molecular weight and fiber diameter on the interfacial behavior in glass fiber/PP composites", *Journal of Applied Polymer Science*, vol. 67, pp. 1191–1197, 1998.

[NAK 72] NAKAMURA K., WATANABE K., KATAYAMA K. *et al.*, "Some aspects of nonisothermal crystallization of polymers. I. Relationship between crystallization temperature, crystallinity and cooling conditions", *Journal of Applied Polymer Science*, vol. 16, no. 5, pp. 1077–1091, 1972.

[NAM 04] NAM Y., SHIM K.B., KIM S.H. *et al.*, "Effect of silica nanoparticle on the quiescent and the shear induced crystallization behaviors of poly(trimethylene terephtalate)", *Polymer Journal*, vol. 36, no. 7, pp. 519–530, 2004.

[NAU 07] NAUDY S., FULCHIRON R., "Static and shear induced crystallization of glass fiber reinforced poly(*m*-xylylene adipamide) with nucleating additives", *Journal of Polymer Science part B: Polymer Physics*, vol. 45, no. 21, pp. 2982–2992, 2007.

[PAN 01] PANTANI R., SPERANZA V., TITOMANLIO G., "Relevance of crystallisation kinetics in the simulation of the injection molding process", *International Polymer Processing*, vol. 16, no. 1, pp. 61–71, 2001.

[PAR 00] PARK C.S., LEE K.J., KIM S.W., "Crystallization kinetics of glass fiber reinforced PBT composites", *Journal of Applied Polymer Science*, vol. 78, no. 3, pp. 576–585, 2000.

[PIO 13] PIORKOWSKA E., "Crystallization in polymer composites and nanocomposites", in PIORKOWSKA G.C., RUTLEDGE (eds), *Handbook of Polymer Crystallization*, Wiley, 2013.

[POG 98] POGODINA N.V., WINTER H.H., "Polypropylene crystallization as a physical gelation process", *Macromolecules*, vol. 31, no. 23, pp. 8164–8172, 1998.

[POG 01] POGODINA N.V., LAVRENKO V.P., SRINIVAS S. *et al.*, "Rheology and structure of isotactic polypropylene near the gel point: quiescent and shear-induced crystallization", *Polymer*, vol. 42, no. 21, pp. 9031–9043, 2001.

[POM 00] POMPE G., MADER E., "Experimental detection of transcrystallinity interphase in glass-fiber/polypropylene composites", *Composite Science and Technology*, vol. 60, pp. 2159–2167, 2000.

[REI 94] REINSCH V.E., REBENFELD L., "Crystallization processes in poly(ethylene terephtalate) as modified by polymer additives and fiber reinforcement", *Journal of Applied Polymer Science*, vol. 52, pp. 649–662, 1994.

[REF 14] REFAA Z., BOUTAOUS M., ROUSSET F. *et al.*, "Crystallization kinetics of poly-(lactic acid) with and without talc: optical microscopy and calorimetric analysis", *AIP Conference Proceedings*, vol. 1593, pp. 342–346, 2014.

[SCH 88] SCHNEIDER W., KÖPPL A., BERGER J., "Non-isothermal crystallization: crystallization of polymers: system of rate equations", *International Polymer Processing*, vol. 3, no. 4, pp. 151–154, 1988.

[SHE 77] SHERWOOD C.H., PRICE F.P., STEIN R.S., "Effect of shear on crystallization kinetics of poly(ethylene oxide) and poly(epsilon-caprolactone) melts", *Journal of Applied Polymer Science*, vol. 63, pp. 77–94, 1977.

[SOM 00] SOMANI H.S., HSIAO B.S., NOGALES A. *et al.*, "Structure development during shear flow-induced crystallization of i-PP: in-situ small-angle X-ray scattering study, *Macromolecules*, vol. 33, no. 25, pp. 9385–9394, 2000.

[TAD 74] TADMOR Z., "Molecular orientation in injection molding", *Journal of Applied Polymer Science*, vol. 18, no. 6, 1753–1772, 1974.

[TAN 03] TANNER R.I., "On the flow of crystallizing polymers I. Linear regime", *Journal of Non-Newtonian Fluid Mechanics*, vol. 112, no. 2–3, pp. 251–268, 2003.

[TAN 05] TANNER R.I., QI F., "A comparison of some models for describing polymer crystallization at low deformation rates", *Journal of Non-Newtonian Fluid Mechanics*, vol. 127, nos. 2–3, pp. 131–141. 2005.

[TIT 97] TITOMANLIO G., SPERANZA V., BRUCATO V., "On the simulation of thermoplastic injection moulding process .2. Relevance of interaction between flow and crystallization", *International Polymer Processing*, vol. 12, no. 1, pp. 45–53, 1997.

[TIT 04] TITOMANLIO G., LAMBERTI G., "Modeling flow induced crystallization in film casting of polypropylene", *Rheologica Acta*, vol. 43, no. 2, pp. 146–158, 2004.

[TRI 96] TRIBOUT C., MONASSE B., HAUDIN J.M., "Experimental study of shear-induced crystallization of an impact polypropylene copolymer", *Colloid Polymer Science*, vol. 274, no. 3, pp. 197–208, 1996.

[ULR 76] ULRICH R.D., PRICE F.P., "Morphology development during shearing of poly(ethylene oxide) melt", *Journal of Applied Polymer Science*, vol. 20, no. 4, pp. 1077–1093, 1976.

[VAR 94] VARGA J., KARGER-KOCSIS J., "The difference between transcrystallization and shear-induced cylindritic crystallization in fiber-reinforced polypropylene composites", *Journal of Material Science Letters*, vol. 13, no. 14, pp. 1069–1071, 1994.

[VER 96] VERHOYEN O., Experimental study and modelization of the crystallization of thermoplastic polymers during injection molding, PhD Thesis, Université Catholique de Louvain, Belgium, 1996.

[VLE 96] VLEESHOUWERS S., MEIJER H.E.H., "A rheological study of shear induced crystallization", *Rheologica Acta*, vol. 35, no. 5 pp. 391–39, 1996.

[WAS 00] WASSNER E., MAIER R.D., "Shear induced crystallization of polypropylene melts", in BINDING D.M. (ed.), *Proceedings of the 13th International Congress on Rheology*, Cambridge, UK, 2000.

[WHI 98] WHITE H.M., BASSETT D.C., "On row structures, secondary nucleation and continuity in alpha-polypropylene", *Polymer*, vol. 39, no. 14, pp. 3211–3219, 1998.

[WOL 78] WOLKOWICZ M.D., "Nucleation and crystal growth in sheared poly(1-butene) melts", *Journal of Polymer Science: Polymer Symposia*, vol. 63, no. 1, pp. 365–382, 1978.

[ZHA 05] ZHANG C.G., HU H.Q., WANG D.J. *et al.*, "In situ optical microscope study of the shear-induced crystallization of isotactic polypropylene", *Polymer*, vol. 46, no. 19, pp. 8157–8161, 2005.

[ZHE 04] ZHENG R., KENNEDY P.K., "A model for post-flow induced crystallization: general equations and predictions", *Journal of Rheology*, vol. 48, no. 4, pp. 823–842, 2004.

[ZIA 88] ZIABICKI A., "The mechanisms of 'neck-like' deformation in high speed melt spinning. 2. Effects of polymer crystallization", *Journal of Non-Newtonian Fluid Mechanics*, vol. 30, pp. 157–168, 1988.

[ZIN 10a] ZINET M., Crystallization modeling in polymer processing: quantification of the thermal and rheological effects, PhD Thesis, INSA Lyon, France, 2010.

[ZIN 10b] ZINET M., EL OTMANI R., BOUTAOUS M. *et al.*, "Numerical modeling of non-isothermal polymer crystallization kinetics: flow and thermal effects", *Polymer Engineering and Science*, vol. 50, pp. 2044–2059, 2010.

[ZUI 01] ZUIDEMA H., PETER S.G., MEIJER H.E.H, "Development and validation of a recoverable strain-based model for flow-induced crystallization of polymers", *Macromolecular Theory and Simulations*, vol. 10, no. 5, pp. 447–460, 2001.

5

From the Characterization and Modeling of Cure-Dependent Properties of Composite Materials to the Simulation of Residual Stresses

5.1. Introduction

The use of composite materials has increased significantly in recent years in aerospace, naval and other high-tech applications. Despite many advantages, such as strength, lightness, etc., over conventional materials, there are certain associated problems. Residual stress is one such problem that arises due to the presence of two or more heterogeneous materials within a composite part [PAR 06, NAW 14]. These stresses lead to the warpage, matrix cracking, delamination, reduced strength and fiber buckling etc., in a composite part [NAW 12, NAW 12].

This chapter includes a discussion on the factors inducing residual stress into the composite parts and its associated problems, methods of determination of composite properties and modeling the residual stress in composite plates using analytical methods.

5.2. Origin of residual stress

Residual stress can be defined as a stress that persists in a material that is free of external forces or temperature gradients. Residual stress influences the physical and

Chapter written by Yasir NAWAB and Frédéric JACQUEMIN.

structural properties of the composite parts significantly [PAR 06], and it is crucial to take it into account while designing and modeling the behavior of composite structures.

Residual stress generates in the resin during the curing process. During the initial stages of curing, thermoset resin is in viscous state and is not able to retain significant amount of stress. But on transformation to rubbery state, it has significant amount of bulk modulus such that hydrostatic stress might arise in three-dimensionally constrained areas due to tooling. Cracks may occur in thick composites in the early stages of curing [ERS 10]. Young's modulus and shear modulus of a resin in glassy state are significantly higher than the modulus values of the same resin in the rubbery state. Therefore, a vitrified resin may be assumed to behave like an elastic solid. Therefore, a strain due to temperature change at this stage is much more significant than the strain caused by chemical shrinkage due to crosslinking reactions.

In general, residual stress is induced in a composite part due to the mismatch of chemical shrinkage, thermal expansion, and hygral behavior of fibers, resin and tool [NAW 12, WIS 06]. Residual stress in a composite part can be divided into three major categories depending upon the source of the stress:

– chemical residual stress;

– thermal residual stress;

– hygral residual stress.

Chemical and thermal residual stress are related to the heterogeneous behavior of the constituents of the composite part during fabrication, whereas hygral stress is the result of mismatched moisture uptake by the fabricated part.

5.2.1. *Mechanical levels of residual stress*

In laminated composites, residual stress is induced at three mechanical levels [PAR 06].

5.2.1.1. *Micromechanical level*

This is the residual stress generated at the level of constituents, i.e. fiber and neighboring matrix. The mismatch in coefficient of thermal expansion and chemical shrinkage between the fibers and the matrix governs this stress. This may lead to matrix cracks and debonding of fiber matrix interface.

5.2.1.2. *Macromechanical residual stress*

This is the stress present on a ply-to-ply scale due to lamina anisotropy and is also called laminate-level residual stress. A composite ply is a thin anisotropic layer of fibrous material filled with a resin. The properties, i.e. thermal coefficients of this ply, are different in longitudinal and transverse direction. In a laminate made by stacking of a certain number of plies at given orientation, the residual stress arises due to a difference in ply coefficients of thermal expansion in the given direction. For example, with cross-ply laminated composites, the 90° fibers impose a mechanical constraint on the 0° fibers during cooling and vice versa, because of differences in thermal shrinkage directions. This stress can be analyzed by classical laminated plate theory.

5.2.1.3. *Global residual stress*

Since fiber reinforced polymer composites are bad conductors of heat, thermal gradients may generate through the thickness of thick composites while curing or cooling [NAW 12]. While cooling a thick laminate, middle plies will have a slower cooling rate in the middle compared to at the surface due to bad conduction properties. Therefore, the surface plies impose a constraint on the shrinkage of the middle plies. A gradient in cooling rate, temperature or moisture conditions may result in a residual stress distribution through the thickness of the laminate. This will often result in a parabolic distribution of compressive residual stresses in the surface plies and tensile stresses in the center plies in a composite part [PAR 06].

5.2.2. *Parameters contributing to the formation of residual stress*

In the following, the major parameters that contribute toward the formation of residual stress in the thermoset composites are presented.

5.2.2.1. *Thermal expansion coefficients*

The expansion coefficient of polymer resin is usually much higher than that of the fibers. Also, the expansion coefficients of many fibers are orthotropic. For example, carbon fibers have very low or slightly negative expansion coefficients in the fiber direction, but higher values in the transverse direction. Mismatch of thermal expansion coefficients is a major source of residual stress generation, especially during cooling of the composite after curing. Due to this mismatch, residual stress is generated at micro, macro and global scales causing shape distortion, matrix crack and fiber buckling [KIM 89].

5.2.2.2. *Cure shrinkage*

Thermoset resins chemical shrinkage occurs due to the formation of complex 3D crosslinking network during curing. This shrinkage is about 5–8% of volume of virgin resin and causes a global volume shrinkage of 2–3% [NAW 12, LI 04]. A significant amount of chemical shrinkage occur in resin before its gel point and has no contribution to residual stress, but after that it can cause similar effects to thermal contraction, with stresses at the micromechanical level, in-plane stresses and distortions in laminates, and changes in curvature in curved plates [NAW 14, NAW 12]. Generally, dilatometric techniques are used for the measurement of volume chemical shrinkage of resin as well as composite [NAW 13]. These techniques include capillary-type dilatometer [NAW 13], plunger-type dilatometer, gravimetric method, rheometer, pycnometer, fiber Bragg grating, thermomechanical analyzer, dynamic mechanical analyzer, ultrasonic waves and by using optical fibers [NAW 13], etc.

5.2.2.3. *Tool–part interaction*

Due to differential strains between the part and mold (tool), on which composite is fabricated, residual stress can arise during the curing. Aluminum or steel tools have much higher expansion coefficients than composite parts, and tend to stretch the parts as they heat up. This can happen as a result of small shear stresses at the tool interface causing tension in the part. Gradients of in-plane stresses through the thickness can arise, causing bending when the stresses are released [TWI 04]. A second tool–part interaction mechanism is due to locking, where the geometry of the part forces it to move with the tool as it expands.

5.2.2.4. *Other mechanisms*

Moisture causes swelling of the matrix which produces similar effects as thermal or chemical volume changes [WU 00]. Volume fraction variations through the thickness can arise, especially where there is resin bleed during the cure. These can produce distortion even in flat unidirectional composites [RAD 95]. Resin rich layers near the tool interface and resin poor regions at the surface tend to have higher and lower thermal expansion coefficients, respectively, leading to convex bowing. Fiber movement during the cure can cause changes in geometry and properties leading to stresses and distortion. Consolidation and resin flow at radii can lead to corner thickening with concave tooling and corner thinning with convex tooling. Wrinkles can also arise at corners, particularly with convex tooling. Differential curing through the thickness due to temperature lags and exotherms can be important in thick parts, and may cause additional residual stresses and gradients through the thickness [BOG 92].

5.2.3. *Problems generated by residual stress*

The following problem/defects may be encountered in composites due to the presence of these stresses:

1) shape distortion;

2) matrix cracking;

3) delamination;

4) reduced strength;

5) fiber buckling, etc.

Some examples of matrix cracking, residual deformation and fiber buckling are given in Figure 5.1. Shape distortions, which occur due to unavoidable residual stress formations during cure, are not easy to overcome. Because of shape distortions, assembling of composite parts becomes difficult and this leads to an increase in manufacturing costs.

a) b) c) (d)

Figure 5.1. *a) Matrix crack, b) deformed composite plate, c) broken composite part, and d) fiber buckling*

Shape distortions can be classified into two major groups: warpage and spring-in. Warpage is described as the distortion of geometry for a manufactured part which is initially thin and flat [NAW 12, WIS 06]. On the other hand, spring-in can be defined as the change in the enclosed angle in a corner-shaped composite part. For a thermoset composite part, change in the enclosed angle of a perpendicular corner is in the order of 1–3 [WIJ 05].

5.3. Determination of composite properties

Determination of composite properties during and after cure is a prerequisite for modeling the residual stress in the composite. In general, modeling of these properties is done from the properties of fibers and resin using various techniques.

However, some techniques for experimental determination of these properties are reported in the literature as well.

5.3.1. *Modeling the mechanical properties of composites*

Self-consistent models based on the mathematical formalism proposed by Kroner [KRO 58] and Eshelby [ESH 57] constitute a reliable method to predict the micromechanical behavior of heterogeneous materials such as composites. The method was initially introduced to treat the case of polycrystalline materials, i.e. duplex steels, aluminum alloys, etc., submitted to purely elastic solicitations. Jacquemin *et al.* [JAC 05] used the similar approach to calculate the effective hygro-elastic properties of composite materials. It was considered that fibers are the ellipsoidal inclusions embedded in an infinite medium called homogeneous effective medium. The latter was supposed to behave as a ply in the composite. The properties of this medium were found by applying homogenization operations on the fibers and matrix. The material was investigated at two different scales for the needs of micromechanical modeling. The properties of the ply at macroscopic level, denoted by using superscript "I", are calculated from properties of fibers and matrix (microscopic level) which are denoted by "f" and "m", respectively [JAC 05].

The microscopic and macroscopic residual stresses generated in the ply due to the temperature change and curing of resin can be found by using the constitutive law. Eshelby demonstrated in his fundamental work that when the elementary inclusions (here the matrix and the fiber) were assumed to have ellipsoidal shapes, stresses and strains were related by the scale transition relation [5.1]:

$$\overline{\overline{\sigma^r}} - \overline{\overline{\sigma^I}} = -\overline{\overline{\overline{\overline{C^I}}}} : \overline{\overline{\overline{\overline{R^I}}}} : \left(\overline{\overline{\varepsilon^r}} - \overline{\overline{\varepsilon^I}} \right) \quad [5.1]$$

Here, "r" stands for the constituent f or m (means for fiber or matrix). $\overline{\overline{\overline{\overline{R^I}}}}$ is the reaction tensor and its value can be found by expression:

$$\overline{\overline{\overline{\overline{R^I}}}} = \left(\overline{\overline{\overline{\overline{C^{I-1}}}}} - \overline{\overline{\overline{\overline{E}}}} \right) : \overline{\overline{\overline{\overline{E^{-1}}}}} \quad [5.2]$$

$\overline{\overline{\overline{\overline{E}}}}$ is the Morris tensor which expresses the dependence of reaction tensor on morphology of constituents. As per the useful relation introduced by Hill [HIL 67], the macroscopic stresses and strains are equal to average of microscopic stresses and strains of constituents and can be expressed as:

$$\left\langle \overline{\overline{\sigma^r}} \right\rangle_{r=f,m} = \overline{\overline{\sigma^I}} \qquad \left\langle \overline{\overline{\varepsilon^r}} \right\rangle_{r=f,m} = \overline{\overline{\varepsilon^I}} \qquad [5.3]$$

The self-consistent model allows determining the macroscopic elasticity tensor, coefficients of thermal expansion and chemical shrinkage of ply from the mechanical, thermal and chemical properties of the constituents.

$$\left\langle \left(\overline{\overline{C^r}} + \overline{\overline{C^I}} : \overline{\overline{R^I}} \right)^{-1} : \left[\overline{\overline{C^r}} : \overline{\mathrm{CTE}^r} \Delta \mathrm{T}^r - \overline{\overline{C^I}} : \overline{\mathrm{CTE}^I} \Delta \mathrm{T}^I + \overline{\overline{C^r}} : \overline{CS^r} \Delta \alpha^r - \overline{\overline{C^I}} : \overline{CS^I} \Delta \alpha^I \right] \right\rangle = 0 \qquad [5.4]$$

The relation [5.5] leads to the classical expression of rigidity:

$$\overline{\overline{C^I}} = \left\langle \left(\overline{\overline{C^r}} + \overline{\overline{C^I}} : \overline{\overline{R^I}} \right)^{-1} : \left(\overline{\overline{C^I}} + \overline{\overline{C^I}} : \overline{\overline{R^I}} \right) : \overline{\overline{C^r}} \right\rangle_{r=f,m} \qquad [5.5]$$

C^I was determined from [5.5] by using iterative method. The determination of homogenized coefficient of chemical shrinkage [MSA 10] is given below:

$$\overline{CS^I} = v^m \overline{\overline{C^{I-1}}} \left\langle \left(\overline{\overline{C^r}} + \overline{\overline{C^I}} : \overline{\overline{R^I}} \right)^{-1} \right\rangle_{r=f,m}^{-1} : \left(\overline{\overline{C^m}} + \overline{\overline{C^I}} : \overline{\overline{R^I}} \right)^{-1} : \overline{\overline{C^m}} : \overline{\mathrm{CS}^m} \qquad [5.6]$$

where v^m is the matrix volume fraction in the considered ply ($v^m + v^f = 1$). Similarly, the coefficient of thermal expansion can be written as:

$$\overline{CTE^I} = \overline{\overline{C^{I-1}}} : \left\langle \left(\overline{\overline{C^r}} + \overline{\overline{C^I}} : \overline{\overline{R^I}} \right)^{-1} \right\rangle_{r=f,m}^{-1} : \left\langle \left(\overline{\overline{C^r}} + \overline{\overline{C^I}} : \overline{\overline{R^I}} \right)^{-1} : \overline{\overline{C^r}} : \overline{\mathrm{CTE}^r} \right\rangle_{r=f,m} \qquad [5.7]$$

For example, Table 5.1 shows the properties of carbon fibers. The indices 1, 2 and 3 are the orientation coefficients, denoting a property along x, y and z-axis, respectively.

Property	Value	Property	Value
E1(GPa)	230	G_{12}(GPa)	27.6
E2, E3(GPa)	20.7	CTE_1 (K^{-1})	-0.38× 10^{-5}
υ12, υ13	0.2	CTE_2 (K^{-1})	20×10^{-6}
G23(GPa)	6.89	CS_1 , CS_2	0

Table 5.1. *Mechanical and thermomechanical properties of carbon fibers*

Similarly, the properties of cured epoxy (M21) resin are presented in Table 5.2.

Property	E_{cured} (GPa)	Y	CTE_{cured} (K-1)	CS	αgel
Value	1.37	0.35	44.9×10-6	0.019	0.52

Table 5.2. *Mechanical and thermomechanical properties of resin M21 [ABO 10]*

In Table 5.3, an example of properties of cured carbon/epoxy composite estimated by self-consistent model is presented.

Property	CTE1 (K-1)	CTE2 (K-1)	CS1 (%)	CS2 (%)	G12 (GPa)	E1 (GPa)	E2 (GPa)	v12
57 vol% fiber	0.21×10–6	29.8× 10–6	–1.17× 10-2	–0.67	6.4	131.2	5.63	0.25

Table 5.3. *Effective properties of cured carbon/epoxy ply*

5.3.2. *Experimental determination of thermomechanical properties of composite*

The direct measurements of thermomechanical properties, i.e. elastic modulus, linear chemical shrinkage, etc., of composites during the full curing cycle are rarely found in the literature. Measurement of these properties results into a more real calculation of residual stresses and considerably lessens the calculation by eliminating the step of modeling the composite properties from the constituents.

Experimental studies of the measurement of tensile modulus and in-plane chemical shrinkage of thermoset composites were also undocumented. These properties are important to define the shape evolution of composite part during curing, and to model the chemical curing stresses. Several authors [NAW 13] proposed techniques to find out the in-plane chemical shrinkage of composites. White and Hahn [WHI 92] placed a 25.4 mm × 152 mm ply of bismaleimide matrix composite on an aluminum caul plate and noted its dimensions manually during curing at room temperature. The authors concluded that the chemical strains were large in magnitude but they were experienced very early in the curing cycle when the mechanical properties were quite low. Manual measurement, curing under atmospheric pressure and experiments on only unidirectional composites are the limitations of this method.

Chemical shrinkage of laminated composite plates was also determined by following warpage [DAN 90] during curing. A certain number of uncured plies were stacked over the fully cured plate having same number of plies and stacking sequence. Due to the mismatch of chemical shrinkage in cured and uncured plies, warpage occurred in the sample which was recorded using the shadow moiré method. This method provided measurements throughout the curing but for the composites with complex stacking sequences, it is very difficult to deconvolute the strains along x, y and z-axes. Moreover, the use of adhesive between uncured and cured portion, flow of resin during curing may also affect the quality of results.

Dynamic mechanical analyzer (DMA) [ERS 10] and thermal mechanical analyzer (TMA) [OLI 06] have also been used for the characterization of shrinkage. Ersoy *et al.* [ERS 10] used DMA for the determination of chemical shrinkage along the thickness direction of partially cured carbon/epoxy laminated composite sample having 25×25×5 mm^3. Since testing is done on partially cured samples, the information obtained on chemical shrinkage is available only after a certain degree of cure, e.g. gel point.

Olivier [OLI 06] adapted TMA to determine the in-plane chemical shrinkage and thermomechanical properties of partially cured laminated composite samples with dimensions of 25 mm × 2.5 mm. The set-up provided good results for samples having only in-plane strain but the presence of out of plane deflection such as in the case of [0/90] can affect the quality of results. Moreover, the sample cannot be subjected to pressure and vacuum during the experiments. To relate the strain with degree of cure, another instrument such as DSC is required.

Strain gauges bonded on the surface or placed among the plies of laminated composite provide the information about strain generated by chemical shrinkage. Research on the measurement of strain induced by chemical as well as thermal expansion using these gauges can be found in the literature [KIM 04]. Strains/shrinkage in more than one direction can be measured simultaneously using this method but results may be affected by the presence of gauges which change the local stiffness of the composite.

Recently, we proposed a new set-up [NAW 15] consisting of an oven and a tensile testing machine. Tests were conducted on mass-scale samples under controlled temperatures and vacuum pressure. Composite strain response was recorded after regular intervals of time by applying a cyclic force. The test gave information on the stiffness of the composite and its length. From these data, the Young's modulus and in-plane chemical shrinkage were calculated. Using this set-up, tests on unidirectional glass/vinylester laminated composite samples were successfully conducted under temperature ramp and pressure conditions. An example of chemical shrinkage versus degree of cure of three glass/vinylester $[90]_8$ composite samples is plotted in Figure 5.2. Dashed curve is added to show the global behavior of chemical shrinkage.

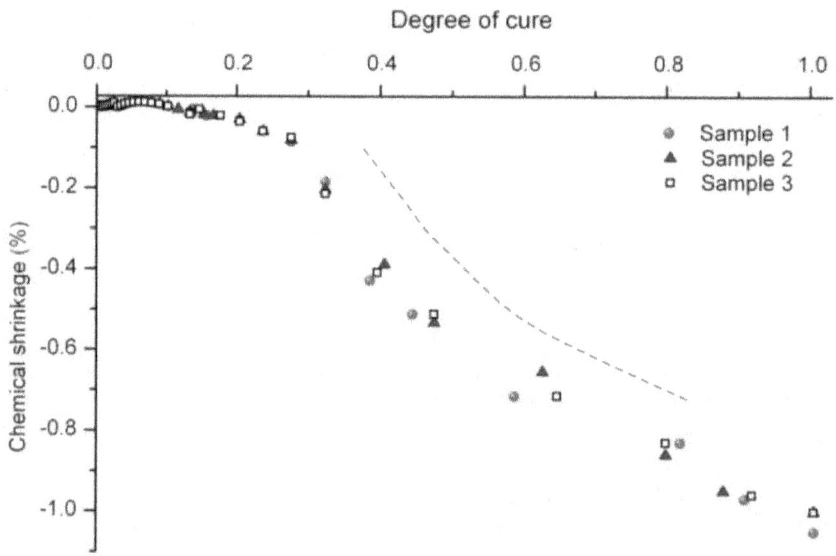

Figure 5.2. *Chemical shrinkage of glass/ vinylester composite $[90]_8$ having V_f=31%. For a color version of the figure, see www.iste.co.uk/boyard/heat.zip*

It can be noted that there was no shrinkage in the composite until $\alpha = 0.1$, which is close to the gel point of vinylester resin. This observation is in agreement with the literature [NAW 12] that chemical shrinkage and mechanical properties of composite are negligible until the gel degree of cure. From this point, the chemical shrinkage increases nonlinearly to reach the maximum value of 1% in fully cured state.

From similar experiments, chemical shrinkage of [0°] composite was found quasi-null. For the same composite, volume chemical shrinkage was found equal to 4.4% [NAW 12]. By comparing the results of volume and linear chemical shrinkage of glass/vinylester composite, it was concluded that in the thin laminated composites most of the shrinkage occurs along their thickness. Young's modulus of carbon/epoxy T700/M21 [NAW 12] was determined as well from the similar experiments, which was started to increase from 5.5±0.2 to 9.2±0.3 GPa in fully cured state. This value is close to the transversal Young's modulus (9.5 GPa) of T700/M21 found in the literature.

5.4. Modeling of residual stress

Several studies [ZHU 01] have been conducted on the modeling of residual stress. This induced stress in a composite part can be divided into three categories depending upon the source of the stress:

– chemical residual stress;

– thermal residual stress;

– hygral residual stress.

Chemical and thermal residual stress are related to the heterogeneous behavior of the constituents of the composite part during fabrication, whereas hygral stress is a result of mismatched moisture uptake in the fabricated part.

Several approaches are found in the literature to find the residual stresses and strains. The thermo-elastic approach was used by many authors [JUN 92] for calculating the thermal residual stresses after the curing. Kim and Hahn [KIM 89] were the first to model the chemical residual stresses along with the thermal residual stresses. Bogetti and Gillespie [BOG 92] used an enhanced form of classical laminate theory to study the residual stress of chemical and thermal origin. The thermo-chemical-elastic approach [MSA 10] has also been used to calculate not only the thermal but also the chemical stresses during the curing cycle. For the same objective, some authors [WEI 82] have used the thermo-visco-elastic approach. This approach was based on the assumption that material behavior is viscoelastic during the thermal changes. Kim [KIM 07] used a viscoelastic model to calculate the chemical residual stresses by considering that the properties of material during curing depend on the cure degree (α) and time (t).

For the calculation of warpage in the laminated plates, two approaches exist: the linear approach, also named classical laminate theory, and the nonlinear approach.

5.4.1. *Linear approach or classical theory of laminates*

The linear approach is based on linear relationship between the displacements and corresponding strains. If the deformation is small, then the bending curvatures (a (along x-axis), b (along y-axis) and/or twisted with twist curvature c (in xy plane)) of the laminated composite plate can be found using:

$$a = -\left(\frac{\partial^2 w^\circ}{\partial x^2}\right), \quad b = -\left(\frac{\partial^2 w^\circ}{\partial y^2}\right), \quad c = -2\left(\frac{\partial w^\circ}{\partial x}\right)\left(\frac{\partial w^\circ}{\partial y}\right)$$

Here, w^o is the out of plane, midplane displacement. Cross ply [0/90] laminates are generally considered for the studies of residual strains and stresses. While cooling this simplest laminated plate from curing temperature, exhibits no twist, so c = 0 and the curvature a and b depend linearly on temperature. If thickness is small and number of 0° and 90° plies are equal, then a = – b and plate deforms in a saddle-like shape.

5.4.2. *Nonlinear approach*

The linear approach was not able to predict correctly the deformation [GIG 04] in laminated plate. For example, room temperature shape of [0/90] unsymmetric laminates predicted by linear theory was saddle-like but in reality, it is cylindrical with axis along the x or y direction.

This difference was explained by assuming a nonlinear strains–displacement relationship [LEE 71]. According to nonlinear theory, the Von Karman simplified equations for nonlinear strains in terms of displacement are written as:

$$\text{Strain along x-axis } \varepsilon_{xx} = \frac{\partial u^\circ}{\partial x} - z\left(\frac{\partial^2 w^\circ}{\partial x^2}\right) + \frac{1}{2}\left(\frac{\partial w^\circ}{\partial x}\right)^2 \quad [5.8]$$

$$\text{Strain along y-axis } \varepsilon_{yy} = \frac{\partial v^\circ}{\partial y} - z\left(\frac{\partial^2 w^\circ}{\partial y^2}\right) + \frac{1}{2}\left(\frac{\partial w^\circ}{\partial y}\right)^2 \quad [5.9]$$

Strain in x-y plane or shear strain:

$$\gamma_{xy} = \frac{\partial u^\circ}{\partial x} + \frac{\partial v^\circ}{\partial y} - 2z\left(\frac{\partial^2 w^\circ}{\partial x \partial y}\right) + \left(\frac{\partial w^\circ}{\partial x}\right)\left(\frac{\partial w^\circ}{\partial y}\right) \qquad [5.10]$$

Here, u^o and v^o are the midplane displacements along x and y-axis, respectively. The factors $\frac{1}{2}\left(\frac{\partial w^\circ}{\partial x}\right)^2, \frac{1}{2}\left(\frac{\partial w^\circ}{\partial y}\right)^2, \left(\frac{\partial w^\circ}{\partial x}\right)\left(\frac{\partial w^\circ}{\partial y}\right)$ represent nonlinearity of strains along x, y-axis and twist. The value of w° is given in [5.11].

$$w^\circ = -\frac{1}{2}(ax^2 + by^2 + cxy) \qquad [5.11]$$

The equations of strains are simplified by many authors [GIG 04, ANT 06]. Gigliotti [GIG 04] solved the displacement equations by using polynomials like the Timoshenko polynomial. It is one of the simplest ways that involves only five unknowns. However, this method provides only the good results for [0/90] laminates. The midplane displacements by this method are:

$$u^\circ = D_1 x - \frac{1}{6}a^2 x^3 - \frac{1}{4}abxy^2 \qquad [5.12]$$

$$v^\circ = D_2 y - \frac{1}{6}b^2 y^3 - \frac{1}{4}abyx^2 \qquad [5.13]$$

Here, D_1 and D_2, a, b and c are the coefficients to be determined. The twist curvature c is zero in the case of cross-ply laminates.

To solve the equations of strains for angle ply as well as cross-ply laminates, Dano and Hyer [DAN 98] used richer equations having 28 coefficients and found the thermally deformed shapes of carbon/epoxy laminates for three groups of plies orientations [-θ/+ θ], [(90-θ)/ θ], [(θ-90)/θ].

5.4.3. *Minimization of energy approach*

The total potential energy of the laminated plate can be found from the known equation:

$$E_{tot} = \frac{1}{2}\sigma_{ij}\varepsilon_{ij} \qquad [5.14]$$

where σ_{ij} is the residual stress field induced by the chemical cure shrinkage and thermal expansion/contraction and ε_{ij} is the strain field. According to the Lagrange–Dirichlet theorem, the object will be in a state of equilibrium if value of first derivative of total potential energy is zero [5.15].

$$\delta E_{tot} = 0 \qquad [5.15]$$

In addition, this equilibrium will be stable if the second derivative of total energy is positive [5.16].

$$\delta^2 E_{tot} \succ 0 \qquad [5.16]$$

Evaluation of total energy at any instant by using equations [5.19] and [5.20] gives the curvatures of laminated plate subjected to chemical shrinkage and thermal expansion/contraction as by-products.

5.4.4. *Application*

Figure 5.3 shows the curvatures developed in 57 vol% fiber carbon/epoxy cross-ply composite plates due to chemical and thermal origin during the full curing cycle versus time [NAW 12]. The temperature and degree of cure of resin are also presented on the same graph. All the short-dotted curves represent the unstable solutions of chemical and thermal curvatures. Chemical deformation zone (from $\alpha = 0.5$ to $\alpha = 1$) is also highlighted on the graph. In this zone, the piece deforms like a saddle in the beginning but after the bifurcation point it converts into a cylinder. The maximum curvature is reached at $\alpha = 1$. It can be noted that after the complete cure of resin (end of chemical deformation zone), there is a constant temperature zone Z (highlighted by the dashed ellipse) before the start of thermal deformation zone. This zone is very important to define the evolution of shape of laminated pieces. It can be described in two possible ways:

1) Zone Z is short enough to prevent any relaxation of stresses. Thus, a chemically deformed piece will maintain itself at maximum curvature up to the beginning of thermal deformation zone. The thermal curvature will appear at the beginning of thermal deformation zone and will result in the increase of curvature of chemically developed cylinder.

2) The chemical stresses in the piece relax in zone Z, so the curvatures caused by chemical shrinkage disappear and the thermal curvatures start from zero. Consequently, the piece deforms again in the saddle-like shape and passing again

through a bifurcation point it becomes a cylinder whose curvature increases with cooling as shown in Figure 5.3.

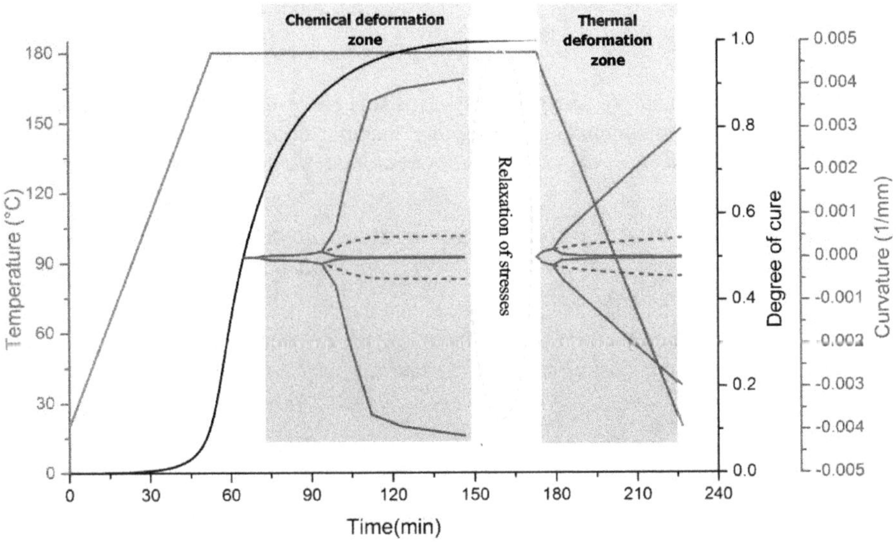

Figure 5.3. *Shape evolution of [0/90] laminate with stress relaxation. For a color version of the figure, see www.iste.co.uk/boyard/heat.zip*

5.5. Conclusion

In this chapter, a review about the determination of thermomechanical properties of composite and experimental and modeling of residual stress is presented. It was observed that for the simulation of shape evolution during curing of cross-ply laminated (carbon/epoxy) composite plates, a nonlinear geometrical approach provides good results. For the determination of composite properties, needed to feed this model, a self-consistent model, such as the model proposed by Eshelby and Kroner, proved to be suitable. Despite a number of simulations found in the literature, the properties of composites (Young's modulus and linear chemical shrinkage) were rarely measured experimentally. A new set-up based on an insulated oven and the tensile testing machine was developed. The results showed that transversal [90°] chemical shrinkage started to appear from the gel point and its evolution versus degree of cure was nonlinear. Chemical shrinkage along fibers direction [0°] was found negligible. Comparison of these results with volume chemical shrinkage of composite leads us to conclude that the magnitude of chemical shrinkage along the thickness was greater than the in-plane chemical shrinkage.

5.6. Bibliography

[ABO 10] ABOU MSALLEM Y., JACQUEMIN F., BOYARD N. *et al.*, "Material characterization and residual stresses simulation during the manufacturing process of epoxy matrix composites", *Composites Part A: Applied Science and Manufacturing*, vol. 41, pp. 108–115, 2010.

[ANT 06] ANTONUCCI V., CUSANO A., GIORDANO M. *et al.*, "Cure-induced residual strain build-up in a thermoset resin", *Composites Part A: Applied Science and Manufacturing*, vol. 37, pp. 592–601, 2006.

[BOG 92] BOGETTI T.A., GILLESPIE J.W., "Process-induced stress and deformation in thick-section thermoset composite laminates", *Journal of Composite Materials*, vol. 26, pp. 626–660, 1992.

[DAN 90] DANIEL I.M., WANG T.M., KARALEKAS D. *et al.*, "Determination of chemical cure shrinkage in composite laminates", *J. Compos. Technol. Res.*, vol. 12, pp. 172–176, 1990.

[DAN 98] DANO M.L., HYER M.W., "Thermally-induced deformation behavior of unsymmetric laminates", *International Journal of Solids and Structures*, vol. 35, pp. 2101–2120, 1998.

[ERS 10] ERSOY N., TUGUTLU M., "Cure kinetics modeling and cure shrinkage behavior of a thermosetting composite", *Polymer Engineering & Science*, vol. 50, pp. 84–92, 2010.

[ESH 57] ESHELBY J.D., "The determination of the elastic field of an ellipsoidal inclusion, and related problems", *Proceedings of the Royal Society of London. Series A Mathematical and Physical Sciences*, vol. 241, no. 1226, pp. 376–396, 1957.

[GIG 04] GIGLIOTTI M., WISNOM M.R., POTTER K.D., "Loss of bifurcation and multiple shapes of thin [0/90] unsymmetric composite plates subject to thermal stress", *Composites Science and Technology*, vol. 64, pp. 109–128, 2004.

[HIL 67] HILL B.R., "The essential structure of constitutive laws for metal composites and polycrystals", *Journal of the Mechanics and Physics of Solids*, vol. 15 pp. 79–95, 1967.

[JAC 05] JACQUEMIN F., FREOUR S., GUILLEN R., "A hygroelastic self-consistent model for fiber-reinforced composites", *Journal of Reinforced Plastics and Composites*, vol. 24, pp. 485–502, 2005.

[JUN 92] JUN W.J., HONG C.S., "Cured shape of unsymmetric laminates with arbitrary lay-up angles", *Journal of Reinforced Plastics and Composites*, vol. 11, pp. 1352–1366, 1992.

[KIM 89] KIM K.S., HAHN H.T., "Residual stress development during processing of graphite/epoxy composites", *Composites Science and Technology*, vol. 36, pp. 121–132, 1989.

[KIM 04] KIM Y.K., "Process-induced residual stress analysis by resin transfer molding", *Journal of Composite Materials*, vol. 38, pp. 959–972, 2004.

[KIM 07] KIM Y.K., WHITE S.R., "Viscoelastic analysis of processing-induced residual stresses in thick composite laminates", *Mechanics of Composite Materials and Structures*, vol. 4, pp. 361–387, 2007.

[KRO 58] KRÖNER E., "Berechnung der elastischen Konstanten des Vielkristalls aus den Konstanten des Einkristalls", *Zeitschrift für Phys*, vol. 151, pp. 504–518, 1958.

[LEE 71] LEE S., "Stability of saddle-like deformed configurations of plates and shallow shells", *International Journal of Non-Linear Mechanics*, vol. 6, pp. 221–236, 1971.

[LI 04] LI C., POTTER K., WISNOM M.R. *et al.*, "In-situ measurement of chemical shrinkage of MY750 epoxy resin by a novel gravimetric method", *Composites Science and Technology*, vol. 64, pp. 55–64, 2004.

[MSA 10] MSALLEM Y.A., JACQUEMIN F., POITOU A., "Residual stresses formation during the manufacturing process of epoxy matrix composites: resin yield stress and anisotropic chemical shrinkage", *International Journal of Material Forming*, vol. 3, pp. 1363–1372, 2010.

[NAW 12a] NAWAB Y., Characterization and modelling of cure dependent properties of thermoset composites – application to the simulation of residual stresses, PhD Thesis, University of Nantes, 2012.

[NAW 12b] NAWAB Y., JACQUEMIN F., CASARI P. *et al.*, "Evolution of chemical and thermal curvatures in thermoset-laminated composite plates during the fabrication process", *Journal of Composite Materials*, vol. 47, pp. 327–339, 2012.

[NAW 12c] NAWAB Y., TARDIF X., BOYARD N. *et al.*, "Determination and modelling of the cure shrinkage of epoxy vinylester resin and associated composites by considering thermal gradients", *Composites Science and Technology*, vol. 73, pp. 81–87, 2012.

[NAW 13] NAWAB Y., SHAHID S., BOYARD N. *et al.*, "Chemical shrinkage characterization techniques for thermoset resins and associated composites", *Journal of Materials Science*, vol. 48, pp. 5387–5409, 2013.

[NAW 14] NAWAB Y., PARK C.H., SAOUAB A. *et al.*, "Shape distortion of carbon/epoxy composite parts during fabrication", *Macromolecular Symposia*, vol. 340, pp. 59–64, 2014.

[NAW 15] NAWAB Y., CASARI P., BOYARD N. *et al.*, "In situ characterisation of in-plane chemical shrinkage of thermoset laminated composite plates using a simple setup", *Journal of Reinforced Plastics and Composites*, vol. 34, pp. 931–938, 2015.

[OLI 06] OLIVIER P., "A note upon the development of residual curing strains in carbon/epoxy laminates. Study by thermomechanical analysis", *Composites Part A: Applied Science and Manufacturing*, vol. 37, pp. 602–616, 2006.

[PAR 06] PARLEVLIET P.P., BERSEE H.E.N., BEUKERS A., "Residual stresses in thermoplastic composites—a study of the literature–Part I: formation of residual stresses", *Composites Part A: Applied Science and Manufacturing*, vol. 37, pp. 1847–1857, 2006.

[RAD 95] RADFORD D.W., "Volume fraction gradient induced warpage in curved composite plates", *Composites Engineering*, vol. 7, pp. 923–934, 1995.

[TWI 04] TWIGG G., POURSARTIP A., FERNLUND G., "Tool–part interaction in composites processing. Part I: experimental investigation and analytical model", *Composites Part A: Applied Science and Manufacturing*, vol. 35, pp. 121–133, 2004.

[WEI 82] WEITSMAN Y., HARPER B., "Optimal cooling of cross-ply composite laminates and adhesive joints", *Journal of Applied Mechanics*, vol. 49, pp. 735–739, 1982.

[WHI 92] WHITE S.R., HAHN H.T., "Process modeling of composite materials: residual stress development during cure. Part I. model formulation", *Journal of Composite Materials*, vol. 26, pp. 2402–2422, 1992.

[WIJ 05] WIJSKAMP S., Shape distortions in composites forming, PhD Thesis, University of Twente, 2005.

[WIS 06] WISNOM M.R., GIGLIOTTI M., ERSOY N. *et al.*, "Mechanisms generating residual stresses and distortion during manufacture of polymer–matrix composite structures", *Composites Part A: Applied Science and Manufacturing*, vol. 37, pp. 522–529, 2006.

[WU 00] WU Y.-J., TAKATOYA T., CHUNG K. *et al.*, "Development of the transient simulated laminate (TSL) methodology for moisture ingression studies using unsymmetric laminates", *Journal of Composite Materials*, vol. 34, pp. 1998–2015, 2000.

[ZHU 01] ZHU Q., GEUBELLE P.H., LI M. *et al.*, "Dimensional accuracy of thermoset composites: simulation of process-induced residual stresses", *Journal of Composite Materials*, vol. 35, pp. 2171–2205, 2001.

6

Heat Transfer in Composite Materials and Porous Media: Multiple-Scale Aspects and Effective Properties

The estimation of effective transport properties for composite materials or porous media is a typical example of multi-scale problems. Starting from the pore-scale transport equations, various macroscale models may be obtained, depending on the physical parameters involved. Upscaling theories are available which can be used to determine the proper form of the macroscale equations. Local or *closure* problems are also available which can be used to calculate macroscale (also called effective) transport properties. This is particularly useful for various industrial objectives: estimation of the properties based on available pore-scale data (computed tomography (CT)-scan images, etc.), optimization of the macroscale properties, etc. This chapter provides a brief introduction to the upscaling of heat transfer in composite materials and porous media. Macroscale equations and effective properties are studied in the case of local equilibrium and non-equilibrium situations, which cover most situations encountered in composite processing and in industrial applications of composite materials.

6.1. Introduction

While composite materials may involve several phases, we restrict the analysis in this chapter to two-phase systems. In Figure 6.1, we present a schematic representation of a two-phase material made up of a β-phase and a σ-phase, typical of a porous medium or a composite material, which emphasizes the multiple-scale features. In particular, we distinguish the micro-scale, or pore-scale, characterized by the length-scales, l_β and l_σ, and the macro-scale of characteristic length-scale L.

Chapter written by Michel QUINTARD.

The macro-scale is in general related to the process under investigation, e.g. mechanical part, heat exchanger, mold, etc. While the pore-scale physics fully determine the macro-scale behavior, there is in general not a particular interest in knowing the l-scale details. Moreover, the details of the pore-scale structure are most of the time unavailable, and, as a result, a direct solution of the pore-scale equations cannot be undertaken. However, macro-scale quantities, e.g. temperature, fluxes, etc., are needed in any application. Therefore, the development of a macro-scale theory is of a particular importance and has received a lot of attention in the literature. Macro-scale theories may be introduced in a heuristic manner. The most prominent example is the introduction of Darcy's law [DAR 56] for describing the relationship between the total flux flowing through a porous column and the pressure drop across the column. The introduction of the generalized Darcy's law for describing at the macro-scale multiphase flow through porous media is also a good example [MUS 49]. While later theoretical investigations did not fully support these multiphase equations, they are still commonly used in engineering practice.

Heuristic laws are certainly very useful. This is not, however, without the risk of erroneous intuitions. Indeed, many questions must be addressed when developing a macro-scale theory:

– what is the meaning of a macro-scale theory? Indeed, if we adopt as a starting point the usual continuum physics, then temperature, fluxes and other pore-scale physical variables have in general a well-defined meaning. On the contrary, it is not necessarily clear what a macro-scale temperature, pressure, etc. is;

– what is the appropriate form of the macro-scale equations? This short introduction to heat transfer in porous media and composite material will clearly illustrate the difficulties generated by such questions;

– what are the *effective properties* to be used when manipulating the macro-scale theory? Indeed, if we consider the temperature field in a two-phase composite material, its pore-scale value will be determined by the complex interplay of phase thermal conductivities, diffusivities and the pore-scale geometry. We will show later that such a problem may be, most of the time but not always, described macroscopically with a heat transfer equation involving a mixture temperature (to be defined later) and an effective thermal conductivity that depends not only on the phase thermal conductivities but also on the pore-scale geometry. Since many modern tools are available today to investigate such pore-scale features (nuclear magnetic resonance (NMR), X-ray scanners, etc.), it is interesting to have a theoretical link between the pore-scale structure and all macro-scale equations and quantity. Indeed, commercial software provides tools to calculate some macro-scale properties from pore-scale tomographic images.

This chapter will try to clarify, in the case of heat transfer, some of these questions.

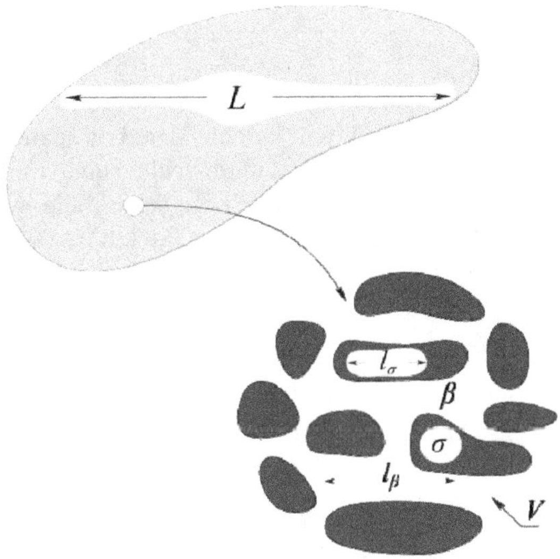

Figure 6.1. *Multiple-scale schematic representation of a two-phase medium*

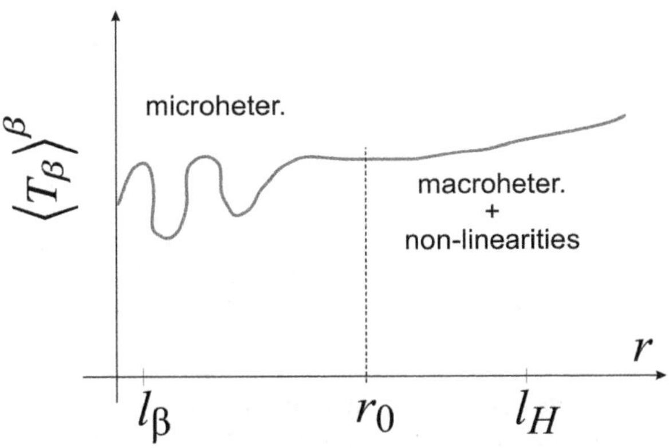

Figure 6.2. *Separation of scales*

6.2. Effective thermal conductivity

We start with a generic and simple heat transfer problem to illustrate the different questions associated with the development of a macro-scale theory. While the original problem, as illustrated in Figure 6.1, clearly features pore-scale interfaces between the different phases, we will work with a continuous but heterogeneous thermal conductivity field. The pore-scale heat transfer equation for temperature T is simply given by:

$$\nabla \cdot (\mathbf{k}(\mathbf{x}) \cdot \nabla T) = 0 \qquad [6.1]$$

where the thermal conductivity, $k(x)$, will be considered as spatially periodic in this short introduction, with a characteristic length-scale equal to l. Typically, in a porous medium or composite material, the pore-scale field will feature l-scale fluctuations as illustrated by the black solid line on the left part of Figure 6.3.

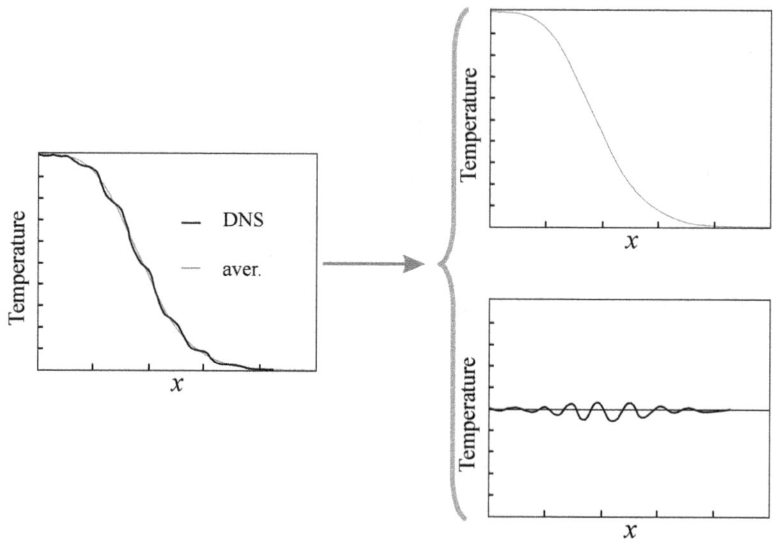

Figure 6.3. *Pore-scale temperature field, deviation field and macro-scale field. For a color version of the figure, see www.iste.co.uk/boyard/heat.zip*

This figure also illustrates the existence of an average temperature signal, which is exactly what a macro-scale theory should be able to predict. We will discuss in what follows a general framework to relate the macro-scale theory giving the average signal to the pore-scale physics.

6.2.1. *Background on upscaling methods*

As outlined above, we want to look at a governing equation for the macro-scale averaged temperature field. Many mathematical definitions may be used to achieve this goal. In this short introduction, we will use spatial averaging to smooth the pore-scale temperature field. The macro-scale temperature will be calculated as:

$$T^* = \langle T \rangle (\mathbf{x}, t) \equiv \frac{1}{V} \int_{\mathbf{r} \in \mathcal{V}(\mathbf{x})} T(\mathbf{r}) \, dV_r \qquad [6.2]$$

where V is the averaging volume of size r illustrated in Figure 6.1. It may be convenient to define an intrinsic phase average, for instance for the temperature T_β in the β-phase, such as:

$$\langle T_\beta \rangle^\beta = \frac{1}{V_\beta} \int_{V_\beta} T_\beta dV \quad ; \quad \langle T_\beta \rangle = \varepsilon_\beta \langle T_\beta \rangle^\beta \qquad [6.3]$$

where V_β is the volume of the β-phase within the averaging volume V and ε_β is the β-phase volume fraction defined by:

$$\varepsilon_\beta = \frac{V_\beta}{V} \qquad [6.4]$$

There are many constraints over the size of the averaging volume in order to get a macro-scale field devoid of l-scale fluctuations. The evolution of the intrinsic temperature field is illustrated in Figure 6.2. The average value sees l-scale fluctuations as r increases then reaches a plateau for a critical value r_0, before experiencing large-scale fluctuations due to media heterogeneities. One constraint of particular importance for obtaining a simple macro-scale theory is the constraint of *separation of scale* expressed by:

$$l_\beta, l_\sigma \ll r_0 \ll L \qquad [6.5]$$

where r_0 defines the size of the minimal representative elementary volume (REV). We will assume that such constraints hold throughout this chapter.

6.2.2. *A simple example: continuous thermal conductivity*

Let us now look at the treatment of the simple example presented in the introduction of this section. It is convenient to look at an approximate solution of equation [6.1] in terms of the following decomposition:

$$T = \langle T \rangle + \tilde{T} \qquad [6.6]$$

where $\tilde{T}$ is called the temperature deviation. The averaged equation reads:

$$\nabla \cdot \left\langle \mathbf{k}(x) \cdot \left(\nabla \langle T \rangle + \nabla \tilde{T}\right)\right\rangle = 0 \quad [6.7]$$

and the micro-scale equation becomes:

$$\nabla \cdot \left(\mathbf{k}(x) \cdot \nabla \tilde{T}\right) = -\nabla \cdot (\mathbf{k}(x) \cdot \nabla \langle T \rangle) \quad [6.8]$$

Now comes some approximation for the obtained coupled problem in order to find a simple solution. Given the linearity of the problem, the assumption of scale separation, an approximate solution can be developed under the form:

$$\tilde{T} = \mathbf{b}.\nabla \langle T \rangle + \ldots \quad [6.9]$$

where the mapping vector, **b**, obeys the following closure problem:

$$\nabla \cdot (\mathbf{k} \cdot \nabla \mathbf{b}) = -\nabla \cdot \tilde{\mathbf{k}} \quad [6.10]$$

with the following additional constraints (the periodicity condition is not essential at this point):

$$\langle \mathrm{b} \rangle = 0; \ \mathrm{b}(\mathrm{X} + 1_{\mathrm{i}}) = \mathrm{b}(\mathrm{X}) \ \mathrm{i} = 1,2,3 \quad [6.11]$$

Equation [6.7], which is the candidate for being the macro-scale equation, can be recast into the following *closed* macro-scale equation (i.e. without pore-scale deviations)

$$\nabla \cdot \left(\mathbf{K}_{eff} \cdot \nabla \langle T \rangle\right) = 0 \quad [6.12]$$

where the effective thermal conductivity tensor is given by:

$$\mathbf{K}_{eff} = \langle \mathbf{k} \cdot (\mathbf{I} + \nabla \mathbf{b}) \rangle \quad [6.13]$$

The upscaling technique not only provides the form of the macro-scale equation but also a way of calculating the effective properties from some knowledge of the pore-scale characteristics. Before looking at some general features of the effective properties, it is important to remind the readers that, technically speaking, many different, upscaling techniques may be designed. All theories have the same goal: deriving macro-scale equations and the associated effective properties. It is beyond the scope of this chapter to provide a comprehensive review. Indications are given in [DAV 13] for a comparison between the volume averaging technique [CAR 84] and the homogenization method [ENE 82, AUR 83]. Other points of views include the stochastic methods, effective-medium theories, etc. [MAT 67, DAG 89, GEL 93].

6.2.3. *Effective thermal conductivity: properties and bounds*

From the aforementioned upscaling theories, several properties can be inferred for the values of the effective thermal conductivity. The first property of importance is the existence of bounds in terms of the thermal conductivity ratio and volume fractions. Let us take a two-phase system containing a β-phase and a σ-phase. It can be shown (see, for instance, [WIE 12, DAG 89, TOR 02]), that the effective thermal conductivity is bounded by the harmonic average and by the arithmetic average, weighted by the phase volume fractions. We have:

$$K_{\perp} \leq K_{eff} \leq K_{\parallel} \tag{6.14}$$

with

$$\frac{1}{K_{\perp}} = \frac{\varepsilon_\sigma}{k_\sigma} + \frac{\varepsilon_\beta}{k_\beta} \quad ; \quad K_{\parallel} = \varepsilon_\sigma k_\sigma + \varepsilon_\beta k_\beta \tag{6.15}$$

Such bounds are useful, in particular if little is known about the pore-scale structure. These bounds, however, do not impose a strong restriction and cannot be used as a real predictive tool. Improvements have been proposed in order to narrow the predictions. Among the popular formulas, one found the result obtained by [HAS 63] and expressed as:

$$\varepsilon_\sigma k_\sigma + \varepsilon_\beta k_\beta - \frac{\varepsilon_\sigma \varepsilon_\beta \left(k_\beta - k_\sigma\right)^2}{\varepsilon_\beta k_\sigma + \varepsilon_\sigma k_\beta + k_\beta (d-1)} \leq K_{eff} \leq \varepsilon_\sigma k_\sigma + \varepsilon_\beta k_\beta - \frac{\varepsilon_\sigma \varepsilon_\beta \left(k_\beta - k_\sigma\right)^2}{\varepsilon_\beta k_\sigma + \varepsilon_\sigma k_\beta + k_\sigma (d-1)} \tag{6.16}$$

where d is the dimension of space and $k_\sigma > k_\beta$.

Estimates based on a few parameters such as the phase volume fractions and the thermal conductivity ratios have been proposed. For low-density inclusions made up of spherical particles (σ-phase), we may use the following estimate due to Maxwell [MAX 73]:

$$K_{eff} = k_\beta \left(1 + \frac{3\varepsilon_\sigma}{\left(\frac{(k_\sigma/k_\beta)+2}{(k_\sigma/k_\beta)-1}\right) - \varepsilon_\sigma} \right) \tag{6.17}$$

The logarithmic mixture law given by:

$$K_{eff} = k_\sigma^{\varepsilon_\sigma} k_\beta^{\varepsilon_\beta} \qquad [6.18]$$

is also popular [LIC 27]. Numerous other estimates have been proposed for various conditions [RAY 92, BRU 35]. While these correlations are useful, they are based on simplistic or idealized representations of the real structure. The effective thermal conductivity may depend strongly on specific geometrical and physical features. This is illustrated in the case of a two-phase system by the results presented in Figures 6.4 and 6.6 [NOZ 85]. In Figure 6.4, the "closure problem" expressed by equations [6.10] and [6.11] is solved for the structure obtained from a tomography image represented in Figure 6.5 for different values of the thermal conductivity ratio k_σ / k_β (the β-phase corresponds to the phase in the pore-space). The resulting effective thermal conductivity K_{eff} (the medium is isotropic to within less than 1%) divided by k_β is plotted versus k_σ / k_β together with various estimates and bounds. Classically, the effective thermal conductivity increases with k_σ / k_β. Bounds or estimates that do not make precise assumptions about the medium real structure, while certainly useful, do not provide in many cases accurate enough predictions. The Maxwell result, for instance, is based on the assumption of non-touching inclusions with a small volume fraction. Wiener's bounds or Hashin and Shtrikman bounds do narrow the expected range of variation but without being truly predictive in this somehow simple medium structure. Indeed, topology plays an important role as is emphasized in the results presented in Figure 6.6.

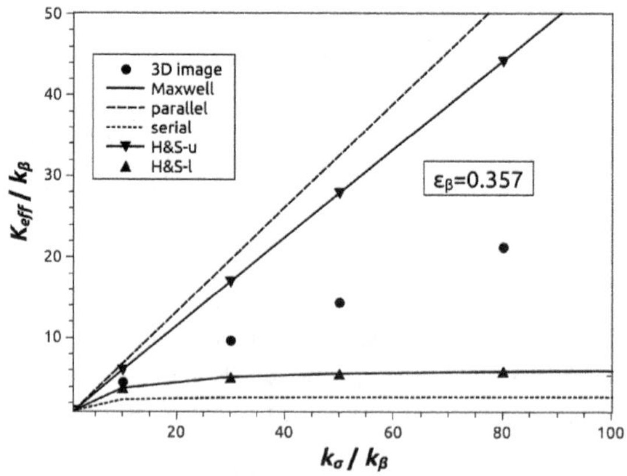

Figure 6.4. *Effective thermal conductivity of a packed bed: ED image = computation from tomography image, estimates and bounds (H&S-u = upper bound from Hashin and Shtrikman; H&S-l = lower bound from Hashin and Shtrikman*

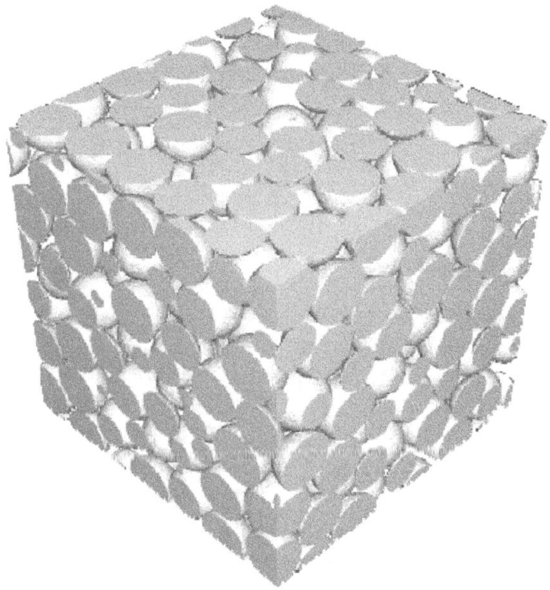

Figure 6.5. *X-ray tomography image of a packed bed made up of spherical grains*

Figure 6.6 represents various experimental and numerical results for a two-phase system, the β-phase filling the pore space. Experimental results referenced in [NOZ 85] correspond to various unconsolidated media. Calculations are for two types of 2D unit cells. One is characterized by solid square inclusions (σ-phase) in a continuous β-phase, while the other features contact points materialized by small solid bridges. Calculations are made for some values of the β-phase volume fraction and the ratio K_{eff}/k_β is plotted as a function of the thermal conductivity ratio, k_σ/k_β. Let us first look at the results for a medium with solid inclusions. Starting from a value of 1 in the homogeneous medium case, the ratio K_{eff}/k_β increases with k_σ/k_β to reach eventually a plateau. Increasing the thermal conductivity of the isolated inclusions does not bring additional effective thermal conductivity after a value of k_σ/k_β roughly equal to 100. This is completely different for the case with contact points [SHO 89]. At first, the behavior is the same for low k_σ/k_β. However, the heat flux goes through the contact point areas at large k_σ/k_β leading to a continuous increase in K_{eff}/k_β. This effect will be particularly dramatic if we work with a metallic composite material. Now, considering the 3D calculations presented in Figure 6.4, we see that the curve is typical of a medium with contact point effects. Is the observed curve really predictive when contact point effects are important? This is difficult to tell since the actual results are very sensitive to the geometry for k_σ/k_β larger than 100 and since the true contact point surface is often below the tomography resolution.

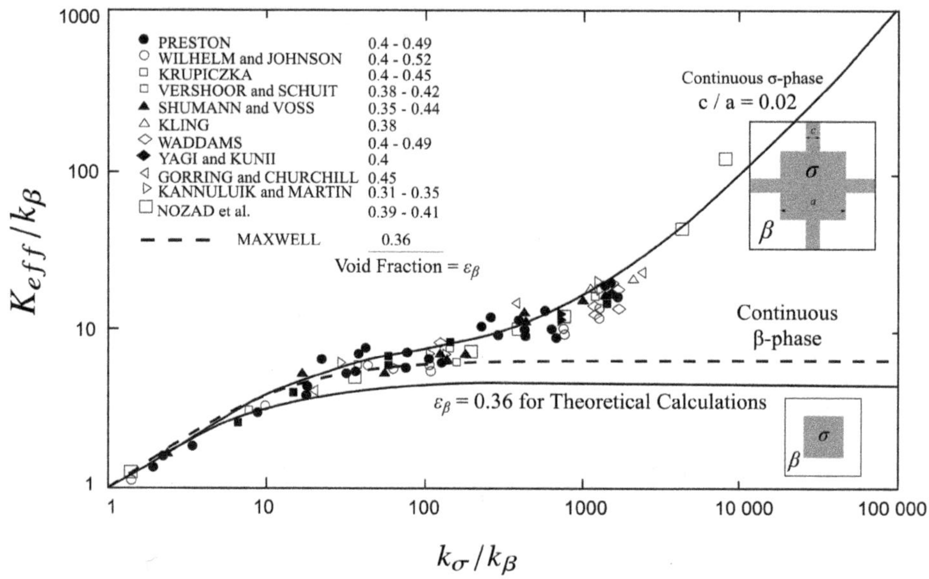

Figure 6.6. *Effective thermal conductivity versus the thermal conductivity ratio (according to [NOZ 85])*

6.3. Local-equilibrium model and thermal dispersion

What is the impact of a flowing fluid over the effective heat transfer equation? To understand what is at stake, we consider below the two-phase problem with β as the flowing phase. The pore-scale problem to be solved is described by the following classical equations:

$$(\rho c_p)_\beta \frac{\partial T_\beta}{\partial t} + (\rho c_p)_\beta \nabla \cdot \left(\mathbf{v}_\beta T_\beta\right) = \nabla \cdot (k_\beta \nabla T_\beta) \ \text{ in } V_\beta \qquad [6.19]$$

$$\text{B.C. 1} \quad T_\beta = T_\sigma \ \text{ at } A_{\beta\sigma} \qquad [6.20]$$

$$\text{B.C. 2} \quad \mathbf{n}_{\beta\sigma} \cdot k_\beta \nabla T_\beta = \mathbf{n}_{\beta\sigma} \cdot k_\sigma \nabla T_\sigma \quad \text{at } A_{\beta\sigma} \tag{6.21}$$

$$(\rho c_p)_\sigma \frac{\partial T_\sigma}{\partial t} = \nabla \cdot (k_\sigma \nabla T_\sigma) \quad \text{in } V_\sigma \tag{6.22}$$

where $(\rho c_p)_\alpha$ is the heat capacity of the α-phase, indices indicate the related phase, v_β, is the pore-scale velocity field. This problem must be completed by the total mass balance equation and the momentum balance equation. Without further discussion (the readers may have a look at the following literature for indications about the upscaling of the Stokes problem [SAN 80, WHI 86, QUI 94, QUI 94a, QUI 94b]), we will assume that the latter two equations may be solved independently from the heat transfer problem, e.g. constant density and viscosity. Therefore, the associated macroscale equation reads:

$$\nabla \cdot \mathbf{V}_\beta = 0 \tag{6.23}$$

for the total mass balance in the case of constant density and Darcy's law for creeping flows written as:

$$\mathbf{V}_\beta = -\frac{1}{\mu_\beta} \mathbf{K}.\left(\nabla P_\beta - \rho_\beta \mathbf{g}\right) \tag{6.24}$$

where the filtration velocity $\mathbf{V}_\beta$, and the intrinsic velocity, $\mathbf{U}_\beta$, are defined by:

$$\mathbf{V}_\beta = \langle \mathbf{v}_\beta \rangle = \varepsilon_\beta \mathbf{U}_\beta \ ; \ \mathbf{U}_\beta = \langle \mathbf{v}_\beta \rangle^\beta \tag{6.25}$$

In terms of the heat transfer problem, does such a system behave like a mixture (or equivalent) phase? The answer is: not necessarily, depending on the pore-scale geometrical features. Indeed, if we consider a transient problem, we may define four main characteristic times for diffusion in the system, two linked to the micro-scale temperature field and two linked to the macro-scale behavior (we will not consider, at this point, the characteristic time due to advection within the pore space, estimates for the macroscale effective conductivity will be discussed later). If these characteristic times are arranged as follows:

$$\left(\frac{(\rho c_p)_\beta l_\beta^2}{k_\beta} \approx \frac{(\rho c_p)_\sigma l_\sigma^2}{k_\sigma} \right) \ll \left(\frac{\varepsilon_\beta (\rho c_p)_\beta L^2}{K_\beta^*} \approx \frac{\varepsilon_\sigma (\rho c_p)_\sigma L^2}{K_\sigma^*} \right) \tag{6.26}$$

i.e. the thermal evolution of the two phases at the micro-scale is quite similar, and relaxes more rapidly than the macro-scale temperatures, then we obtain a macro-scale local equilibrium behavior characterized by the approximation:

$$\langle T_\beta \rangle^\beta = \langle T_\sigma \rangle^\sigma = \langle T \rangle \qquad [6.27]$$

and, therefore, the heat may be calculated as:

$$\varepsilon_\beta \left(\rho c_p\right)_\beta \langle T_\beta \rangle^\beta + \varepsilon_\sigma \left(\rho c_p\right)_\sigma \langle T_\sigma \rangle^\sigma = \left(\rho c_p\right)^* \langle T \rangle \qquad [6.28]$$

where we have the effective heat capacity given by:

$$\left(\rho c_p\right)^* = \varepsilon_\beta \left(\rho c_p\right)_\beta + \varepsilon_\sigma \left(\rho c_p\right)_\sigma \qquad [6.29]$$

It must be emphasized that the *mixture temperature* in the local equilibrium case is also approximately equal to the intrinsic phase averages. Under these circumstances, upscaling theories provide the following local-equilibrium thermal dispersion equation:

$$\left(\rho c_p\right)^* \frac{\partial \langle T \rangle}{\partial t} + \varepsilon_\beta \left(\rho c_p\right)_\beta \mathbf{U}_\beta \cdot \nabla \langle T \rangle = \nabla \cdot \left(\mathbf{K}_{eq}^* \cdot \nabla \langle T \rangle\right) \qquad [6.30]$$

where K_{eq}^* is the thermal dispersion tensor [ENE 82, CAR 84, WHI 99]. Its typical behavior is represented in Figure 6.7 which plots the ratio of K_{eq}^* / k_β (longitudinal component, i.e. in the flow direction) versus the particle Péclet number given by:

$$Pe = \frac{\left(\rho C_p\right)_\beta \langle v_\beta \rangle d_p}{k_\beta} \qquad [6.31]$$

for a system made up of glass beads and with flowing air in the x-direction. This curve is typical of dispersion mechanisms. The longitudinal thermal conductivity tends to the value K_{eff} in the diffusive regime at low Péclet numbers. Then, after a transition that depends on the thermal conductivity ratio, the dispersion curve scales like Pe^n ($n<2$). We may notice that the transition occurs at Pe numbers larger than 1, i.e. larger than the usual transition value for the case of solute dispersion (solute dispersion would correspond here to the limit k_σ / k_β tends to zero).

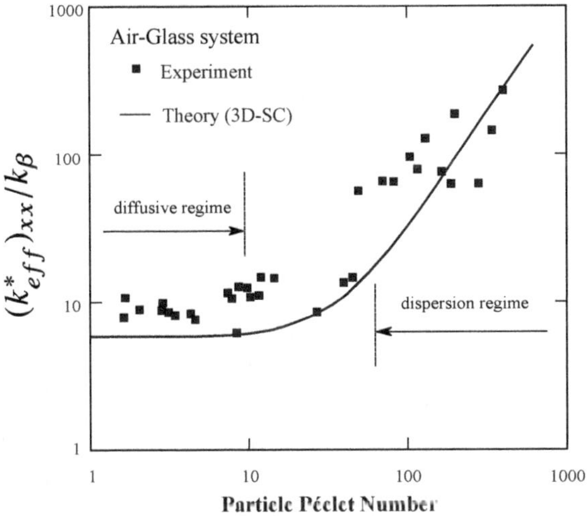

Figure 6.7. *Longitudinal thermal dispersion curve for air and packed glass beads. Experiments from [YAG 60, GUN 74] and numerical calculations from [QUI 97]*

6.4. Local equilibrium versus local non-equilibrium models

The constraints for having a local equilibrium situation may be too restrictive if the characteristics of the two phases are rather different. For instance, if the thermal diffusivity in the solid phase is rather small compared to its β-phase counterpart, the average σ-phase temperature field will lag behind thus producing a *local non-equilibrium* situation (one example will be provided later in Figure 6.10) characterized by, in general, $\langle T_\beta\rangle^\beta \neq \langle T_\sigma\rangle^\sigma$. In this case, several theories have been proposed to handle such non-equilibrium situations. They are briefly introduced below.

6.4.1. *The two-equation model*

The simplest model used in practice is based on the following two-equation model [SCH 75, GLA 88]:

$$\varepsilon_\beta \left(\rho c_p\right)_\beta \frac{\partial \langle T_\beta\rangle^\beta}{\partial t} + \left(\rho c_p\right)_\beta \langle \mathbf{v}_\beta\rangle . \nabla \langle T_\beta\rangle^\beta = \nabla . \left(\mathbf{K}^*_\beta . \nabla \langle T_\beta\rangle^\beta\right) - h\left(\langle T_\beta\rangle^\beta - \langle T_\sigma\rangle^\sigma\right) \qquad [6.32]$$

$$\varepsilon_\sigma \left(\rho c_p\right)_\sigma \frac{\partial \langle T_\sigma\rangle^\sigma}{\partial t} = \nabla . \left(\mathbf{K}^*_\sigma . \nabla \langle T_\sigma\rangle^\sigma\right) - h\left(\langle T_\sigma\rangle^\sigma - \langle T_\beta\rangle^\beta\right) \qquad [6.33]$$

where h is a volume heat exchange coefficient and where phase effective thermal dispersion tensors have been introduced. If h is large enough, this two-equation model will give back the local equilibrium model with:

$$\mathbf{K}_{eq} = \mathbf{K}_\beta^* + \mathbf{K}_\sigma^* \tag{6.34}$$

The heat exchange coefficient is not intrinsic to the pore-scale boundary condition, as it is often introduced in some heuristic models, but instead depends on the flow conditions and thus is a *result* of the upscaling process. One example of such a coefficient determination is shown in Figure 6.8. The experimental device, described more thoroughly in [GRA 88], is made up of glass beads arranged in a hexagonal unit cell. The figure reproduces the values of the dimensionless number hl_β^2 / k_β as a function of the unit cell Péclet number for two couples of phases. Experimental values are taken from [GRA 88], numerical solutions of the closure problems giving the mass exchange coefficient are taken from [QUI 97] and correspond to 2D and 3D hexagonal arrangements of circles and spheres, respectively. There is a relatively good agreement between theoretical predictions and experimental values for the low thermal conductivity ratio, i.e. the nylon-water system, while there is a significant difference for the large thermal conductivity ratio, i.e. the brass-water system, for which contact point effects are likely to play a role.

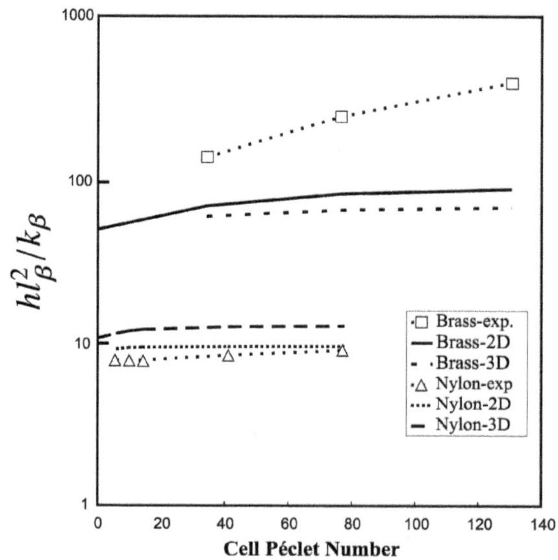

Figure 6.8. *Experimental and numerical determination of the heat exchange coefficient for brass-water* ($k_\sigma / k_\beta = 180$) *and nylon-water systems* ($k_\sigma / k_\beta = 0.46$) *[QUI 97]*

It is interesting to look at the evolution of the volume heat exchange coefficient with the pore-scale physical characteristics. From the dimensionless analysis as exemplified in Figure 6.8, we see that, first, h is inversely proportional to the square of a pore-scale characteristic length. Therefore, local non-equilibrium conditions are likely to occur in media with large pore-scale dimensions. This occurs in some heat storage devices, fractured media, to cite a few examples. However, this also depends on the thermal characteristics and therefore this may be also important, even for media with relatively small characteristic lengths, if the process characteristic time is small. This might be the case, for instance, when modeling the fate of composite materials during the firing of rocket boosters, as discussed in [FLO 91, PUR 04]. In these papers, it is shown that the position of the pyrolysis front is affected by the choice of a local equilibrium or local non-equilibrium model. This is beyond the scope of this chapter to discuss in details the conditions under which we may assume local equilibrium or not. This point has received a lot of attention in the literature [WHI 91, QUI 93, AMI 94, QUI 95, MIN 99, VAD 05]. Simple estimates of the various time and length-scales are enough to decide whether a non-equilibrium analysis is needed or not if differences are clear (for instance, conditions such as in equation [6.26] clearly defined by differences of several orders of magnitude). If this is not the case, it has been observed in the literature that the range of validity depends on the geometry and topology of the unit cell [QUI 93], the boundary value problem solved [VAF 90, VAD 05], the processes involved, for instance natural convection [REE 05], phase change [SOE 90, DUV 04], the coupling with reactive transport [FAT 97, OLI 01], etc. The readers may find further information in [VAF 90, QUI 93, DAV 10].

6.4.2. *Further discussion*

This section provides the readers with an extended discussion and some opening toward more sophisticated models. No details about the mathematical developments are provided and the readers are referred to the cited literature for a better understanding.

Upscaling theories provide generalized forms of the simple two-temperature model given by equations [6.32] and [6.33]. They contain additional terms in the partial derivative equations such as cross-diffusion terms and various pseudo-convective terms [CAR 84, ZAN 84, LEV 85, QUI 93, QUI 97]. All these models, however, suffer from the same restriction that the heat exchange term is only able to reproduce a single characteristic time for the exchange process between the two phases, determined by the value of the heat exchange coefficient h. Indeed, it can be shown that a more accurate macro-scale equation would involve spatial and time convolutions, as it is suggested in [MOY 97]. It has also been demonstrated (see [LEV 85, DAV 12]) that the two-equation model has an asymptotic behavior for some initial boundary value problems (mainly heat transfer in a semi-infinite, homogeneous medium with simple inlet boundary conditions, e.g. Dirichlet condition). This asymptotic behavior may be represented by a one-temperature model for the mixture temperature $\langle T \rangle$ (but without the assumption that $\langle T_\beta \rangle^\beta \approx \langle T_\sigma \rangle^\sigma$) mathematically equivalent to equation [6.30] but with an effective thermal dispersion tensor, K_∞^*, given by:

$$\mathbf{K}_\infty^* = \mathbf{K}_{eq}^* + \frac{1}{h}\frac{\left(\varepsilon_\beta \left(\rho c_p\right)_\beta \varepsilon_\sigma \left(\rho c_p\right)_\sigma\right)^2}{\left(\varepsilon_\beta \left(\rho c_p\right)_\beta + \varepsilon_\sigma \left(\rho c_p\right)_\sigma\right)^2}\mathbf{U}_\beta \mathbf{U}_\beta \qquad [6.35]$$

From this expression, we see that this asymptotic value, which is representative of the dynamic of the non-equilibrium situation, is *larger* than the local equilibrium value. Its value can be calculated by solving the closure problems associated with the two-temperature model [QUI 97] or it may be obtained by a direct upscaling with a specific procedure involving deviations to the mixture temperature, without making the assumption expressed by equation [6.27]. Such a model is often called a non-equilibrium one-temperature model. The readers will find additional information on this subject in [MOY 00, QUI 01, DAV 10]. We see from this partial discussion that several models may be developed to account for the complex interplay of the various physical phenomena involved. Indeed, this has triggered the proposition of various additional models, in addition to the three discussed here. For instance, other non-equilibrium one-equation models have been proposed in the literature such as dual-phase-lagging heat conduction models [WAN 08]. Other mathematical forms may be used to reproduce some of the features of the non-equilibrium situations, for instance equations with fractional derivatives which have already been used for dealing with some problems of dispersion in porous media [SAB 07].

At this point, we see that several models can be used, as summarized in Figure 6.9 according to [DAV 15]. The pore-scale problem is associated with direct numerical simulation (DNS). Assuming the macro-scale problem is somehow one-dimensional, the palette of various models comprises one-equation local equilibrium models, one-equation non-equilibrium models of various sorts, *N*-equation models and a variety of so-called *mixed* models which couples macro-scale equations with pore-scale equations where the upscaling process is difficult to carry on. Some of the models have not been discussed in this chapter, and we refer the readers to [DAV 15] for further discussion.

Which model should we use? This has been a source of much confusion in the literature. Of course, if we *a priori* know that the system will behave close to local equilibrium, the use of the one-equation equilibrium model is preferable. If we solve a 1D heat transfer problem over a semi-infinite homogeneous medium, it might be tempting to use the one-equation non-equilibrium model, therefore with an effective dispersion tensor larger than the one for the equilibrium model, etc. It is important to understand that, in many real cases, there are different stages during the evolution of the temperature field, which may require one model different from the model necessary at another stage. This underestimated effect has been illustrated in [DAV 10, DAV 15] in the case of a porous medium made of an array of cylinders and subject to a steady-state Stokes flow with Dirichlet conditions for temperature at both ends. The resulting temperature field at the macro-scale is 1D, as is illustrated in Figure 6.10. Symbols represent the computed averaged temperatures for the β-phase and the σ-phase coming from DNS, while the solid line corresponds to the macro-scale local equilibrium temperature field (additional comparisons with the two-equation model and the one-equation non-equilibrium models are provided in [DAV 15]). A close look at the various macro-scale fields allows us to draw the following picture for the various stages encountered during the simulation:

– after a short period for which the temperature fields are localized in one or two unit cells (and which corresponds to non-homogenizable situations), the solution calls for a local non-equilibrium treatment, for instance a two-equation model;

– since the medium has a large longitudinal extension, the asymptotic behavior is later observed, characterized by a larger thermal conductivity (larger spreading of the thermal plume) than the one-equation equilibrium model;

– because of the boundary condition at the exit, we finally obtain a steady-state solution and we see that the conditions *revert to local equilibrium.* An error would be made if we would attempt to predict this final temperature field with the asymptotic thermal conductivity value which is larger than K_{eq}^{*} as discussed before.

In conclusion, we have to be very careful about the choice of the model we choose to handle a specific case. For the example discussed here, the choice of a two-equation model was preferable since it covers the two limiting cases: local-equilibrium and asymptotic behavior, of course at the expense of some complexity and more effective parameters to be determined.

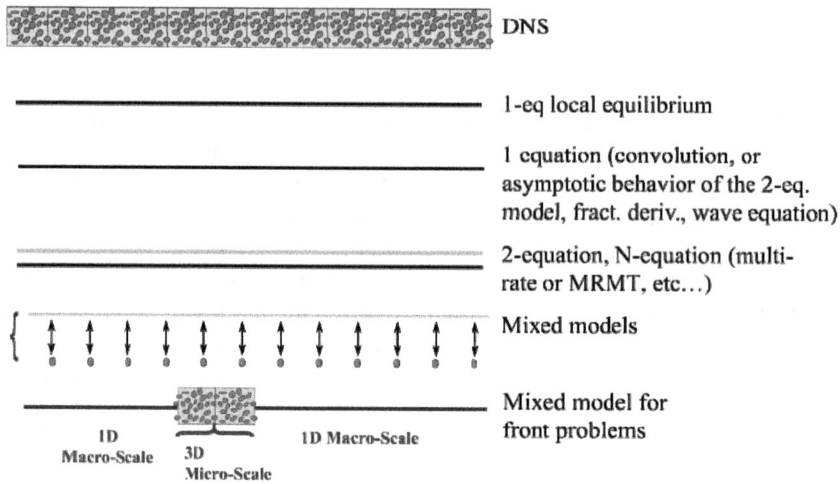

Figure 6.9. *Schematic representation of the various models that can be used*

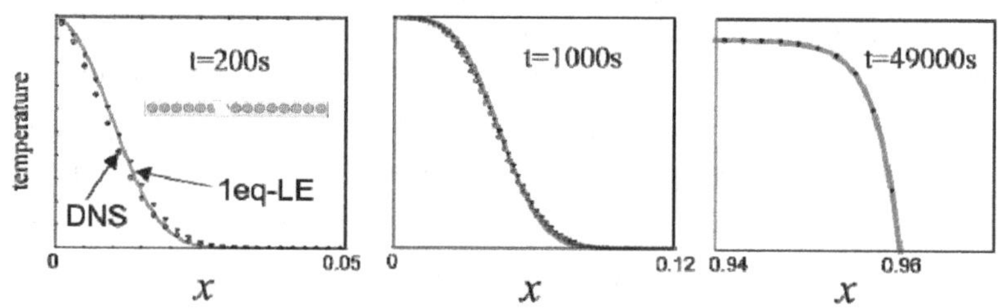

Figure 6.10. *Transition between different models during a single experiment [DAV 10]. For a color version of the figure, see www.iste.co.uk/boyard/heat.zip*

6.5. Various extensions

So far, we have considered the classical two-phase heat transfer problem. Many developments have been made to incorporate in the upscaling studies additional physical mechanisms, such as chemical reaction [FAT 97, OLI 01], drying and boiling [WHI 77, DUV 04], Soret effect [LAC 02, DAV 10], radiative heat transfer [SIN 92, KAV 95, JEA 12, LER 01], superfluid flow [ALL 10], etc. In this section, we focus on two cases which are often encountered in many industrial applications involving composite materials.

6.5.1. *Effect of homogeneous and heterogeneous thermal sources*

This section deals with the case of heat transfer with homogeneous and constant heat sources, for instance a volume heat source in the σ-phase denoted by Φ_σ, or a heterogeneous heat source, denoted by Ω, created at the β–σ interface. These sources may come from various mechanisms, including radiative heat transfer, chemical reactions, microwave or electromagnetic induction heating, etc.

The case of local-equilibrium is relatively straightforward and the resulting macro-scale equation is simply:

$$\left(\rho c_p\right)^* \frac{\partial \langle T \rangle}{\partial t} + \varepsilon_\beta \left(\rho c_p\right)_\beta \mathbf{U}_\beta \cdot \nabla \langle T \rangle = \nabla \cdot \left(\mathbf{K}^*_{eq} \cdot \nabla \langle T \rangle\right) + a_v \Omega + \varepsilon_\sigma \Phi_\sigma \qquad [6.36]$$

where a_v is the specific area of the porous medium [QUI 00, DAV 15].

The case of local non-equilibrium is more complicated. The homogeneous heat source appears in the related phase equation while the heterogeneous heat source must be distributed between the two equations. This problem has been addressed theoretically in [QUI 00,]. The two-equation model is slightly modified to introduce a distribution coefficient which partitions Ω between the two macro-scale equations. We have:

$$\begin{aligned} \varepsilon_\beta \left(\rho c_p\right)_\beta \frac{\partial \left\langle T_\beta \right\rangle^\beta}{\partial t} + \left(\rho c_p\right)_\beta \left\langle \mathbf{v}_\beta \right\rangle . \nabla \left\langle T_\beta \right\rangle^\beta - \mathbf{u}_{\beta\beta} . \nabla \left\langle T_\beta \right\rangle^\beta - \mathbf{u}_{\beta\sigma} . \nabla \langle T_\sigma \rangle^\sigma = \\ \nabla . \left(\mathbf{K}^*_{\beta\beta} . \nabla \left\langle T_\beta \right\rangle^\beta + \mathbf{K}^*_{\beta\sigma} . \nabla \langle T_\sigma \rangle^\sigma \right) - h\left(\left\langle T_\beta \right\rangle^\beta - \langle T_\sigma \rangle^\sigma \right) + a_v \xi \Omega \end{aligned} \qquad [6.37]$$

$$\begin{aligned} \varepsilon_\sigma \left(\rho c_p\right)_\sigma \frac{\partial \langle T_\sigma \rangle^\sigma}{\partial t} - \mathbf{u}_{\sigma\beta} . \nabla \left\langle T_\beta \right\rangle^\beta - \mathbf{u}_{\sigma\sigma} . \nabla \langle T_\sigma \rangle^\sigma = \nabla . \left(\mathbf{K}^*_{\sigma\beta} . \nabla \left\langle T_\beta \right\rangle^\beta + \mathbf{K}^*_{\sigma\sigma} . \nabla \langle T_\sigma \rangle^\sigma \right) \\ -h\left(\langle T_\sigma \rangle^\sigma - \left\langle T_\beta \right\rangle^\beta \right) + a_v (1-\xi) \Omega + \varepsilon_\sigma \Phi_\sigma \end{aligned} \qquad [6.38]$$

The theoretical results show that:

$$\begin{aligned} &\xi = 1 \quad if \quad k_\beta >> k_\sigma \\ &\xi = 0 \quad if \quad k_\beta << k_\sigma \end{aligned} \qquad [6.39]$$

In other words, the heterogeneous heat source is distributed in the phase with the *higher diffusivity*.

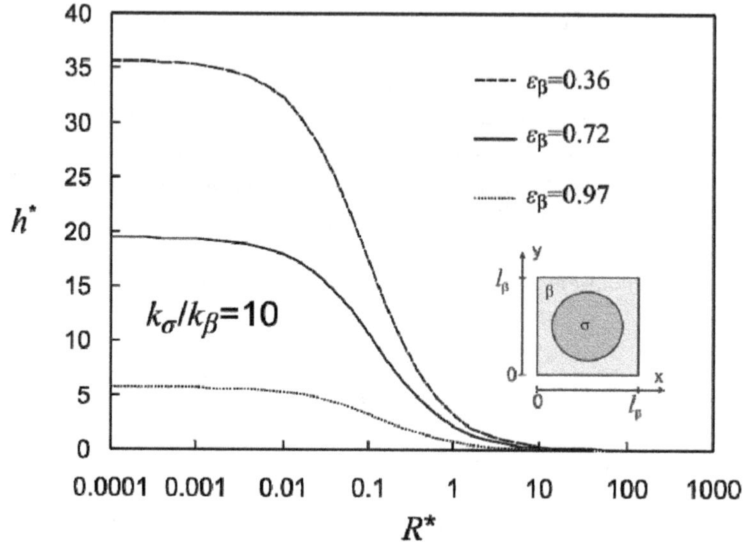

Figure 6.11. *Evolution of the exchange coefficient versus the thermal resistance* ($h^* = \frac{hl_\beta^2}{k_\beta}$, $R^* = \frac{Rk_\beta}{l_\beta}$) *extracted from [GOB 98]*

6.5.2. *Interfacial thermal resistance*

The case of heat transfer with an interfacial thermal resistance is very important for composite material applications [BHA 90] because of the dramatic effect it may have on the effective thermal conductivity and other effective parameters, as shown theoretically in [HAS 87, AUR 09, GOB 98]. This time, the boundary conditions at the β-α interface are replaced by:

$$\text{B.C.2} \quad -\mathbf{n}_{\beta\sigma}.k_\beta \nabla T_\beta = -\mathbf{n}_{\beta\sigma}.k_\sigma \nabla T_\sigma = \frac{1}{R}\left(T_\beta - T_\sigma\right) \text{ at } A_{\beta\sigma} \qquad [6.40]$$

where R is the interfacial thermal resistance. We see that a very small value will make this problem similar to the problem with a continuous temperature discussed in the previous sections. On the contrary, a large value of R will have a tendency to increase the temperature difference between the two phases, thus promoting local non-equilibrium situations. We understand that the resulting macro-scale models will have their effective parameters affected by the value of the interfacial thermal resistance. As an illustration, we represent in Figure 6.11 the evolution of a dimensionless heat exchange coefficient, $h^* = \frac{hl_\beta^2}{k_\beta}$, coming from the two-equation model versus a dimensionless number, $R^* = \frac{Rk_\beta}{l_\beta}$, characterizing the interfacial thermal resistance. The medium unit cell is an array of cylinders, schematically representing a fibrous composite material. Three calculations with different volume fractions have been carried out with the thermal conductivity ratio kept at a value $k_\sigma / k_\beta = 10$. We see that, for small values of R^*, we recover the classical two-equation model while the exchange coefficient becomes dramatically very small when R^* increases, thus favoring local non-equilibrium behaviors.

6.6. Conclusions

Heat transfer in multi-phase, multi-scale systems is a complex phenomenon characterized by various spatial and time scales. This results in different macro-scale models. The major classes are represented by:

– local equilibrium models for a mixture temperature close to each intrinsic phase average temperature;

– non-equilibrium models that relax the constraint of local equilibrium. Several forms have been proposed in the literature among which the classical two-equation model is the most used in practice, while it does have limitations if we reach points far from equilibrium. Other candidates are telegraph equations, equations with convolution terms, mixed models involving a coupling between macro-scale and micro-scale transfers, etc.

Remarkably, transitions between the models may occur during a given process. In this case, it is recommended to use models that embed several others, like the two-equation model.

All these models require the determination of several effective parameters. Bounds and estimates are available for the most important: the effective thermal conductivity in the one-equation case. If these estimates are not accurate enough, we have to rely on experimental determinations or direct calculations from the pore-scale geometry.

Coupling with other transport mechanisms (phase change, chemical reaction, etc.) and their impact on the form of the macro-scale equations and the effective properties is still an open research problem, while several models are already available and known to work relatively well.

In this chapter, we used a notation involving explicit averages in an attempt to explain the relationship between porescale physics and the macro-scale description. This notation may be replaced by a simpler notation for macro-scale temperatures if the underlying multi-scale features are no longer needed.

6.7. Bibliography

[ALL 10] ALLAIN H., QUINTARD M., PRAT M. *et al.*, "Upscaling of superfluid helium flow in porous media", *International Journal of Heat and Mass Transfer*, vol. 53, pp. 4852–4864, 2010.

[AMI 94] AMIRI A., VAFAI K., "Analysis of dispersion effects and non-thermal equilibrium, non-darcian, variable porosity incompressible flow through porous media", *Journal of Heat and Mass Transfer*, vol. 37, no. 6, pp. 939–954, 1994.

[AUR 83] AURIAULT J.-L., "Effective macroscopic description for heat conduction in periodic composites", *International Journal of Heat and Mass Transfer*, vol. 26, no. 6, pp. 861–869, 1983.

[AUR 94] AURIAULT J.-L., ENE H.I., "Macroscopic modelling of heat transfer in composites with interfacial thermal barrier", *Journal of Heat and Mass Transfer*, vol. 37, no. 18, pp. 2885–2892, 1994.

[BHA 90] BHATT H., DONALDSON K., HASSELMAN D.P. *et al.*, "Role of the interfacial thermal barrier in the effective thermal diffusivity/conductivity of sic-fiber-reinforced reaction-bonded silicon nitride", *Journal of the American Ceramic Society*, vol. 73, no. 2, pp. 312–316, 1990.

[BRU 35] BRUGGEMAN D.A.G., "Berechnung verschiedener physikalischer konstanten von heterogenen substanzen", *Annalen der Physik*, vol. 24, pp. 636–679, 1935.

[CAR 84] CARBONELL R.G., WHITAKER S., "Heat and mass transfer in porous media", in BEAR J., CORAPCIOGLU M.Y. (eds), *Fundamentals of Transport Phenomena in Porous Media*, Martinus Nijhof Publ., 1984.

[DAG 89] DAGAN G., *Flow and Transport in Porous Formations*, Springer-Verlag, Heidelberg Berlin, New York, 1989.

[DAR 56] DARCY H., *Les fontaines publiques de la ville de Dijon*, Dalmont, Paris, 1856.

[DAV 10a] DAVARZANI H., MARCOUX M., QUINTARD M., "Theoretical predictions of the effective thermodiffusion coefficients in porous media", *International Journal of Heat and Mass Transfer*, vol. 53, pp. 1514–1528, 2010.

[DAV 10b] DAVIT Y., QUINTARD M., DEBENEST G., "Equivalence between volume averaging and moments matching techniques for mass transport models in porous media", *International Journal of Heat and Mass Transfer*, vol. 53, no. 21–22, pp. 4985–4993, 2010.

[DAV 12] DAVIT Y., WOOD B., DEBENEST G. *et al.*, "Correspondence between one- and two-equation models for solute transport in two-region heterogeneous porous media", *Transport in Porous Media*, vol. 95, pp. 213–238, 2012.

[DAV 13] DAVIT Y., BELL C.G., BYRNE H.M. *et al.*, "Homogenization via formal multiscale asymptotics and volume averaging: how do the two techniques compare?", *Advances in Water Resources*, vol. 62, pp. 178–206, 2013.

[DAV 15] DAVIT Y., QUINTARD M., "Theoretical analysis of transport in porous media", in HADIM H., VAFAI K. (eds), *Handbook of Porous Media*, Taylor & Francis, 2015.

[DUV 04] DUVAL F., FICHOT F., QUINTARD M., "A local thermal non-equilibrium model for two-phase flows with phase-change in porous media", *International Journal of Heat and Mass Transfer*, vol. 47, no. 3, pp. 613–639, 2004.

[ENE 82] ENE H.I., SANCHEZ-PALENCIA E., "On thermal equation for flow in porous media", *International Journal of Engineering Science*, vol. 20, no. 5, pp. 623–630, 1982.

[FAT 97] FATEHI M., KAVIANY M., "Role of gas-phase reaction and gas-solid thermal nonequilibrium in reverse combustion", *International Journal of Heat and Mass Transfer*, vol. 40, no. 11, pp. 2607–2620, 1997.

[FLO 91] FLORIO J., HENDERSON J.B., TEST F.L. *et al.*, "A study of the effects of the assumption of local-thermal equilibrium on the overall thermally-induced response of a decomposing, glass-filled polymer composite", *International Journal of Heat and Mass Transfer*, vol. 34, pp. 135–147, 1991.

[GEL 93] GELHAR L.W., *Stochastic Subsurface Hydrology*, Prentice Hall, Englewood Cliffs, NJ, 1993.

[GLA 88] GLATZMAIER G.C., RAMIREZ W.F., "Use of volume averaging for the modeling of thermal properties of porous material", *Chemical Engineering Science*, vol. 43, pp. 3157–3169, 1988.

[GOB 98] GOBBÉ C., RAMOND L., QUINTARD M., "Macroscopic description of unsteady heat transfer in heterogeneous media with consideration of interfacial thermal barriers", *High Temperature-High Pressure*, vol. 30, no. 3, pp. 365–372, 1998.

[GRA 88] GRANGEOT G., Description des transferts de chaleur en milieu poreux, PhD Thesis, University of Bordeaux I, 1988.

[GUN 74] GUNN D.J., DE SOUZA J.F.C., "Heat transfer and axial dispersion in packed beds", *Chemical Engineering Science*, vol. 29, pp. 1363–1371, 1974.

[HAS 63] HASHIN Z., SHTRIKMAN S., "A variational approach to the theory of the elastic behaviour of multiphase materials", *Journal of the Mechanics and Physics of Solids*, vol. 11, no. 2, pp. 127–140, 1963.

[HAS 87] HASSELMAN D.P.H., LLOYD F., "Effective thermal conductivity of composites with interfacial thermal barrier resistance", *Journal of Composite Materials*, vol. 21, no. 6, pp. 508–515, 1987.

[KAV 95] KAVIANY M., *Principles of Heat Transfer in Porous Media*, Springer, New York, 1995.

[LAC 02] LACABANNE B., BLANCHER S., CREFF R. *et al.*, "Soret effect in multicomponent flow through porous media: local study and upscaling process", *Thermal Nonequilibrium Phenomena in Fluid Mixtures*, Springer, pp. 448–465, 2002.

[LER 13] LEROY V., GOYEAU B., TAINE J., "Coupled upscaling approaches for conduction, convection, and radiation in porous media: theoretical developments", *Transport in Porous Media*, vol. 98, no. 2 pp. 323–347, 2013.

[LEV 85] LEVEC J., CARBONELL R.G., "Longitudinal and lateral thermal dispersion in packed beds - 1: Theory", *AIChE Journal*, vol. 31, pp. 581–590, 1985.

[LIC 27] LICHTENECKER K., "The common root of the logarithmic mixture law and the beginning of the entrophy function", *Physikalische Zeitschrift*, vol. 28, pp. 417–418, 1927.

[MAT 67] MATHERON G., *Eléments pour une théorie des milieux poreux,* Masson, Paris, 1967.

[MAX 73] MAXWELL J.C., *A Treatise on Electricity and Magnetism*, vol. 1, Clarendon Press, London, 1873.

[MIN 99] MINKOWYCZ W.J., HAJI-SHEIKH A., VAFAI K., "On departure from local thermal equilibrium in porous media due to a rapidly changing heat source: the sparrow number", *International Journal of Heat and Mass Transfer*, vol. 42, no. 18, pp. 3373–3385, 1999.

[MOY 97] MOYNE C., "Two-equation model for a diffusive process in porous media using the volume averaging method with an unsteady state closure", *Advances in Water Resources*, vol. 20, nos. 2–3, pp. 63–76, 1997.

[MOY 00] MOYNE C., DIDIERJEAN S., AMARAL SOUTO H.P. *et al.*, "Thermal dispersion in porous media: one-equation model", *International Journal of Heat and Mass Transfer*, vol. 43, no. 20, pp. 3853–3867, 2000.

[MUS 49] MUSKAT M., *Physical Principles of Oil Production*, McGraw-Hill, New York, 1949.

[NOZ 85] NOZAD I., CARBONELL R.G., WHITAKER S., "Heat conduction in multiphase systems I: theory and experiment for two-phase systems", *Chemical Engineering Science*, vol. 40, pp. 843–855, 1985.

[OLI 01] OLIVEIRA A.A.M., KAVIANY M., "Non-equilibrium in the transport of heat and reactants in combustion in porous media", *Progress in Energy and Combustion Science*, vol. 27, no. 5, pp. 523–545, 2001.

[PUR 04] PUIROUX N., PRAT M., QUINTARD M., "Non-equilibrium theories for macroscale heat transfer: ablative composite layer systems", *International Journal of Thermal Sciences*, vol. 43, no. 6, pp. 541–554, 2004.

[QUI 93] QUINTARD M., WHITAKER S., "One- and two-equation models for transient diffusion processes in two-phase systems", *Advances in Heat Transfer*, Academic Press, New York, pp. 369–464, 1993.

[QUI 94a] QUINTARD M., WHITAKER S., "Transport in ordered and disordered porous media I: The cellular average and the use of weighting functions", *Transport in Porous Media*, vol. 14, pp. 163–177, 1994.

[QUI 94b] QUINTARD M., WHITAKER S., "Transport in ordered and disordered porous media II: Generalized volume averaging", *Transport in Porous Media*, vol. 14, pp. 179–206, 1994.

[QUI 94c] QUINTARD M., WHITAKER S., "Transport in ordered and disordered porous media III: Closure and comparison between theory and experiment", *Transport in Porous Media*, vol. 15, pp. 31–49, 1994.

[QUI 95] QUINTARD M., WHITAKER S., "Local thermal equilibrium for transient heat conduction: Theory and comparison with numerical experiments", *International Journal of Heat and Mass Transfer*, vol. 38, no. 15, pp. 2779–2796, 1995.

[QUI 97] QUINTARD M., KAVIANY M., WHITAKER S., "Two-medium treatment of heat transfer in porous media: Numerical results for effective properties", *Advances in Water Resources*, vol. 20, no. 2–3, pp. 77–94, 1997.

[QUI 00] QUINTARD M., WHITAKE S., "Theoretical analysis of transport in porous media", in VAFAI K. (ed.), *Handbook of Heat Transfer in Porous Media*, Marcel Decker Inc., N.Y., 2000.

[QUI 01] QUINTARD M., CHERBLANC F., WHITAKER S., "Dispersion in heterogeneous porous media : one-equation non-equilibrium model", *Transport in Porous Media*, vol. 44, no. 1, pp. 181–203, 2001.

[RAY 92] RAYLEIGH R.S., "On the influence of abstacles arranged in rectangular order upon the properties of the medium", *Philosophical Magazine*, vol. 34, pp. 481–489, 1892.

[REE 05] REES D.A.S., POP. I., "Local thermal non-equilibrium in porous medium convection", in INGHAM D.B., POP I. (eds), *Transport Phenomena in Porous Media {III}*, Pergamon, Oxford, 2005.

[SAB 07] SABATIER J., PRAKASH AGRAWAL O., TENREIRO J.A., *Advances in Fractional Calculus: Theoretical Developments and Applications in Physics and Engineering*, Springer, 2007.

[SAN 80] SANCHEZ-PALENCIA E., *Non-Homogeneous Media and Vibration Theory*, vol. 127, Springer, New York, 1980.

[SCH 75] SCHLÜNDER E.U., "Equivalence of one- and two-phase models for heat transfer processes in packed beds: one-dimensional theory", *Chemical Engineering Science*, vol. 30, pp. 449–452, 1975.

[SHO 89] SHONNARD D.R., WHITAKER S., "The effective thermal conductivity for a point-contact porous medium: an experimental study", *International Journal Heat and Mass Transfer*, vol. 32, pp. 503–512, 1989.

[SIN 92] SINGH B.P., MASSOUD KAVIANY, "Modelling radiative heat transfer in packed beds", *International Journal of Heat and Mass Transfer*, vol. 35, no. 6, pp. 1397–1405, 1992.

[SOE 90] SOEZEN M., VAFAI K., "Analysis of the non-thermal equilibrium condensing flow of a gas through a packed bed", *International Journal Heat and Mass Transfer*, vol. 33, pp. 1247–1261, 1990.

[TAI 12] TAINE J., IACONA E., "Upscaling statistical methodology for radiative transfer in porous media: new trends", *Journal of Heat Transfer*, vol. 134, no. 3, 2012.

[TOR 02] TORQUATO S., *Random Heterogeneous Materials: Microstructure and Macroscopic Properties*, Springer-Verlag, New York, 2002.

[VAD 05] VADASZ P., "Explicit conditions for local thermal equilibrium in porous media heat conduction", *Transport in Porous Media*, vol. 59, no. 3, pp. 341–355, 2005.

[VAF 90] VAFAI K., SOZEN M., "Analysis of energy and momentum transport for fluid flow through a porous bed", *Journal of Heat Transfer*, vol. 112, pp. 690–699, 1990.

[WAN 08] WANG L., WEI X., "Equivalence between dual-phase-lagging and two-phase-system heat conduction processes", *International Journal of Heat and Mass Transfer*, vol. 51, pp. 1751–1756, 2008.

[WHI 77] WHITAKER S., "Toward a diffusion theory of drying", *Industrial & Engineering Chemistry Fundamentals*, vol. 16, pp. 408–414, 1977.

[WHI 86] WHITAKER S., "Flow in porous media I: a theoretical derivation of Darcy's law", *Transport in Porous Media*, vol. 1, pp. 3–35, 1986.

[WHI 91] WHITAKER S., "Improved constraints for the principle of local thermal equilibrium", *Industrial & Engineering Chemistry Fundamentals*, vol. 30, pp. 983–997, 1991.

[WHI 99] WHITAKER S., *The Method of Volume Averaging*, Kluwer Academic Publishers, Dordrecht, 1999.

[WIE 12] WIENER O., "Die theorie des mischkörpers für das feld der stationären strömung. erste abhandlung: Die mittelwertsätze für kraft, polarisation und energie", *Abhandlungen der Mathematisch Physischen Klasse der Königlischen-Sachsischen Gesellschaft der Wissenschaften*, vol. 32, pp. 509–604, 1912.

[YAG 60] YAGI S., KUNII D., WAKAO N., "Studies on axial effective thermal conductivities in packed beds", *AIChE Journal*, vol. 6, no. 4, pp. 543–546, 1960.

[ZAN 84] ZANOTTI F., CARBONELL R.G., "Development of transport equations for multiphase systems - I: General development for two-phase systems", *Chemical Engineering Science*, vol. 39, pp. 263–278, 1984.

7

Thermal Optimization of Forming Processes

This chapter is devoted to the optimization of forming processes through the thermal control of tools. Two applications are detailed. The first application addresses the problem optimal temperature control for the curing of an aeronautical part molded by resin transfer molding (RTM) process. The second application targets the development of a design methodology of cooling channels in thermoplastic injection molds. Before discussing the results of these two studies, a contextualization of optimization problems is proposed.

7.1. Context of optimization

The optimization in forming processes aims for two major objectives. The first objective concerns the improvement of the quality of the produced parts and the second objective concerns the increase in production rates. The problem lies in the simultaneous achievement of these two incompatible objectives.

Short cycle times induce rapid changes inside the part and increase the risk of creation of inhomogeneities and trapping of residual stresses in the parts. Conversely, the absence of gradients (thermal, transformations, etc.) is consistent with quality; but inevitably results in longer cycles. Compromises are to be found to satisfy these two constraints or favor one of the two aspects. In all cases, the optimization of the part quality and/or the production rate requires to consider the whole forming process, the thermal control of the part being dependent on the thermal control of the tools.

Chapter written by Vincent SOBOTKA.

Thus, even when the optimal thermal cycle for the part is known, the one to impose to the mold can be very different and may involve solving another specific optimization problem. Furthermore, each process has its own constraints. For example, we cannot reach high production rates using a mold having a large thermal inertia. Various issues must be thus addressed before optimization procedure can be set. As just mentioned, the first of them concerns the objective, which will lead to the formulation of a mathematical objective function to be minimized. The control parameters contained in this function depend on the process. The processes can thus be classified according to these parameters. It is necessary to discuss the modeling of the problem, including the scale of description and the coupling between the different physics.

Finally, we discuss the question of the choice of the minimization algorithm that must provide a good compromise between quality of the identification and efficiency in terms of computation time. These aspects are detailed in the following sections.

7.2. Heat transfer: optimization lever

Heat transfer is a controllable lever on almost all forming processes. An action on this lever allows action of other properties such as chemical reaction. First, we propose to gather the processes together according to their thermal specificities, i.e. according to the involved transfer modes and according to the steady or transient nature of the thermal setpoint. This classification makes it possible to easily show the action parameters and the type of optimization that can be considered. The first classification leads to distinguish processes in closed molds and those in opened molds (Table 7.1).

For closed mold processes, the main transfer mode is conduction that occurs in the mold and into the part. In this case, the inertia of the molds is the major factor that controls the cycle. We must therefore act on this parameter to obtain a significant gain on the production rate. In the case of opened mold processes, the three modes can be encountered as for example during the autoclave-curing of prepregs. Transfers are conductive in the mold and into the part; but are coupled with convective or convective-radiative heat transfers between the mold and heating system. The control parameters can be in this case the flow rate and temperature of the fluid, the position of the blowing nozzles or infrared panels, the supply of additional heat sources. In the case of multiple loadings in autoclaves, positions of the tools relative to each other are also parameters that can be optimized.

Conduction	Conduction/convection/radiation
Injection molding: RTM, CRTM, RIM, BMC, TP Compression molding: BMC, SMC Extrusion Pultrusion	Prepregs cure Infusion Filament winding Thermoforming Tape placement

Table 7.1. *Classification of processes according to the transfer mode*

For the second classification, the distinction is made between constant and time-dependent thermal setpoints. The first category relates generally to cyclic processes in which cycle times are relatively slow and the second category relates to cyclic processes at a high rate, as well as continuous processes.

In the first case, the optimization consists in determining the optimal temperature cycle in the process. In the second case, the goal is to determine the spatial distribution of heat sources and sinks. Some combinations are shown in Table 7.2.

Processes associated with constant tool temperature		
	Cyclic processes in closed molds	Continuous processes in opened molds
Process	Compression Injection Transfer Materials: – Thermosets (BMC-SMC): hot tools – Thermoplastics: cold tools	Filament winding Tape placement
Heat transfer mode	Conduction	Conduction – Convection – Radiation
Control parameters	– Tool temperature – Location of heat sources	– Mandrel temperature – Rotation or placement speed – Radiative / Convective power
Processes with time-dependant setpoints		
	Cyclic processes in closed molds	Continuous processes in opened molds
Process	RTM (thermoset or thermoplastic)	Prepregs in autoclave (Thermoset or thermoplastic) Thermoset infusion
Heat transfer mode	Conduction	Conduction – Convection – Radiation
Control parameters	– Setpoint temperature – Location of heat sources	– Temperature cycle of the oven / autoclave – Complementary heat sources – Pressure

Table 7.2. *List of processes according to the proposed classification*

As well as the processes classified according to criteria related to heat transfer, we propose to differentiate the optimization problems in two subgroups:

– the first one concerns the *control problems* for which we search the optimal setpoint temperature or heat flux using control parameters already present in the process in order to reach a given criterion in the part. This category of problems is illustrated by the example detailed in section 7.6;

– the second concerns the *design problems* that consist of designing in an optimal way the entire thermal control system. The control parameters are much more numerous since they include those of the previous point, as well as concern the positions, forms and materials of the tools. An example for the design of a thermoplastic injection mold is proposed in section 7.7.

7.3. Definition of the optimization criterion

The definition of the objective is a prerequisite for any optimization procedure. This is a major issue because it corresponds to the mathematical translation of what is expected as a quality criterion for the produced parts.

It must also be compatible with the constraints (time, thermal inertia, etc.) engendered by the considered molding process. It can be single or multiple. The main objectives [HAS 10, QIA 06, LAM 04, LIN 02] encountered in the literature concern:

– obtaining a uniform thermal history in the part during heating and cooling phases;

– obtaining a desired temperature level at the end of the cycle to be able to start a new cycle in the same conditions;

– minimizing the cycle time;

– obtaining uniform transformation (crosslinking and crystallization).

These objectives are translated by criteria which evaluate:

– the difference between the part temperature and the target temperature over the entire surface of the part [BAR 82, PAR 87];

– the deviation from the average temperature of the part [QIA 06, LAM 04];

– the difference between maximum and minimum temperature in the part [MAT 04, HUA 01];

– the transformation gradients.

Often, authors combine the objective functions by introducing a weighting factor between them [TAN 98, MAT 04, HUA 01]. Multi-objective methods also help to address this problem.

7.4. Problem modeling

In this section, we address the following three points to describe the problem; the answers are guided by the objectives:

– what is the spatial scale for the process?

– what are the step(s) of the process?

– what are the coupling(s)?

7.4.1. *Spatial scale*

This spatial scale spreads over several orders of magnitude since it extends from about 10 µm for the filament of a yarn to several meters for a mold. However, in the case of process optimization, the minimum considered scale is that of the composite part. This latter is thus considered as an equivalent homogeneous material with effective properties.

7.4.2. *Time scale: process steps*

Almost all of the composite forming processes can be subdivided into two steps: the impregnation of the reinforcement with the matrix and the consolidation. It is therefore necessary to perform a sensitivity analysis of each of these steps on the desired criterion. For example, in conventional processes, the time dedicated to the consolidation is much more important than that dedicated to impregnation. Therefore, in the case of a composite part molded by RTM, it is likely that the impregnation step will have little effect on the crosslinking reaction if this latter is slow. On the contrary, in the case of a rapid cycle, e.g. thermoplastic RTMs having a cycle time of less than 2 min, it becomes necessary to take into account the evolution of heat transfer during filling in the optimization process to determine the final properties of the part.

7.4.3. *Multi-physical aspects*

Alongside the issue of scale of description of phenomena is the question of coupling between thermal, chemical, mechanical and rheological phenomena that make mathematical and numerical resolutions more complex. In this chapter, we will limit the coupling between nonlinear heat transfers with transformation kinetics. These couplings can be strong, especially in the case where the transformations are fast and highly exothermic. Another class of couplings concerns the boundary conditions between the part and the surface of the tool. The analysis of this coupling is of primary importance because it helps answer the question of the description scale. In the case of strong coupling, it is mandatory to jointly describe the transfers in the tool and in the parts. When couplings remain weak between the part and the mold, it is possible to successively solve two optimization problems: the first one in the part and the second one in the mold [BAI 03, BAI 96, BAI 97].

7.5. Numerical optimization methods

Two main optimization methods exist to minimize the objective function: stochastic methods [HUA 01, ABO 00] and deterministic methods [QIA 06, BAR 85,MAT 04, TAN 98]. The first class of methods moves toward the global minimum and uses a random selection of combinations of the parameters on the solutions space. The second class of methods aims to achieve a local minimum of the objective function usually using the calculation of gradients to determine optimal parameters. These methods require the calculation of the functional and its derivatives as a function of the estimated parameters. Within the gradient methods, there are methods which require the calculation of the first derivative of the function [FAV 02, JAR 91, ABO 01, MAI 13] and those for example the methods of quasi-Newton, Levenberg–Marquardt, Gauss–Newton, which require the calculation of the Jacobian matrix of the functional J as well as of the Hessian function or at least an approximation. These methods are very effective, particularly in the vicinity of the minimum, however, the calculation of the Jacobian can be very costly in terms of computation time.

For all the works presented in this chapter, the conjugate gradient method [BAR 82] was used. An algorithm of this method is summarized in Figure 7.1. From an initial choice of parameters, the direct problem, described by the system of equations $E(T,\beta) = 0$ is solved and the objective function J is evaluated. If the functional verifies the inequality $\|J\| < \varepsilon$, it is considered that optimal parameters are achieved and the procedure is stopped. Otherwise, the unknowns' vector β is corrected using the descent depth ρ and descent direction w. The method of Fletcher–Reeves directions is used to calculate w.

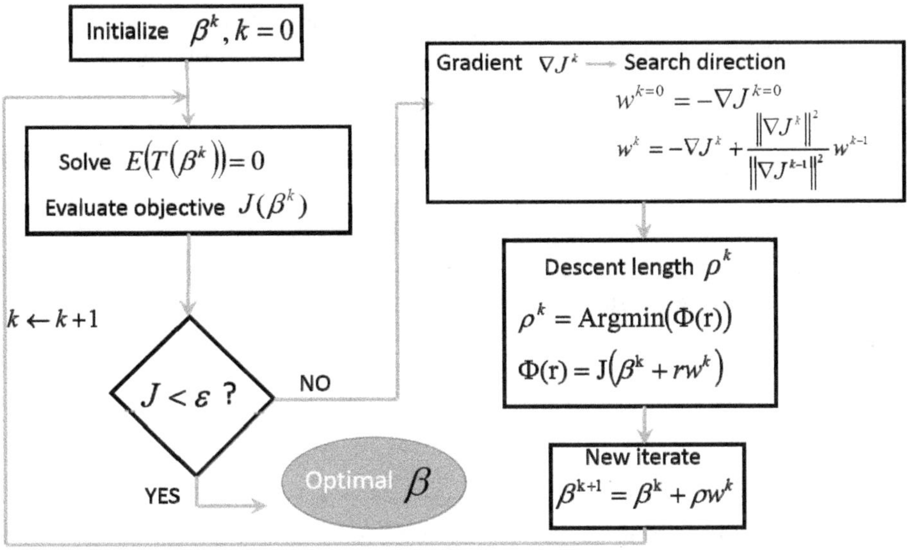

Figure 7.1. *Conjugate gradient algorithm*

This step requires the calculation of the components of the gradient of the function J with respect to the parameters to be estimated. However, there is no explicit relationship between the functional and the parameters to be assessed. To evaluate these components, we can use a numerical calculation which consists in calculating the following difference (equation [7.1]) to determine the gradient components with respect to the parameter z_i:

$$\frac{\partial J}{\partial z_i} \approx \frac{J(z_1,...,z_i+\delta z_i,...,z_p) - J(z_1,...,z_i,...,z_p)}{\delta z_i} \qquad [7.1]$$

where δz_i stands for an infinitesimal increment of the ith component of the vector z.

This technique is not satisfactory in the case where many parameters are to be estimated, because at each optimization iteration, we must solve the direct problem $E(T,\beta) = 0$ as many times as there are parameters p in the vector z. Moreover, the choice of the increment δz_i remains delicate because if it is too small, the value of $\frac{\partial J}{\partial z_i}$ is indeterminate and, on the contrary, if it is too large, the linear approximation is no longer verified. An alternative is to calculate the gradient using a Lagrangian approach [HEN 91, ALI 94] which requires the formulation of the adjoint problem, and potentially the sensitivity problem.

7.5.1. *The adjoint problem*

Let us introduce the Lagrange multiplier ψ [ABO 01, WOO 03] and the Lagrangian $L(\psi, T, \beta)$ associated with the optimization problem defined by the criterion J and the constraint corresponding to the direct problem E(T,β) = 0. The Lagrangian is written:

$$L(T,\beta,\psi) = J(T) - \langle \psi, E(T,\beta) \rangle \qquad [7.2]$$

This problem consists of finding the stationary points (ψ, T, β) of the Lagrangian L. For these points, dL = 0. Not all stationary points are minima. They may be maxima points of inflexion or saddle points. But, the procedure is to find all the stationary points of L and then inspect these to find the ones at which L is minimum. The procedure guarantees that, at all stationary points found, the constraint will be satisfied; hence, only those that minimize J will be considered.

In this method, the parameters T, β and Ψ are assumed to be independent.

$$dL = \frac{\partial L}{\partial T}\delta T + \frac{\partial L}{\partial \beta}\delta\beta + \frac{\partial L}{\partial \psi}\delta\psi \qquad [7.3]$$

By setting the adjoint variable Ψ at an arbitrary value, the differential of L is equal to:

$$\delta L = \frac{\partial L}{\partial T}\delta T + \frac{\partial L}{\partial \beta}\delta\beta \qquad [7.4]$$

We choose the Lagrange multiplier so as to verify the following condition:

$$\frac{\partial L}{\partial T}\delta T = 0, \forall \delta T \qquad [7.5]$$

This condition leads to find the Lagrange multiplier so as to be the solution of a set of equations called adjoint equations. When the adjoint variable is determined, the differential of L is equal to $\delta L = \frac{\partial L}{\partial \beta}\delta\beta$. Furthermore, when the temperature field T verifies the equations of the direct problem E(T,β) = 0 and we obtain $\delta L = \delta J$. The components of the gradient of J can then be determined.

7.5.2. *Practical setting of the method*

The Lagrangian is explicitly calculated: $L(T,\beta,\psi) = J(T) - \langle \psi, E(T,\beta) \rangle$

Its differential is also calculated: $dL = \frac{\partial L}{\partial T}\delta T + \frac{\partial L}{\partial \beta}\delta\beta$.

ψ is searched such as $\frac{\partial L}{\partial T} = 0$, $\forall \delta T$ which leads to the solving of the set of adjoint equations. We then obtain the explicit expression of the components of the gradient of J which depend on T and ψ.

From a numerical point of view, first the direct problem is solved, and then the adjoint problem. We can then compute the components of the gradient; these latter depending only on the temperature field and adjoint variable field.

This method remains robust to solve optimization problems linear or not, coupled or not, in 2D or 3D. It is interesting in the case where the parameter vector has many components and, as such, is very well suited for optimization problems in forming processes. It can be implemented easily with commercial software like Comsol Multiphysics®. This method will be used to deal with the two examples presented in this chapter and which consist of an optimal control problem and optimal design problem.

7.6. Example of process optimization: determination of optimal heat flux setpoint

The study described in this section was conducted as part of a project called *RTM Région* (2004–2008) funded by the Pays de la Loire. The ambition of the project was the development of a stand-alone aeronautical RTM tool of large dimensions integrating its own thermal regulation system. The major challenge was related to the reduction of cost of molds and their environment by using less expensive materials and manufacturing technique. Regarding the thermal aspects, the objective was: to determine the setpoint temperature allowing us to reach a temperature cycle in the composite part, imposed by the specifications.

7.6.1. *Experimental setup and constraint*

The composite part to manufacture is an H-beam (Figure 7.2) of 9 m long and whose thickness varies from 5 to 9 mm. It is composed of carbon reinforcement (G1151) and epoxy matrix (RTM6). The mold (Figure 7.4) comprises several removable metal elements (Figure 7.3). The components of this tool are made up of aluminum (3.05 t) and steel (4.7 t).

Figure 7.2. *Composite part*

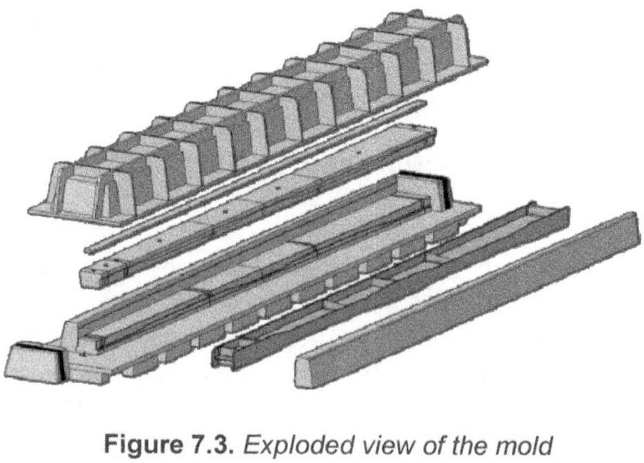

Figure 7.3. *Exploded view of the mold*

The desired molding cycle that must be reached in the part consists of four steps:

1) Heating the carbon preform from ambient temperature to 120°C at 2 K/min followed by temperature dwell at 120°C.

2) Injecting the resin in the preform at this temperature level.

3) Heating the part again to reach 180°C at 2 K/min followed by temperature dwell at 180°C so as to polymerize the resin.

4) Cooling the composite part down to ambient temperature before removing it from the mold.

Figure 7.4. *RTM mold*

The thermal design led to a total power of 167 kW to meet the specifications of 2 K/min in the part. Oil heated by eight thermal control units circulates in 45 plates located on external surface of the mold (Figure 7.4). The problem to solve is to determine the setpoint temperature or heat flux at the interface between the mold and the heating plates which allows us to reach the desired temperature cycle in the composite part.

As the elements are made up of different materials, different thermal expansions occur and the contact between them does not remain constant according to temperature. Heat transfer being strongly dependent of these contacts, they must be quantified. Thermal contact resistances (TCRs) are then used to model the evolution of the contacts and their influence on the temperature field in the mold. As a result, the optimal temperature setpoint cannot be estimated without preliminary knowledge of the most influential TCRs. Thus, there are two problems to solve: the first one is to determine the TCRs and the second one is to estimate the optimal setpoint.

7.6.2. *Instrumentation of the mold and the preform*

The geometry of the mold as well as the location of the heat exchangers leads to 2D heat transfer on a cross-section of the mold, as the one depicted in Figure 7.5. For this reason, the sensors used to record temperature are placed in a transverse section located in the middle of the length of the mold.

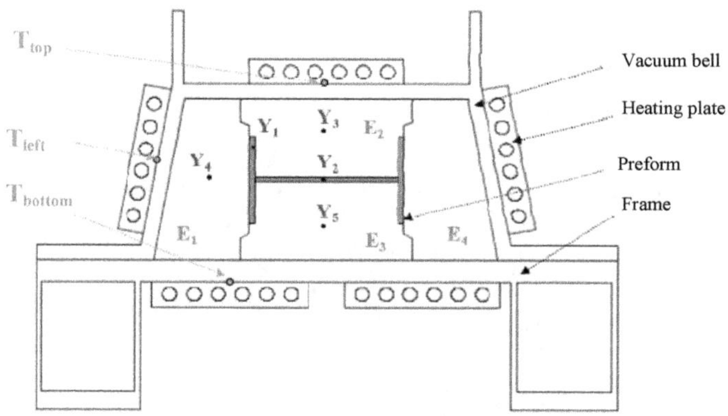

Figure 7.5. *2D cut view of the instrumented mold*

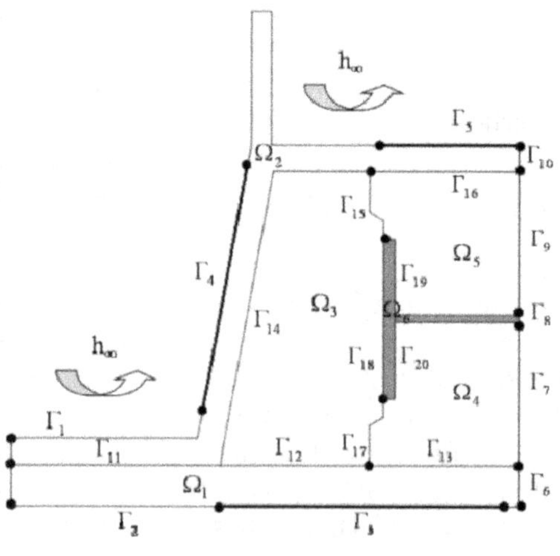

Figure 7.6. *2D domain used for the numerical solving*

Five K-type thermocouples are inserted at different locations in the carbon preform and in the mold. In the following, the readings of these sensors will be denoted by $(Y_i)_{i=1,5}$. Three additional thermocouples denoted by T_{left}, T_{bottom} and T_{top} in Figure 7.5 are placed under each heating plate.

7.6.3. *Thermal modeling*

The symmetry of the problem allows the modeling of only one-half of the initial domain as shown in Figure 7.6. Heat transfers are only conductive. Non-linearity is introduced in the model through the presence of temperature-dependent TCRs. The source term related to the crosslinking of the resin is not taken into account; this latter leads to a very weak coupling and its impact on heat transfer remains negligible.

For practical reasons, only heat transfer equations of the subdomain Ω_2 are presented (equation [7.6]). These equations can easily be generalized to the whole domain. In the following equations, q(t) is the heat flux density, h_∞ is the heat transfer coefficient with the surroundings, T_∞ is the ambient temperature and T_{left} and T_{top} are the temperatures measured by the thermocouples on boundaries Γ_4 and Γ_5.

$$\rho Cp\frac{\partial T_{\Omega_2}}{\partial t} = \nabla.\left(\lambda \nabla T_{\Omega_2}\right) \text{ in } \Omega_2, t>0 \qquad -\lambda\frac{\partial T_{\Omega_2}}{\partial n}\bigg|_{\Gamma_{14}} = \frac{1}{RTC_1}\left(T_{\Omega_3}-T_{\Omega_2}\right) \text{ on } \Gamma_{14}, t>0$$

$$-\lambda\frac{\partial T_{\Omega_2}}{\partial n}\bigg|_{\Gamma_{16}} = \frac{1}{RTC_2}\left(T_{\Omega_5}-T_{\Omega_2}\right) \text{ on } \Gamma_{16}, t>0 \qquad -\lambda\frac{\partial T_{\Omega_2}}{\partial n}\bigg|_{\Gamma_1} = h_\infty\left(T_\infty-T_{\Omega_2}\right) \text{ on } \Gamma_1, t>0$$

$$-\lambda\frac{\partial T_{\Omega_2}}{\partial n}\bigg|_{\Gamma_{10}} = 0 \text{ on } \Gamma_{10}, t>0 \qquad -\lambda\frac{\partial T_{\Omega_2}}{\partial n}\bigg|_{\Gamma_4\cup\Gamma_5} = q(t) \text{ or } \begin{cases} T_{\Omega_2}\big|_{\Gamma_4} = T_{left}(t) \\ T_{\Omega_2}\big|_{\Gamma_5} = T_{top}(t) \end{cases}$$

$$\text{on } \Gamma_4 \cup \Gamma_5, t>0 \qquad [7.6]$$

$$T_{\Omega_2} = T_0 \text{ in } \Omega_2, t=0$$

The heat conduction problem was solved by using the finite element solver Comsol® which is efficient for solving multi-domain heat conduction problems with non-perfect thermal contact.

These two inverse problems are solved by using the classical conjugate gradient algorithm and consist of finding the optimal value β_{opt} which minimizes the objective function:

$$J(\beta)=\sum_{j}\int_{0}^{t_f}\left\|Y_j(t)-T\left(x_j,y_j,t;\beta\right)\right\|^2 dt \qquad [7.7]$$

where β is the unknown vector, Y_j are the target temperatures, either the experimental temperatures or the desired temperatures and T are the computed temperatures.

7.6.3.1. *TCR identification problem*

A sensitivity analysis showed that only the thermal contacts at the internal boundaries Γ_{14}, Γ_{16} and Γ_{13} impact heat transfer. Thus, only these three significant TCRs are estimated and for convenient purposes of mathematical derivation, the estimated parameters are rewritten as their inverses: $h_i=\left(TCR_i^{-1}\right)_{i=1,2,3}$, respectively, on the boundaries. They are evaluated as piecewise linear functions versus time. The correlation time/temperature is a second time after the estimation.

In this problem, the temperatures T_{left}, T_{top} and T_{bottom} are imposed as boundary conditions, and the experimental temperatures $\left(Y_i\right)_{i=3,4,5}$ are considered as target temperatures in the criterion J.

7.6.3.2. *Control problem*

Once the evolutions of these TCRs are estimated, the optimal setpoint temperature can be determined. The heat flux density on the boundaries $\left(\Gamma_i\right)_{i=3,4,5}$ is considered as the unknown of the control problem. In this problem, no experimental data are required for the estimation of q(t). Target temperatures $\left(Y_i\right)_{i=1,2}$ correspond to the desired temperature cycle inside the composite part.

7.6.3.3. *The solving algorithm*

The objective functions are minimized using the conjugate gradient algorithm coupled to the adjoint variable technique. This condition leads to find the Lagrange multiplier so as to be the solution of the following set of adjoint equations. As for the direct problem, only the adjoint equations concerning the subdomain Ω_2 are presented.

$$\rho Cp\frac{\partial \psi_{\Omega_2}}{\partial t}+\nabla.\left(\lambda\nabla\psi_{\Omega_2}\right)=\sum_j (Y_j-T)\delta\left(x-x_j\right)\delta\left(y-y_j\right) \quad \text{in}\,\Omega_2, t>0 \qquad -\lambda\left.\frac{\partial \psi_{\Omega_2}}{\partial n}\right|_{\Gamma_{14}}=h_1\left(\psi_{\Omega_3}-\psi_{\Omega_2}\right) \text{ on}\,\Gamma_{14}, t>0$$

$$-\lambda\left.\frac{\partial \psi_{\Omega_2}}{\partial n}\right|_{\Gamma_{16}}=h_2\left(\psi_{\Omega_5}-\psi_{\Omega_2}\right) \text{ on}\,\Gamma_{16}, t>0 \qquad -\lambda\left.\frac{\partial \psi_{\Omega_2}}{\partial n}\right|_{\Gamma_1}+h_\infty T_{\Omega_2}=0 \text{ on}\,\Gamma_1, t>0$$

$$-\lambda\left.\frac{\partial \psi_{\Omega_2}}{\partial n}\right|_{\Gamma_{10}}=0 \text{ on}\,\Gamma_{10}, t>0 \qquad -\lambda\left.\frac{\partial T_{\Omega_2}}{\partial n}\right|_{\Gamma_4\cup\Gamma_5}=0$$

$$\text{or}\; \psi_{\Omega_2}\Big|_{\Gamma_4\cup\Gamma_5}=0 \text{ on } \Gamma_4\cup\Gamma_5, t>0 \qquad [7.8]$$

$$\psi_{\Omega_2}=0 \text{ in}\,\Omega_2, t=t_f$$

Note that the adjoint problem remains the same for both problems, this one and the one of the optimal setpoint. The numerical solution of this adjoint problem (equation [7.8]) is also performed with Comsol® on the same mesh as the direct problem.

7.6.3.4. *The gradient components*

– Identification problem: $\beta=\left(h_i\right)_{i=1,2,3}$.

Each coefficient $\left(h_i\right)_{i=1,2,3}$ is considered as a function of time. It is then approximated in the form $h_i\left(t\right)=\sum_{k=1}^{N_p} h_{ik}\sigma_k\left(t\right)$, where σ_k is a given set of N_p basis functions over the time interval. The gradient components are then equal to:

$$\frac{\partial J}{\partial h_{ik}}=\int_{t=0}^{t_f}\int_{\Gamma_{h_i}}\left(\psi_m-\psi_n\right)\left(T_n-T_m\right)\sigma_k d\Gamma dt \qquad [7.9]$$

where Γ_{h_i} corresponds to the boundary on which the parameter h_i is applied. ψ_m and ψ_n represent the adjoint fields on the domains Ω_m and Ω_n delimited by the boundary Γ_{h_i}. Thus, the vector size of β is $3\times N_p$.

– Control problem: $\beta = (q_i)_{i=1,N_q}$

The heat flux q(t) is approximated in the form $q(t) = \sum_{k=1}^{N_q} q_k \sigma_k(t)$ where σ_k is a given set of N_q basis functions over the time interval. The gradient components are then equal to:

$$\frac{\partial J}{\partial q_k} = \int_{t=0}^{t_f} \int_{\Gamma_3 \cup \Gamma_4 \cup \Gamma_5} \psi_i \sigma_k d\Gamma dt \qquad [7.10]$$

7.6.4. *Experimental data from a composite part molding*

The molding of a composite part was performed. Recordings from the different thermocouples located at the external surface of the mold and inside the mold are shown in Figure 7.7.

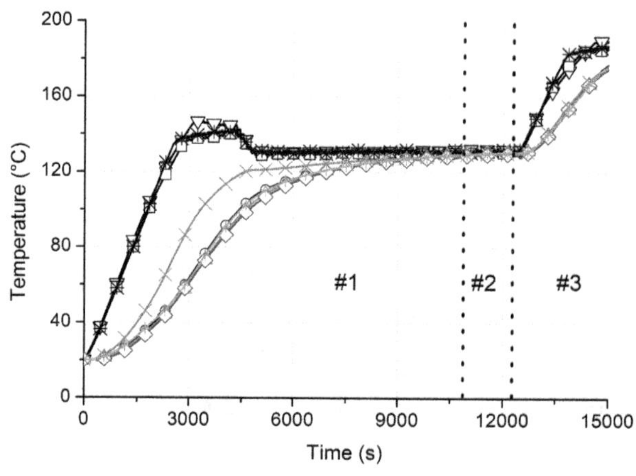

Figure 7.7. *Surface temperature of the mold (× T_{bottom}, O T_{left}, + T_{top}) and internal mold and preform temperatures $(Y_i)_{i=1,5}$ (O Y_1, Δ Y_2, ◇ Y_3, × Y_4, +Y_5) . For a color version of the figure, see www.iste.co.uk/boyard/heat.zip*

Injection (phase #2) occurs between 11,200 and 12,200 s. During the second temperature rise (phase #3), the temperatures are uniform in the mold and the preform. Two phenomena explain this evolution: the thermal expansion of the mold elements and the presence of resin of the interstitial zones between the elements of the mold. The increase in temperature causes an expansion of the elements of the mold. The design of the mold ensures a good contact between the elements at the maximum temperature, i.e. at 180°C. Since resin is injected at 120°C, there are at this temperature some interstices between the elements of the mold. During the injection, the resin flows in these interstices, improving the contact between the elements of the mold and thus improving heat transfer.

7.6.5. *Estimation of the thermal contact resistances*

The TCRs are estimated from the previous measurements. Measurements provided by thermocouples T_{bottom}, T_{left}, T_{top} (Figure 7.5) are used as boundary conditions in the direct problem (equation [7.6]). Recordings of thermocouples located in the metallic elements $(Y_i)_{i=3,4,5}$ are used as target values to solve this problem.

The estimated values of TCR_1, TCR_2 and TCR_3 are then plotted in Figure 7.8 as temperature functions, respectively, of the elements E_1, E_2 and E_3 (Figure 7.5). The evolutions of TCR can be divided into two stages: one before and one after the injection temperature. Before the injection, values of TCR decrease according to temperature. This evolution due to thermal expansion of metallic elements is in compliance with the previous explanations. After the injection, the TCR remains quasi-constant with low values, due to the presence of resin between the elements of the mold.

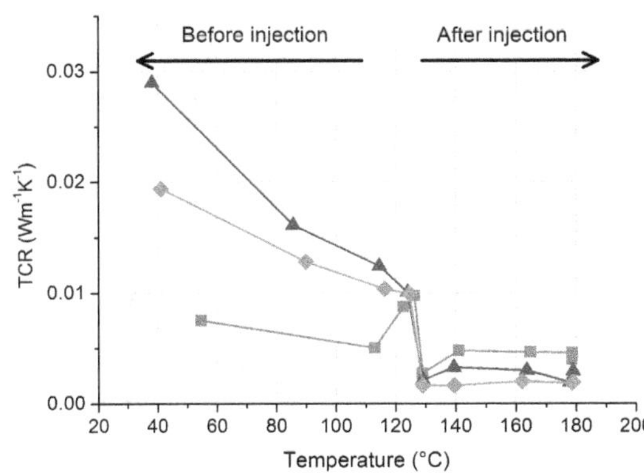

Figure 7.8. *Estimated TCR (TCR_1■, TCR_2▲, TCR_3♦). For a color version of the figure, see www.iste.co.uk/boyard/heat.zip*

7.6.6. Determination of the optimal setpoint

Once the evolution of these TCRs is estimated, their values are used in the conduction model during the determination of the optimal setpoint heat flux on the boundaries $(\Gamma_i)_{i=3,4,5}$. Unfortunately, due to the cost of molding such composite part, no experiment was carried out to validate the computed optimal setpoint temperature cycle.

However, temperature histories $(Y_i)_{i=1,2}$ were recorded in the preform during non-optimized moldings. We decided then to use the data as target temperature in the objective function (equation [7.7]) to solve the problem of optimal setpoint.

This is a key-point of the methodology to validate the determination of the mold setpoint temperature.

The heat flux density was searched as a linear piecewise function defined by $q(t)=\sum_{k=1}^{N_q} q_k\sigma_k(t)$. A set of Nq = 70 basis functions was used to describe the heat flux on the time interval [0s; 30,000s]. The result of this estimation is plotted in Figure 7.9. This figure shows target temperatures in the preform and computed ones after convergence of the method. The very good agreement between both temperatures over the whole cycle demonstrates the robustness of the method.

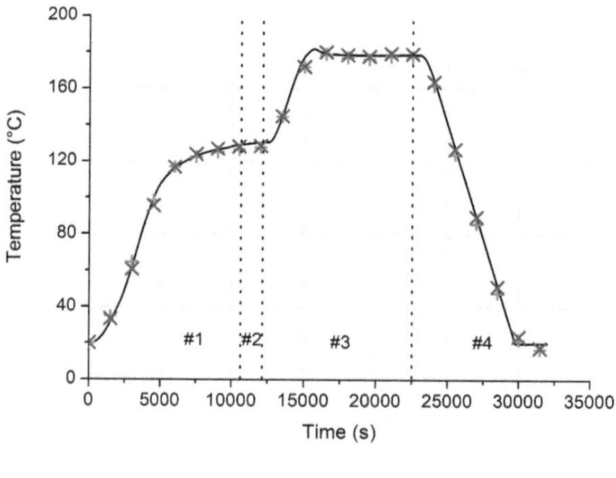

Figure 7.9. *Target (—) and computed (+ Y_1, × Y_2) temperatures. For a color version of the figure, see www.iste.co.uk/boyard/heat.zip*

Figure 7.10 shows the heat flux determined by the algorithm to reach the target temperature. This evolution can be divided into three parts. The first part corresponds to the first heating rise and the first plateau at 120°C (between 0 and 12,000 s). During this stage, the heat flux increases rapidly to attain 10,000 W/m^2 at about 2,500 s. It decreases then linearly to reach a value which remains quasi-constant between 6,000 and 12,000 s. This evolution permits the temperature of the preform to increase linearly at the right rate. The decrease in the heat flux from 2,500 s prevents the temperature of the preform to heat over the desired temperature of the first plateau. The heat flux anticipates, therefore, the evolution of the temperature so as to compensate the thermal inertia of the mold and the preform. The same kind of evolution occurs during the second heating ramp and then during the cooling. In particular, the heat flux remains constant and positive during the two plateaus at 120 and 180°C to compensate the heat losses with the surroundings even if the mold is covered by an insulating material. At the end of the cooling, the heat flux equals zero indicating an isothermal state at ambient temperature. The composite part can then be removed from the mold.

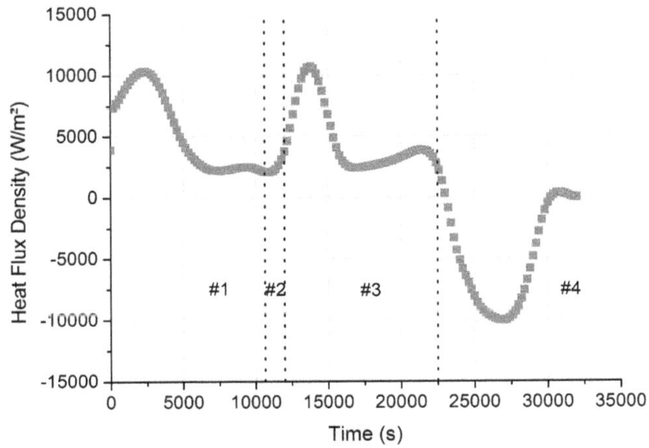

Figure 7.10. *Optimal heat flux setpoint*

These results confirm that, even in the case of a part undergoing a slow thermal cycle, the determination of optimal setpoint is not obvious since the structure of the mold becomes massive and complex. The use of a numerical optimization is required; the functions to determine being non-intuitive.

7.7. Optimal design of molds

7.7.1. *OSOTO project*

Contrary to what was presented in the previous section, when thermal optimization is performed during the mold design, it offers greater choice of control parameters and leads to better efficiency of the heat transfer control in the molded part. The results gathered in this section come from the Osoto program (2007–2010) whose objective was to set an optimal design method of thermal control systems in thermoplastic injection molds in order to minimize cycle time and maximize quality of produced parts.

7.7.2. *The considered thermoplastic part*

The 3D studied part is a plastic box (Figure 7.11). To be representative, the hot sprue is also added in the model.

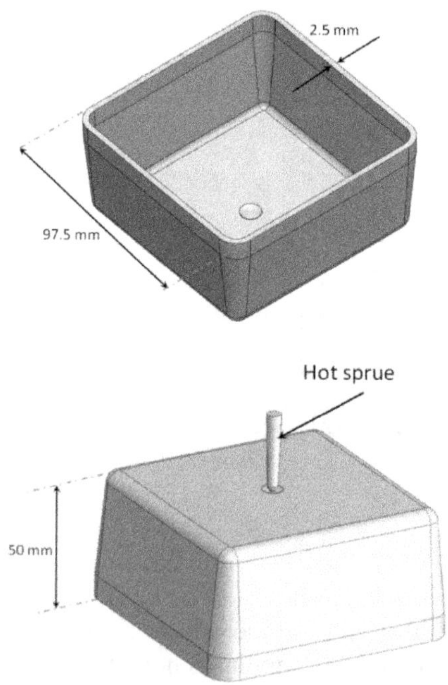

Figure 7.11. *Geometry of the part. Left: top view, right: bottom view*

The injected polymer is a semi-crystalline PA 66 Technyl A218 and the mold is made of maraging steel. Thermophysical properties are assumed to be constant in the considered temperature range except the specific heat capacity that integrates the crystallization enthalpy term. This simplification allows a significant computation time saving, but can be discussed.

7.7.3. *General methodology*

The approach used in this work is divided into two successive stages and aims to determine:

– in a first step, the spatial distribution of the temperature T_∞ of the cooling fluid, from a given configuration;

– in a second step, the number, position and shape of the cooling channels in the mold, from the results of the previous stage.

7.7.4. *Conformal cooling approach*

The concept of conformal cooling surface was considered, i.e. channels that follow the shape of the molding cavity. The first step consists of determining the geometry of this conformal surface. Consequently, morphological concepts are introduced. They consist of the definition of two surfaces named, respectively, as the erosion and the dilation of the molding cavity. The dilation of a set K by a structuring element B corresponds to the Minkowski addition of the set K by the transpose element $\breve{B}$.

$$K \oplus \breve{B} = \left\{ x \in \Re^n, K \cap B_x \neq \varnothing \right\} \quad [7.12]$$

The erosion corresponds to the Minkowski subtraction of the set K by the transpose element $\breve{B}$.

$$K - \breve{B} = \left\{ x \in \Re^n, B_x \subset K \neq \varnothing \right\} \quad [7.13]$$

In our case, both surfaces are built with a ball as structuring element to build both surfaces [SER 82]. The surfaces obtained with this technique for the plastic box are shown in Figure 7.12. For symmetry reason, only one quarter of the part is considered for the numerical model.

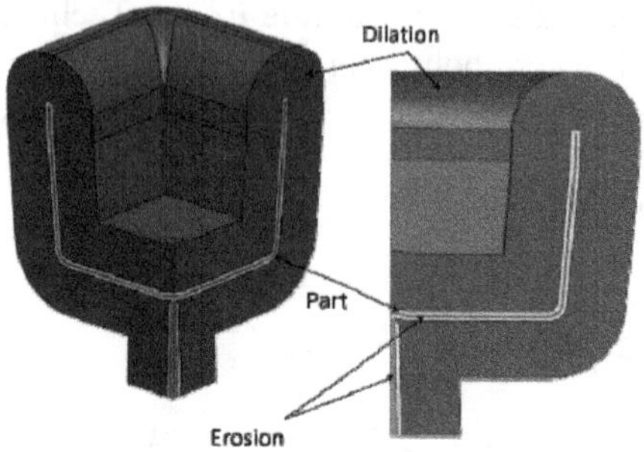

Figure 7.12. *¼ of the part with the erosion and dilation. For a color version of the figure, see www.iste.co.uk/boyard/heat.zip*

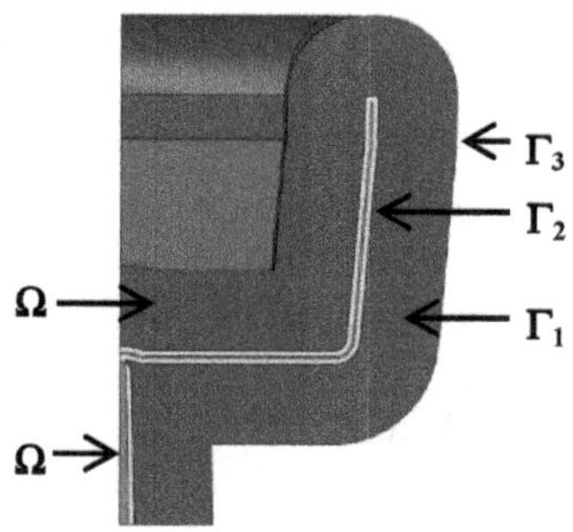

Figure 7.13. *Domain used for the numerical simulation. For a color version of the figure, see www.iste.co.uk/boyard/heat.zip*

Both surfaces are useful. The optimization method consists of determining the temperature along the dilation that allows reaching a target temperature along the erosion. The actual cooling channels will be built *a posteriori* from these results. This will represent the second step of the method.

7.7.5. *Heat transfer model in the process*

The study of heat conduction in the plastic part and in the mold is a transient cyclic nonlinear problem. At each cycle, a molten polymer at the temperature T_{inj} is injected into the cold molding cavity and cooled down until the ejection temperature T_{ejec} is reached. The part is then ejected and the next cycle begins. Heat is extracted from the molding cavity by a conjugated conductive-convective process through the mold and the coolant fluid. In a first approach, the injection phase can be modeled as the instantaneous non-perfect contact between two media at two different temperatures; the filling phase being neglected [SOB 13]. Considering these hypotheses, the heat transfer during the cooling of the injection molding process is described by the following set of equations, over the time interval $0^+ < t < t_c$ for the domain shown in Figure 7.13.

– Equations in the polymer Ω_1 and in the mold Ω_2:

$$\rho_1 C_{p1} (T_1) \frac{\partial T_1}{\partial t} = \nabla.(\lambda_1 \nabla T_1) \quad \text{in} \quad \Omega_1 , \forall t \in 0 < t < t_f$$

$$\rho_2 C_{p2} \frac{\partial T_2}{\partial t} = \nabla.(\lambda_2 \nabla T_2) \quad \text{in} \quad \Omega_2 , \forall t \in 0 < t < t_c \qquad [7.14]$$

– Boundary conditions:

$$-\lambda_1 \left.\frac{\partial T_1}{\partial n}\right|_{\Gamma_2} = \frac{T_1 - T_2}{TCR} \quad \text{on} \quad \Gamma_2 , \forall t \in 0 < t < t_f$$

$$-\lambda_2 \left.\frac{\partial T_2}{\partial n}\right|_{\Gamma_2} = \begin{cases} \dfrac{T_2 - T_1}{TCR} & \text{on} \quad \Gamma_2, \forall t \in 0 < t < t_f \\ h_{eq} (T_2 - T_{out}) & \text{on} \quad \Gamma_2, \forall t \in t_f < t < t_c \end{cases} \qquad [7.15]$$

$$-\lambda_2 \left.\frac{\partial T_2}{\partial n}\right|_{\Gamma_3} = h\left(T_2 - T_\infty(s)\right) \quad \text{on} \quad \Gamma_3, \forall t \in t_f < t < t_c$$

– Initial conditions:

$$T_1(t = 0^+) = T_{inj} \qquad [7.16]$$
$$T_2(t = 0^+) = T_2(t = t_c^-)$$

This set of equations is denoted by $E(T_1,T_2,T_\infty)$. It corresponds to one injection cycle lasting t_c. The cyclic aspect is taken into account through the periodic condition (equation [7.16]). In this simplified model, the ejection phase is related to $t_f<t<t_c$, the molding cavity is empty. Heat losses through Γ_2 are given by equation [7.15b] which models heat transfer by natural convection with the ambient air and by thermal radiation. The polymer is in contact with the mold only during the interval $0<t<t_f$. Its temperature is ignored over $t_f<t<t_c$. In this example, the ejection time is set to 5 s. $T\infty(s)$ stands for the distribution of the coolant fluid temperature on

Γ_3, s being the curvilinear abscissa and h being the convective heat transfer coefficient between the fluid and the mold due to forced convection. $T_2(t = tc^-)$ corresponds to the temperature field in the mold at the end of the previous cycle (Figure 7.14). The injection temperature of the polymer T_{inj} remains unchanged at each new cycle.

Figure 7.14 illustrates the evolution of the molding cavity surface temperature for several cycles. After several cycles, a periodic steady-state is reached. The average mold interface temperature remains constant and oscillations become independent of the cycle number. Industrial production occurs during this state. Thus, to be representative, heat transfer analysis and optimization of cooling are considered during this regime.

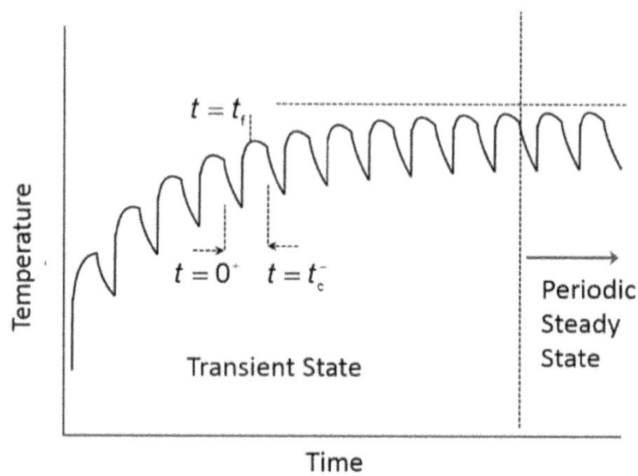

Figure 7.14. *Evolution of the mold surface temperature during cycles*

After the definition of the numerical domain, the choice of the control parameters, the modeling of the direct problem, the next step is the definition of minimization criterion.

7.7.6. *Objective function*

As indicated in the previous section, a dual objective is considered since it consists of solidifying the plastic part in a given/minimum time while increasing the quality of the part. To address this problem, we define an objective function J (equation [7.17]) composed of two terms. The goal of the first term J_1 is to reach the temperature level T_{ejec} along the erosion of the part Γ_1 at the end of the cycle t_f. This term is related to the efficiency of the process. The smaller this term, the lower the temperature of the part is. The second term, J_2, used in [MAT 04] and [TAN 98], aims to homogenize the temperature distribution on the surface of the part Γ_2 and therefore reduces the components of thermal gradient along the surface. This term is linked to the quality of the part. The objective function is computed over the subinterval $t_f - \tau < t < t_f$

$$J(T_\infty) = \underbrace{\int_0^{t_f}\int_{\Omega_1}\left(\frac{T_{ejec} - T_1}{T_{inj} - T_{ejec}}\right)^2 \sigma\eta \, d\Omega \, dt}_{J_1} + \underbrace{\int_0^{t_f}\int_{\Gamma_2}\left(\frac{\overline{T}_1 - T_1}{\Delta T_{ref}}\right)^2 \sigma \, d\Gamma \, dt}_{J_2} \qquad [7.17]$$

$$\overline{T}_1 = \frac{1}{\Gamma_2}\int_{\Gamma_2} T_1 \, d\Gamma \qquad [7.18]$$

$$\sigma = \begin{cases} 0 \text{ if } 0 < t < t_f - \tau \\ 1 \text{ if } t_f - \tau < t < t_f \end{cases} \qquad [7.19]$$

$$\eta = \begin{cases} 1 \text{ on } \Gamma_1 \\ 0 \text{ elsewhere} \end{cases} \qquad [7.20]$$

These two terms are inconsistent. Indeed, making uniform the temperature profile at the interface meanwhile extracting the total heat flux needed to obtain a desired level of temperature in the part becomes antagonistic problem. The best solution is a compromise between quality and efficiency. This compromise is achieved by introducing a weighting parameter between the two terms of the objective function [AGA 10]. For convenience, it is introduced in the form of a "temperature parameter", named ΔT_{ref}[K], and can be compared to the temperature range ($T_{inj}-T_{ejec}$) of the process. The inverse problem statement consists of determining the optimal distribution $T_\infty^*(s)$, $s\in\Gamma_3$, which minimizes the objective function. As indicated previously, the conjugate gradient method is used. The gradient of the objective function is computed through the adjoint technique [BOT 98]. The development being similar to the one used in the section related to RTM mold, we do not detail it for this study.

7.7.7. *Minimization of the functional J*

The simulation is performed with Comsol Multiphysics®, v.4.2a. For our case, we choose the value of ΔT_{ref} equal to twice the difference between T_{inj} and T_{ejec}.

This choice means that the priority is given to the achievement of the ejection temperature in the part at the end of the cycle, at the expense of thermal gradients generated in the part.

After convergence, the average temperature obtained along the erosion of the part (Γ_1) at the end of the cooling time is 489.6 K for a desired level of temperature of 488 K. Along the part surface Γ_2, more than 70% of calculated temperatures are in the range $[T_{average}-\beta;T_{average}+\beta]$ where $T_{average} = 386.4$ K stands for the average temperature along the part surface at the end of the cooling time and $\beta = 6.5$ K is the standard deviation along the part surface.

The temperature field obtained with the optimal fluid temperature distribution is plotted in Figure 7.15. Several cold zones can be distinguished in this figure. These "cold" areas are not affected by the thermal variations during a cycle. They are located in the thermal steady-state zone.

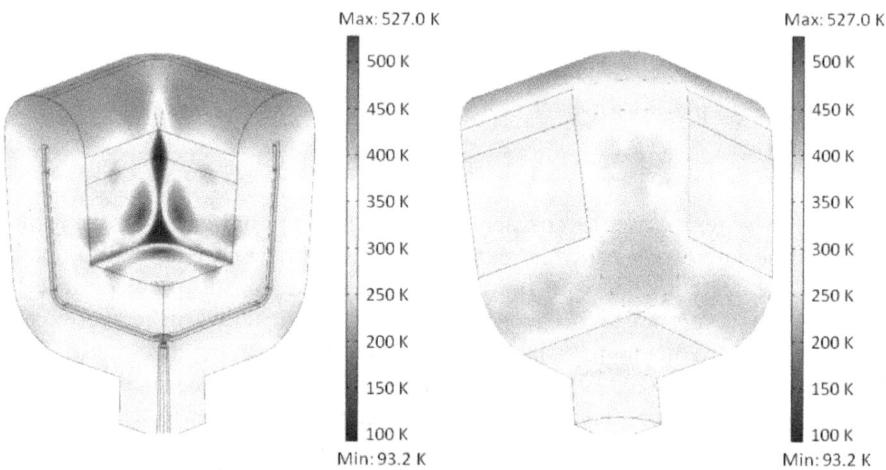

Figure 7.15. *Temperature field in the mold at the end of the optimal cooling time after convergence. Left: front view. Right: back view. For a color version of the figure, see www.iste.co.uk/boyard/heat.zip*

7.7.8. Cooling channel design

The cooling surface obtained along the dilation is, of course, not feasible in practice. The next step consists, from the temperature profile obtained on the cooling surface, of defining discrete channels that can be machined. The proposed methodology consists of giving to the discrete cooling channels, the same shape as some isotherms located in the stationary zone. Considering a turbulent flow regime, the temperature of the coolant fluid is assumed to be very close to the value of the estimated isotherm. Several isotherms can be selected to build the channels. Additional factors (mechanical rigidity, acceptable temperature level in the mold and distribution of the channels all around the part) must then be taken into account to help the designer to choose the most appropriate isotherms. In our configuration, the isotherm 332 K encounters all these criteria. Isotherms are then exported from Comsol to CAD software in order to build cooling channels (ArtCam® and Solidworks®). Figure 7.16 shows the first design for realistic cooling channels.

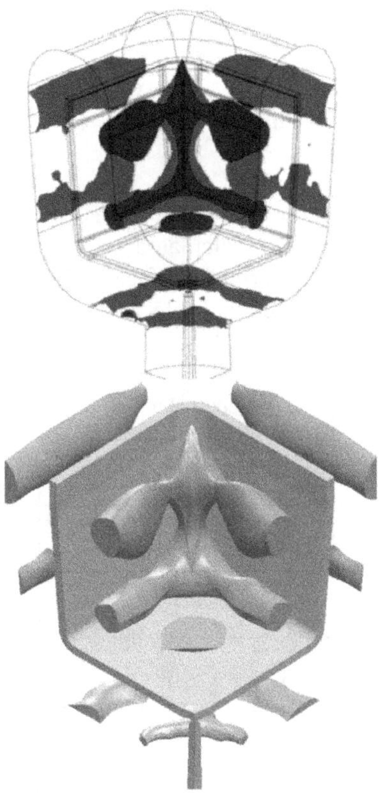

Figure 7.16. *Left: isotherms at 332 K. Right: cooling channels based on the shape of isotherms. For a color version of the figure, see www.iste.co.uk/boyard/heat.zip*

Nevertheless, the shape of the channels in Figure 7.16 generates too large pressure drops to be effective. This design should be modified to improve the flow. The solution proposed for the mold with real channels is shown in Figure 7.17. This development is carried out *a posteriori* by a designer.

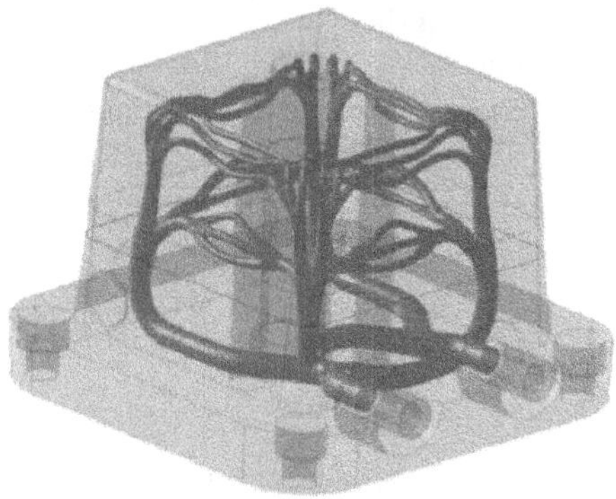

Figure 7.17. *Final design*

To quantify the impact of these modifications on the cooling of the part, the norms of the objective function J calculated on a quarter of the part using the solution obtained with the conformal surface (Figure 7.15) and with discrete channels (Figure 7.17) are compared in Table 7.3.

	Discrete channels	Conformal surface
$J(m^2 \cdot s)$	$2.55 \cdot 10^{-3}$	$1.27 \cdot 10^{-5}$
$J_1(m^2 \cdot s)$	$2.54 \cdot 10^{-3}$	$7.9 \cdot 10^{-6}$
$J_2(m^2 \cdot s)$	$1.79 \cdot 10^{-5}$	$4.8 \cdot 10^{-6}$

Table 7.3. *Comparison of norms of J for solutions obtained with the conformal surface and with discrete channels*

From this, we note that the value of J is lower with the cooling obtained from the cooling surface regulation than with the cooling channels. The transition from a continuous cooling surface to discrete cooling channels designed for laser sintering has lowered the level of temperature in the part at the end of the cooling. The final solution is not as efficient as the regulation by the cooling surface which is due to the transition from continuous surface regulation to discrete cooling channels. However, this approach allows the design of cooling channels by a rigorous methodology and does not require the experience of a designer, no assumption being made on the number, shape and location of the channels.

From the result of Figure 7.17, it is possible to reuse the method proposed in the case of the RTM part so as to recalculate the optimal setpoint using the discrete channels given by the final design (section 7.7).

7.8. Conclusions and outlook

In this chapter, the objective was to point out the requirement to consider not only the polymer/composite part in the optimization method but also the whole process and notably the tools. Indeed, heat transfer regulation is performed through the tools. The control of heat transfer in the parts thus passes first through the control of heat transfer in the tools. The coupling of transfers between the molded part and the tool can be more or less important depending on the contact conditions between them. Nevertheless, even in the case where it is possible to deal with both problems separately, the optimization problem of the tool is not trivial and must be addressed.

In addition, taking into account coupling terms, related in particular to the transformation of the material (crystallization/crosslinking), may lead to an increase the computation times. This is particularly important in the case where it is not possible to sequentially perform the optimizations in the part and in the mold. A sensitivity analysis of these couplings must be carried out on the solution. In the case of low sensitivity, the use of approximate models may be efficient. The choice of the solving algorithm must also arise. Proper generalized decomposition (PGD) methods [GHN 11, GHN 12], which have appeared in recent years, are to be considered principally in the case of linear problems and weak couplings.

Regardless of the coupling, the proposed methodology should be part of a design process for which the heat remains central. In the example presented in this chapter, the method aims to “automatically” build cooling channels in a mold. Other criteria such as the use of multi-materials allowing us to design areas more or less conductive and thus to guide heat in the tools may be considered. This constitutes a new design approach. A generalization of this approach aims at adapting the methodology to processes involving other heat transfer modes. This would lead to an extension of the optimization approach, which could be developed to each process.

7.9. Bibliography

[ABO 00] ABOU KHACHFE R., Résolution numérique de problèmes inverses 2D non linéaires de conduction de la chaleur par la méthode des éléments finis et l'algorithme du gradient conjugué – Validation expérimentale, PhD thesis, University of Nantes, 2000.

[ABO 01] ABOU KHACHFE R., JARNY Y., "Determination of heat sources and heat transfer coefficient for two-dimensional heat flow – numerical and experimental study", *International Journal of Heat and Mass Transfer*, vol. 44, no. 7, pp. 1309–1322, 2001.

[AGA 10] AGAZZI A., SOBOTKA V., LE GOFF R. *et al.*, "A methodology for the design of effective cooling system in injection moulding", *International Journal of Material Forming*, vol. 3, no. 1, pp. 13–16, 2010.

[ALI 94] ALIFANOV O. M., *Inverse Heat Transfer Problems*, Springer-Verlag, 1994.

[BAI 96] BAILLEUL J.-L., DELAUNAY D., JARNY Y., "Determination of temperature variable properties of composite materials: methodology and experimental results", *Journal of Reinforced Plastic Composites*, vol. 15, no. 5, pp. 479–496, May 1996.

[BAI 97] BAILLEUL J.-L., Optimisation du cycle de cuisson de pièces épaisses en matériau composite. Application à un préimprégné résine epoxyde/fibres de verre, PhD thesis, University of Nantes, 1997.

[BAI 03] BAILLEUL J.L., SOBOTKA V., DELAUNAY D. *et al.*, "Inverse algorithm for optimal processing of composite materials", *Composites Part A: Applied Science Manufacturing*, vol. 34, no. 8, pp. 695–708, 2003.

[BAR 82] BARONE M.R., CAULK D.A., "Optimal arrangement of holes in a two-dimensional heat conductor by a special boundary integral method", *International Journal for Numerical Methods in Engineering*, vol. 18, no. 5, pp. 675–685, 1982.

[BOT 98] LE BOT P., Comportement thermique des semi-cristallins injectés. Application à la prédiction des retraits, PhD thesis, University of Nantes, 1998.

[FAV 02] FAVENNEC Y.J., Modélisation numérique en chauffage par induction : Analyse inverse et optimisation, PhD thesis, Ecole nationale supérieure des mines de Paris, 2002.

[GHN 11] GHNATIOS C., CHINESTA F., CUETO E. *et al.*, "Methodological approach to efficient modeling and optimization of thermal processes taking place in a die: Application to pultrusion", *Composites Part A: Applied Science Manufacturing*, vol. 42, no. 9, pp. 1169–1178, 2011.

[GHN 12] GHNATIOS C., MASSON F., HUERTA A. *et al.*, "Proper generalized decomposition based dynamic data-driven control of thermal processes", *Computer Methods in Applied Mechanics and Engineering*, vol. 213–216, pp. 29–41, 2012.

[HAS 10] HASSAN H., REGNIER N., PUJOS C. *et al.*, "Modeling the effect of cooling system on the shrinkage and temperature of the polymer by injection molding", *Applied Thermal Engineering*, vol. 30, no. 13, pp. 1547–1557, 2010.

[HEN 91] HENSEL E., *Inverse Theory and Applications for Engineers*, Prentice Hall, 1991.

[HUA 01] HUANG J., FADEL G., "Bi-objective optimization design of heterogeneous injection mold cooling systems", *Journal of Mechanical Design*, vol. 123, no. 2, pp. 226–239, 2001.

[JAR 91] JARNY Y., OZISIK M.N., ANDARDON J.P.B, "A general optimization method using adjoint equation for solving multidimensional inverse heat-conduction", *International Journal of Heat and Mass Transfer*, vol. 34, no. 11, pp. 2911–2919, 1991.

[LAC 01] LACRAMPE M.F., PABIOT J., "Défauts d'aspect des pièces injectées," *Caoutchoucs & Plastiques*, no. 797, pp. 30–37, 2001.

[LAM 04] LAM Y.C., ZHAI L.Y., TAI K. *et al.*, "An evolutionary approach for cooling system optimization in plastic injection moulding", *International Journal of Production Research*, vol. 42, no. 10, pp. 2047–2061, 2004.

[LEC 99] LECOINTE D., Caractérisation et simulation des processus de transferts lors d'injection de résine pour le procédé RTM, PhD thesis, University of Nantes, 1999.

[LIN 02] LIN J., "Optimum cooling system design of a free-form injection mold using an abductive network", *Journal of Materials Processing Technology*, vol. 120, nos. 1–3, pp. 226–236, 2002.

[MAI 13] MAILLET D., JARNY Y., PETIT D., "Problèmes inverses en diffusion thermique Outils spécifiques de conduction inverse et de régularisation", *Techniques de l'Ingénieur, Transferts thermiques*, vol. ref.: be8267, 2013.

[MAT 04] MATHEY E., PENAZZI L., SCHMIDT F. *et al.*, "Automatic optimization of the cooling of injection mold based on the boundary element method", *Proceedings of the 8th International Conference on Numerical Methods in Industrial Forming Processes*, vol. 712, no. 1, pp. 222–227, 2004.

[MOL 12] MOLDFLOW, "Moldflow material database", available at: www.moldflow.eu, 2012.

[PAR 87] PARANG M., ARIMILLI R.V., KETKAR S.P., "Optimal positionning of tubes in arbitrary two-dimensional regions using a special boundary integral method", *Journal of Heat Transfer*, vol. 109, pp. 826–830, 1987.

[QIA 06] QIAO H., "A systematic computer-aided approach to cooling system optimal design in plastic injection molding", *International Journal of Mechanical Sciences*, vol. 48, no. 4, pp. 430–439, April 2006.

[SER 82] SERRA J., *Image Analysis and Mathematical Morphology*, Academic Press, Orlando, FL, 1982.

[SOB 13] SOBOTKA V., AGAZZI A., BOYARD N. *et al.*, "Parametric model for the analytical determination of the solidification and cooling times of semi-crystalline polymers", *Applied Thermal Engineering*, vol. 50, no. 1, pp. 416–421, 2013.

[TAN 98] TANG L., POCHIRAJU K., CHASSAPIS C. *et al.*, "A computer-aided optimization approach for the design of injection mold cooling systems", *Journal of Mechanical Design*, vol. 120, no. 2, pp. 165–174, 1998.

[WOO 03] WOODBURY K.A., *Inverse Engineering Handbook-The Adjoint Method to compute the Numerical Solutions of Inverse Problems-Jarny*, CRC Press, 2003.

8

Modeling of Thermoplastic Welding

Welding of thermoplastic materials or thermoplastic composite parts is a valuable process as it allows designing parts with quite simple forms and assembling them without adding material. Nevertheless, the mechanical properties around the welded line are often significantly lower than the ones of bulk materials. Optimizing the mechanical properties requires the understanding of the physical mechanisms involved in welding processes. Intimate contact and macromolecular diffusion, which are the two main physical phenomena that govern healing of interfaces, are presented in details. As these phenomena depend a lot on temperature and welding processes are highly anisothermal, thermal modeling of processes or accurate local temperature measurements are necessary to record temperature histories around welded lines in order to assess interface healing. The welding of thermoplastic composite tapes is modeled for two types of heating: hot nitrogen torches and ultrasounds. We show how the phenomena are coupled in the thermo-physical simulations. Moreover, heating can degrade the polymer, therefore modify the molecular mass and then directly interact with macromolecular diffusion phenomena.

8.1. Introduction

8.1.1. *Polymer welding processes*

Welding is a well-established technology in the thermoplastic industry where the efficiency of welded joints can approach the properties of bulk materials. Thermoplastics or thermoplastic matrix composites can be assembled in such a way by heating them at a temperature above glass transition for an amorphous matrix and

Chapter written by Gilles REGNIER and Steven LE CORRE.

above fusion temperature for a semi-crystalline matrix. Thermoplastic welding technologies can be classified according to the technology introducing the type of heating [GRI 90, AGE 01]. We proposed to classify them into three classes: bulk heating, surface heating and mechanical energy dissipation techniques.

Bulk heating techniques (compression molding, electromagnetic heating techniques such as microwaves and dielectric welding [LEE 84], autoclaving and induction for some composites) are available for performing a joining method with no added weight to the final structure. For multi-layer composites, which are excellent shields in the microwave range [YAR 98], microwave welding is poorly suitable especially when composites are reinforced by carbon fibers. An induction heating technique is available if thermoplastics are reinforced by conductive materials (ferromagnetic materials): iron particles, micrometer-sized particles of iron oxide, stainless steel, specific ceramic, ferrite or graphite, acting as susceptors [WIL 90]. However, in all cases, the entire part is brought to the melting temperature and a complex tooling is generally needed to maintain pressure on the entire part in order to prevent deconsolidation.

Hot melt thermoplastic adhesive films [VIN 89] or amorphous films involving co-molding of an amorphous thermoplastic to a semi-crystalline matrix laminate prior to bonding may also be inserted between the welded surfaces to improve filling of part mismatch [COG 89].

In surface substrate heating techniques (hot plate welding, hot gas welding, radiant welding and laser welding), the heating device should be removed from the surfaces between the stages of heating and welding, except for resistive implant welding, also called electro-fusion welding [ATK 89]. Heating times may be long because of the low thermal conduction of the polymer matrix. Between heating and welding stages, surface temperature may drop, where the region having the maximum temperature is located below the skin of the part [TAY 88]. Thus, the pressure required to consolidate the joined surface may cause warpage/flow in the higher temperature inner region [BEN 86]. As the whole welding surface must be heated rather uniformly, these processes involve limitations on size and geometry of the parts.

Heating by mechanical dissipation techniques (ultrasonic welding, vibration welding and spin welding) has been widely used in the plastics industry [GRI 95]. Vibration welding and spin welding are less appropriate to the joining of thermoplastic composites as the motion of the substrates may cause deterioration or modification of the local microstructure [MAG 89]. Ultrasonic welding is certainly one of the most commonly used welding processes for thermoplastics with many applications in automotive parts [GRI 95]. To dissipate mechanical energy, small asperities called energy directors generally in the form of triangular protrusions are molded onto the part to initiate melting. Nevertheless, ultrasonic welding applications to flexible polymers or semi-crystalline polymers with a rubbery amorphous phase are limited since they absorb most of energy in bulk [BEN 86]. Kenney [KEN 92] insisted to consider the requirements of ultrasonic welding at the early stage of the design of a part.

8.1.2. *Healing mechanisms of polymer interfaces*

For all these welding processes, the healing of the polymer/polymer interface was described by Wool *et al.* [WOO 81] who identified three main sequential stages, namely: (1) surface rearrangement and surface approach, (2) wetting, and (3) diffusion. In stage (1), the interface has no mechanical properties as the two distinct faces still exist. The completion of the wetting stage marks the achievement of intimate contact between the two surfaces. Potential barriers associated with inhomogeneities or porosities at the interface have disappeared, and molecular chains are free to move across the interface in a process of interdiffusion also called autohesion, which ultimately leads to the collapse of the interface.

Wool *et al.* [WOO 81] proposed to model the wetting stage by a phenomenological approach of nucleation and radial growth of wetting surface, which led an expression of Avrami's type. Nevertheless, the time scale of wetting phenomenon is very difficult to quantify experimentally. It is also difficult to separate this phenomenon from surface rearrangement. As it seems that the time scale of wetting phenomenon is one order of magnitude less than surface rearrangement and molecular diffusion, most of the researchers gather stages (1) and (2) into one stage generally called intimate contact. Modeling of intimate contact and macromolecular diffusion stages are presented in the next section.

8.2. Physics of thermoplastic welding

8.2.1. *Intimate contact at interface*

Whatever the elaboration technique of polymer or composite parts, surfaces always present a certain roughness. Microasperities and microvalleys as well as more macroscopic surface waves lead to a non-perfect contact between the two parts to assemble. As commonly admitted in thermoplastic welding, intimate contact is the first step in forming a bond between two thermoplastic surfaces that have been brought together under both heat and pressure [BUT 98, SCH 09, TSA 09]. As visible from Figure 8.1, surface roughness may be very important, leading to a rather low initial area of contact between the two adjacent surfaces.

a)

b)

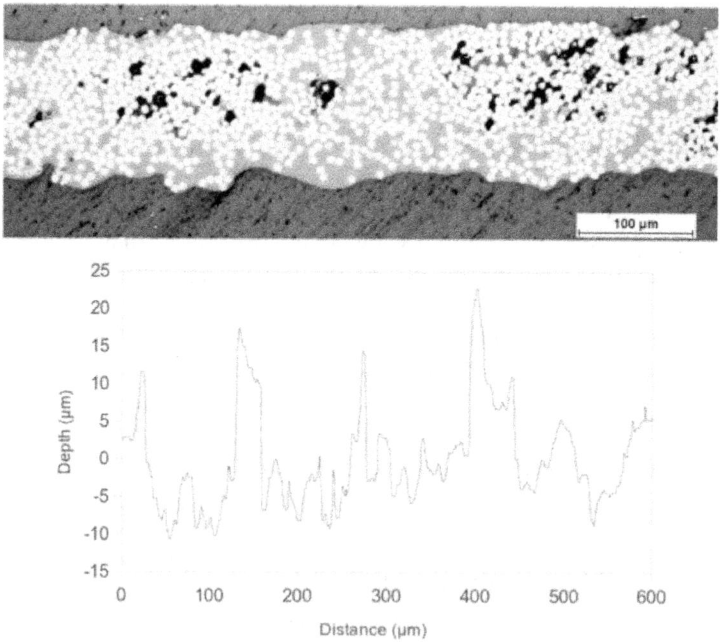

Figure 8.1. *Cross-section of an APC2 carbon/PEEK composite tape a) and associated roughness profile b) [LEV 14]*

The quality of the contact can be described through the concept of *degree of intimate contact* D_{ic}, as introduced by Lee and Springer [LEE 87]. It is a macroscopic descriptor of the contact that can be defined as the ratio of the real contact area A_{cont} to the total area of the surfaces to be welded A_{tot}:

$$D_{ic} = \frac{A_{cont}}{A_{tot}} \qquad [8.1]$$

In thermal studies, it is known as the *real rate of* contact. To describe this variable, we may use optical profilometers that nowadays enable us to characterize solid surfaces from nanometric to millimetric scales. They easily provide statistic descriptors such as the arithmetic roughness R_a, the quadratic one R_m or the average slope of the profile. Automated devices are also able to produce full surface reconstructions, suited for finer modeling [GRU 12].

Once the surface topography is obtained, the main problem is the prediction of its evolution while submitted to the welding pressure. This question has been the subject of several models during the last couple of decades. Dara and Loos [DAR 85] developed a model representing the surface of a thermoplastic as a distribution of rectangles of different sizes. Lee and Springer [LEE 87] simplified this model by using rectangular elements of the same size (Figure 8.1). Their approach, owing to its great simplicity and efficiency, has now become the most widely used.

In the representation of Figure 8.2, the degree of intimate contact can be defined as:

$$D_{ic} = \frac{b}{b_0+w_0} \qquad [8.2]$$

where the geometrical parameters b, b_0 and w_0 are defined in Figure 8.2(a). In this very idealized picture, the surface topography is described by its average roughness, which may be finally attributed to the height a of rectangles, as proposed in [MAN 92, YAN 01, YAN 03].

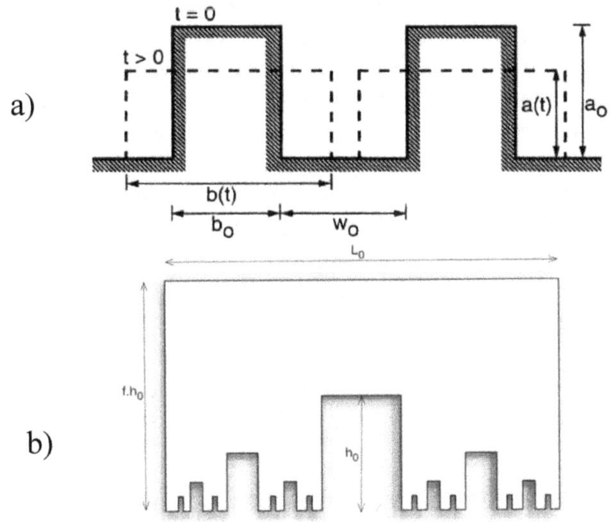

Figure 8.2. *a) Idealization of surface asperities by periodic rectangular elements, before and after squeezing [LEE 87]; b) fractal Cantor set description of the adherents surface [LEV 14]*

Thus, using the volume conservation of rectangles, we can write:

$$D_{ic} = \frac{a/a_0}{1+w_0/b_0} \quad [8.3]$$

The squeezing kinematics is then modeled by a so-called "squeeze flow" model for a Newtonian fluid, and leads to the following differential form [MAN 92]:

$$\dot{D}_{ic} = \left(\frac{D_{ic,0}}{D_{ic}}\right)^4 \left(\frac{a_0}{b_0}\right)^2 \frac{P}{\mu(T(t))} \quad [8.4]$$

where P is the applied pressure and where the viscosity of the thermoplastic μ can be defined by an Arrhenius-type law. It reduces to the Lee and Springer's model in an isothermal case:

$$D_{ic} = \frac{1}{1+w_0/b_0}\left[1 + \frac{5P}{\mu(T)}\left(1 + \frac{w_0}{b_0}\right)\left(\frac{a_0}{b_0}\right)^2 t\right] \quad [8.5]$$

These expressions clearly underline the role of time, pressure and temperature – through its direct influence on the viscosity – as well as thermoplastic initial surface quality, through rudimental parameters a_0, b_0 and w_0 [BUT 98, YAN 03, YAN 01].

In order to improve such a rough description of the surface profile, the *fractal Cantor set-based model* was proposed by Yang and Pitchumani [YAN 01, YAN 01]. Parameters a_0, b_0 and w_0 cannot be determined physically and must therefore be fitted using experimental results. Their approach consists of describing the surface morphology as a Cantor set fractal surface (Figure 8.2(b)), where each geometrical parameter can be obtained from a surface profile measurement. The squeezing kinetics of this morphology and the resulting evolution of D_{ic} are then calculated by the Lee and Springer rectangular model [LEE 87], applied iteratively from the smallest to the biggest rectangles of the fractal geometry [YAN 01]. As illustrated in the parametric study of Figure 8.3, the model is sensitive to the fractal dimension D and capable of predicting more subtle effect than the idealized rectangle approach. One of the main benefits of this method is that it can be based on direct surface measurements as illustrated in several implementations [YAN 01, YAN 03, WAR 96 YAN 02, CHE 10]. As the fractal Cantor model is scale-invariant, the precise accuracy of a particular measuring method is not of significant importance [CHE 10].

More recently, a model using a more realistic description of the initial surface has been proposed [GRU 12]. The surface geometry was measured by a profilometer and meshed. The squeezing step was then computed using Polyflow® software. The authors showed a good correlation of this new model with the Lee and Springer model.

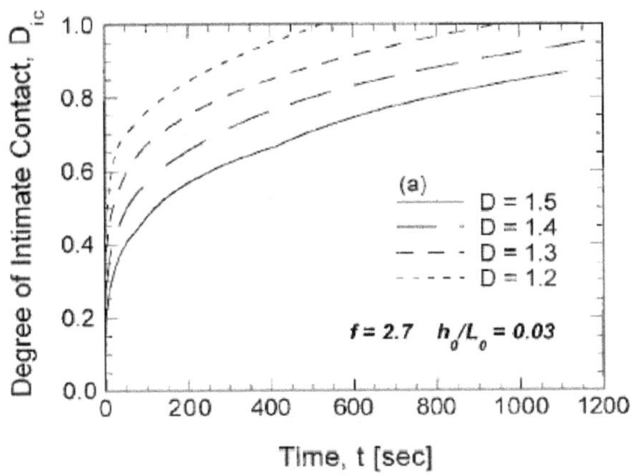

Figure 8.3. *Parametric effects of the fractal surface parameters on the development of the D_{ic} contact [YAN 01]*

Despite the different attempts to reach a finer geometrical description of surface roughness, the simple idealized rectangle model still seems to remain the most simple and efficient as the numerous successful applications of it show [AGE 01, MAN 92, SON 97, NIC 05, LAM 05, TIE 06]. Nevertheless, the underlying assumptions of this approach are very restrictive (2D model, Newtonian squeeze flow) and further improvements are expected in the future, especially for an optimal control of interlaminar adhesion of thermoplastic composite parts. Furthermore, experimental measurements of the D_{ic} during processing, which are generally assumed impossible, can be done through thermal measurements [LEV 14]. Finer validations of these intimate contact models should also be possible.

8.2.2. *Macromolecular diffusion*

8.2.2.1. *Modeling of a polymer chain*

At equilibrium in a melted state, a polymer chain takes the configuration of a statistical pellet [FLO 49] with a given gyration radius *Rg* and a chain end-to-end distance *R*. Many authors [FLO 49 RUB 03, FET 07] have modeled this statistical pellet as a set of *n* freely jointed rigid segments having a length *l* (Figure 8.4), therefore the length of the stretched chain is equal to nl. It is shown that the mean square values of gyration radius Rg and chain end-to-end distance R are:

$$\langle R^2 \rangle = nl^2 \text{ and } \langle R_g^2 \rangle = \langle R^2 \rangle / 6 \quad [8.6]$$

In a real polymer chain, the segments are not fully free in all directions, then $<R^2>$ and $<Rg^2>$ increase. Flory defined [FLO 49] the chain characteristic ratio C_∞ as the ratio of the real polymer mean-square end-to-end distance $<R^2>_0$ and that of a freely jointed chain:

$$C_\infty = \frac{< R^2 >_0}{nl^2} \quad [8.7]$$

The chain appears to be more flexible when C_∞ decreases. Values of C_∞ are given for some polymers [FET 07]: C_∞ is equal to 8.26 for PE and to 6.15 for iPP.

Another way to express the fact that chain segments are not fully free in all directions is to consider the Kuhn length defined as the effective monomer size for the equivalent freely jointed chain (N Kuhn monomers of length b). The length of the considered chain is equal to $nl = Nb$. The real polymer mean-square end-to-end distance $<R^2>_0$ is then equal to:

$$< R^2 >_0 = C_\infty nl^2 = Nb^2 \quad [8.8]$$

From equation [8.8], it can be deduced that $b = C_\infty l$

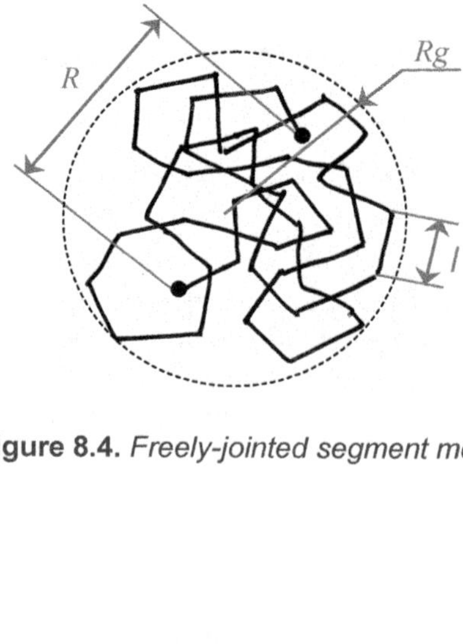

Figure 8.4. *Freely-jointed segment model*

8.2.2.2. *Modeling of macromolecular diffusion*

A large amount of works exists on the study of polymer chain movements, most of them are based on the previous chain modeling. Only the most significant for the topic are presented.

Rouse [ROU 71] proposed to model the polymer chain by a set of beads freely-joined by springs. He showed that the time τ_{Rouse} needed to have an overall movement of the chain is equal to:

$$\tau_{Rouse} = \frac{\xi b N^2}{3\pi^2 k_B T} \tag{8.9}$$

where ξ is a friction coefficient, k_b is the Boltzmann constant, b and N are the Kuhn parameters and T is the temperature. This Rouse time is proportional to the square of the chain length or the molecular mass. It turns out that this model applies for polymer chains or chain branches having a molecular mass less than the critical molecular mass Mc, which denotes the transition in the melt viscosity versus molecular mass relation as the exponent changes from 1 to 3.4. The ratio of M_c to molecular mass between entanglements M_e has long taken to be 2 and thus independently to species. However, an empirical compilation [FET 99] has shown that this ratio varies from 1.4 to 3.5.

For polymer chains having a molecular mass larger than M_c, i.e. for entangled polymer chains, molecular movements can be divided into two types:

– movements at a small scale which do not affect the topology of entanglements. The scale time of these movements is Rouse time;

– movements at a larger scale which modify the topology of entanglements. These movements obey to another scale time because chains interact with neighboring.

For the last type of molecular movement, de Gennes [GEN 71] proposed the well-accepted reptation theory in which the polymer chain is confined in a tube, which is supposed to model the constraints of neighboring chains. The chain has a Brownian movement and can only exit the tube by its two extremities. The tube disappears and appears gradually (Figure 8.5) as the chain moves as a snake in the tube, hence the reptation theory. The time required for the chain to completely exit the initial tube and then to totally loose the memory of its initial configuration is:

$$\tau_{Reptation} = \frac{\xi b^4 N^3}{3\pi^2 k_B T a^3} \tag{8.10}$$

where a is the diameter of the tube, which can be approximated to the mean square distance between entanglements:

$$a^2 = b^2 N_e / 6 \tag{8.11}$$

with N_e equal to the number of Kuhn segments between entanglements. Therefore, the reptation time is proportional to the cube of chain length or molecular mass. Then, the ratio between reptation time and Rouse time is:

$$\frac{\tau_{Reptation}}{\tau_{Rouse}} = 2\frac{N}{N_e} = 2\frac{M}{M_e} \tag{8.12}$$

For example, the ratio between reptation time and Rouse time is nearly two decades for an injection molding PE with a molecular mass of 60 kg/mol, knowing that the mass between entanglements for PE is equal to 1.4 kg/mol [FET 07].

Figure 8.5. *Reptation theory of de Gennes*

8.2.2.3. *Macromolecular diffusion and polymer interface healing*

8.2.2.3.1. Amorphous polymers

Numerous theoretical and experimental studies have been done on interface healing of amorphous polymers. Most of them are based on chain reptation mechanism. De Gennes [GEN 79] considered that healing is driven by the density of chain bridges at the interface, Kausch [KAU 89] supposed that healing is related to the number of entanglements formed at the interface and Wool [WOO 83] who has most certainly worked on this topic assessed that healing is controlled by the length of chain interpenetration. All of them found that the ratio of fracture toughness for a healing time t less than the reptation time $G_{IC}(t)$ to the polymer bulk fracture toughness $G_{IC\infty}$ is proportional to the ratio of time t to reptation time at the power 0.5:

$$\frac{G_{IC}(t)}{G_{IC\infty}} \propto \left(\frac{t}{\tau_{Reptation}}\right)^{\frac{1}{2}} \tag{8.13}$$

Thus, the induced stress necessary to fracture the interface σ(t) varies as:

$$\frac{\sigma(t)}{\sigma_\infty} \propto \left(\frac{t}{\tau_{Reptation}}\right)^{\frac{1}{4}} \qquad [8.14]$$

Nevertheless, the dependence of stress and fracture toughness with molecular mass varies with modeling and rupture mode. First, it was observed that the bulk fracture toughness is reached for a reptation time corresponding to a polymer of 8 M_c and becomes independent of molecular mass [WOO 89]. In this case, the fracture mechanism is chain rupture and agrees with scaling laws of de Gennes and Kausch [GEN 79, KAU 89]:

$$G_{ic}(t) \propto t^{\frac{1}{2}} M^{-\frac{3}{2}} \text{ and } G_{IC\infty} \text{ independant of molar mass} \qquad [8.15]$$

For PE, 8 M_c corresponds to a molecular mass of 40 kg/mol. Yang and Pitchumani [YAN 02] supposed it is sufficient that the chains come out of the tube along a distance corresponding to 8 M_c to have a good healing.

For a polymer with a molecular mass between M_c and 8 M_c, the fracture mechanism should be chain pull out mechanism (disentanglement) and Wool modeling [WOO 89] seems to agree with experimental tests:

$$G_{ic}(t) \propto t^{\frac{1}{2}} M^{-\frac{1}{2}} \text{ and } G_{IC\infty} \propto M \text{ or } \frac{G_{IC}(t)}{G_{IC\infty}} \propto t^{\frac{1}{2}} M^{-\frac{3}{2}} \qquad [8.16]$$

$$\sigma(t) \propto t^{\frac{1}{4}} M^{-\frac{1}{4}} \text{ and } \sigma_\infty \propto M^{\frac{1}{2}} \text{ or } \frac{\sigma(t)}{\sigma_\infty} \propto t^{\frac{1}{4}} M^{-\frac{1}{4}} \qquad [8.17]$$

8.2.2.3.2. Semi-crystalline polymers

In semi-crystalline polymers, crystallites constituted of chains from each side of the interface can be formed through the interface during the cooling [XUE 00]. This co-crystallization phenomenon occurs with incomplete chain diffusion at the interface and takes part in interface consolidation [TSA 09, YUA 90, GEN 97]. Rupture mechanisms at interface in semi-crystalline polymers are therefore more complex and certainly need further investigations to precisely model them.

When the amorphous phase in a semi-crystalline polymer is glassy at test temperature, rupture mechanisms are probably close to the ones occurring in amorphous polymers having a molecular mass higher than 8 M_c, i.e. by chain rupture.

For semi-crystalline polymers with a rubbery amorphous phase (like PE or PP at ambient temperature), the high mobility of the amorphous phase favors chain pull out and disentanglement. In this case, Fayolle *et al.* [FAY 03] showed that the chain pull out phenomenon occurs for semi-crystalline polymers having a molar mass until 50 M_c.

8.3. Linear viscoelasticity to quantify the macromolecular diffusion

Linear viscoelasticity (LVE) precisely reflects the distribution of relaxation times in a polymer and is therefore strongly correlated to the molecular structure, i.e. the molecular weight, molecular weight distribution and molecular architecture [GRA 74, FER 80]. Hence, LVE is a powerful tool, which provides fundamental insights about the link between dynamics and polymer structure [VAN 07].

From linear viscoelastic frequency sweep tests, two characteristic relaxation times can be determined from the terminal relaxation zone. Firstly, the number-average relaxation time determined as $\tau_n = \eta_0 / G_N^0$ can be experimentally obtained from the intersection of the G'' terminal regime with G_N^0, where G_N^0 is the plateau modulus and η_0 the zero-shear viscosity. Secondly, the intersection frequency ω_w of G' and G'' terminal regimes gives the weight-average terminal relaxation time τ_w as follows: $\tau_w = 1/\omega_w = J_e^0 \eta_0$, where J_e^0 is the steady-state recoverable shear compliance. Therefore, the breadth of relaxation time distribution can be assessed to $\tau_w / \tau_n = J_e^0 \eta_0$ and varies in the typical range 2 to 3 for nearly monodisperse polymers or polymers with a very narrow molecular mass distribution [FER 80].

For monodisperse polymers, the reptation time τ_R of tube models is closely linked to the terminal peak position of G" [GRA 80] and should be intermediate between τ_w and τ_n. No simple response can be given for polydisperse polymers as obstacles in tube modeling are not really fixed due to smaller chains.

It is convenient to model the linear viscoelastic response of a polymer by rheological models. The simplest one is certainly the Maxwell model, which has only one relaxation time. Then, all relaxation times that can be determined from a frequency sweep response are identical: τ_w, τ_n, maximum of *G"*, *G'* and *G"* crossing, viscosity passage from Newtonian to rheothinning regimes (Figure 8.6). Thus, this model cannot precisely model the linear viscoelastic response of a monodisperse polymer since $\tau_w/\tau_n \sim 2$ But a multimode Maxwell model can either fit the linear viscoelastic response of a monodisperse or a polydisperse polymer, depending on the number of parallel branches.

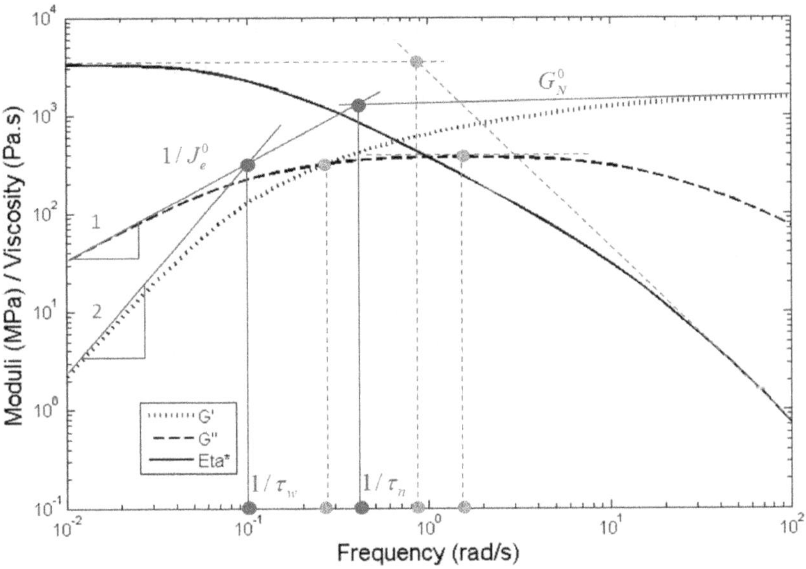

Figure 8.6. *Frequency sweep response of a Maxwell multimode model Relaxation times (0.03/0.1/0.3/1/3/10 s) – respective associated moduli (100/300/300/300/300/200 MPa). For a color version of the figure, see www.iste.co.uk/boyard/heat.zip*

The question is: which relaxation time should be considered to have a healed interface? Let us first consider an amorphous polymer. For a commercial polymer, i.e. a polydisperse polymer, the answer is not simple in a general case because molecular movements are complex and it is not so easy to link dynamics and complex polymer structure. In all cases, interface healing surely needs complete polymer chain reorganization, thus long relaxation times of the terminal behavior should be considered and τ_w is without any doubt a good assessment of a complete polymer diffusion through the interface.

For semi-crystalline polymers, the response is still more complicated because of co-crystallization through the interface. Although co-crystallation could improve interface healing, it is safer to consider that interface is healed when the chain diffusion process through the interface is achieved like for amorphous polymers.

For amorphous and semi-crystalline thermoplastics, the determination of the terminal relaxation time is often not so easy because wide frequency range master curves should be determined for highly polydisperse polymers. Zanetto [ZAN 01] and Régnier *et al.* [REG 07] considered the frequency viscosity passage from Newtonian to rheothinning regimes. Nevertheless, they verified that this relaxation time was not too far from τ_w for one temperature, this was certainly true because the molecular mass distribution of, respectively, PA12 and PEEK was not too large, which is generally not the case for polyolefins.

8.4. Application to continuous welding of composite tape

8.4.1. *Process description*

Continuous carbon fiber-reinforced tapes are used in the aircraft industry to build structural parts by welding them one to another to constitute the desired part. A two-steps procedure was needed in the past: first, a manual or automated tape placement, second, a consolidation in autoclave. The aim of this study was to investigate a new process, named "Drapcocot" [COI 05], developed by EADS, Dassault Aviation and Eurocopter, in which placement and consolidation are performed simultaneously in order to increase the productivity.

The process principle can be briefly described as follows (Figure 8.7): the first layers are supposed to be already laid down. Then, the tape under consideration is placed in its position using the tow guide. Two torches blow hot nitrogen in order to melt the material at the interface between the plies to weld. A roller applies a normal force on the tape in order to improve the adhesion between the plies. The process parameters that can be controlled are torches temperature, head velocity, roller temperature and roller force.

The material was the well-known APC-2™, preimpregnated UD carbon-PEEK composite supplied by Cytec®.

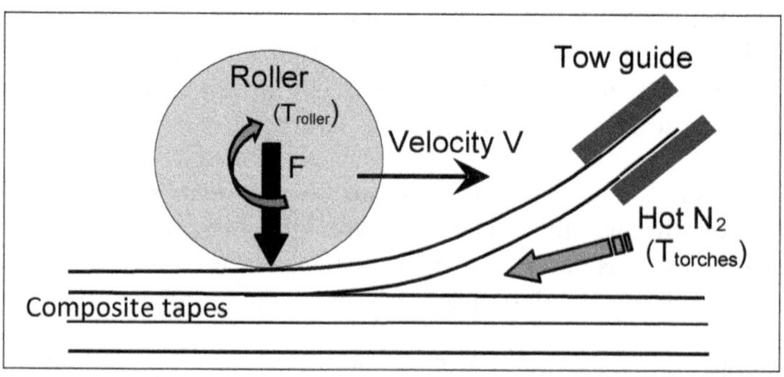

Figure 8.7. *Schematic principle of "Drapcocot" process*

8.4.2. *Influence of processing conditions on interfacial strength*

The weld quality was characterized using peel tests on two layers samples, which differed by their processing conditions: torches temperature and lay-down velocity [REG 07]. The applied roller force was kept constant and high enough to make the intimate contact time [NIC 05] negligible, compared to the welding time. The peel energy (Figure 8.8) displays a maximum and it appears thus that adhesion is under the dependence of two antagonist and thermally activated phenomena. The role of velocity puts in evidence the importance of kinetics aspects. In the low temperature torches domain (< 700 C), adhesion is an increasing function of temperature and a decreasing function of velocity, which suggests a thermally activated diffusional process. The decrease of adhesion at high temperature could be explained by thermal aging, which adversely affects the welding quality. Other phenomena, for instance co-crystallization at interface before complete molecular diffusion, may also affect the mechanical performance of welded joints [COI 05].

According to these assumptions, the model elaboration involves three steps: (1) determination of kinetic laws for chain diffusion; (2) identification of thermal aging processes susceptible to influence welding and determination of their kinetic parameters and (3) thermokinetic mapping of the process to determine the material thermal history at every point.

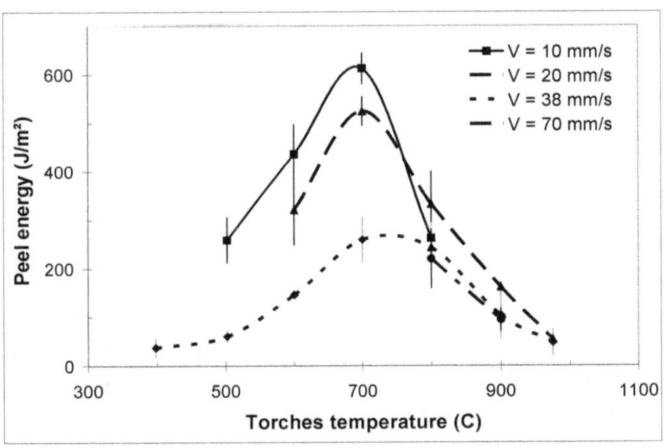

Figure 8.8. *Influence of parameters on interfacial adhesion*

8.4.3. *Modeling of macromolecular diffusion*

Once the intimate contact between plies is established, the interdiffusion of macromolecular chains at the interface occurs as long as the matrix is melted, i.e.

before crystallization quenches long range chain motions. Unfortunately, the PEEK grade used for APC-2™ was not available and thus rheometric measurements were made on a pure PEEK grade (Victrex PEEK G450) which was supposed to be the highest boundary of the APC-2™ matrix in terms of molecular weight.

As it was difficult to experimentally determine the terminal relaxation time, the diffusion time was assessed by determining the passage between Newtonian and rheothinning domains of complex viscosity η* (see section 8.3). The measurements were performed on an Ares™ Rheometric Scientific rheometer using parallel plates geometry under nitrogen in the 310–410 C temperature range at 2% strain amplitude with angular frequency ranging from 0.01 to 100 rad.s-1. It can be recalled that the main temperature transitions of PEEK are glass transition temperature Tg = 140°C, melting point Tm= 330°C and crystallization temperature, which begins at about 280°C in a programmed temperature decreasing experiment.

Characteristic relaxation time versus temperature λ(T) (Figure 8.9) was identified from Carreau law:

$$\eta^*(T) = \eta_0(T)\left[1+\left(\lambda(T)\omega\right)^{\alpha}\right]^{(m-1)/\alpha} \quad [8.18]$$

It obeys an Arrhenius law with a consistent activation energy of 59.7 KJ.mol^{-1} that allows extrapolating it at higher temperatures inaccessible to experiments.

To quantify the extent of reptation in an anisotherm process, it is possible to solve the diffusion equation in the nonisothermal case [WOO 81, SAI 89]. Yang *et al.* [YAN 02] proposed to consider a diffusion criterion D defined by:

$$D = \int_{t_{melting\,\&\,contact}}^{t_{crystallisation}} \frac{dt}{\lambda\left(T(t)\right)} \quad [8.19]$$

where D that depends on material thermal history and assessed diffusion time describes the probability for a chain to cross the interface. D = 1 means that the processing time is equivalent to the reptation time in an isothermal experiment, therefore D has to be larger than 1 to consider the interface healing achieved. D was calculated for various sets of processing parameters taking the origin of time when interfaces are in contact and reach melting point and end of time when crystallization began. For this, a non-isothermal crystallization kinetic was identified and implemented in the model [NIC 05]. Thus, it was possible to plot the envelope of points where D = 1 in a process window (see section 8.4.5).

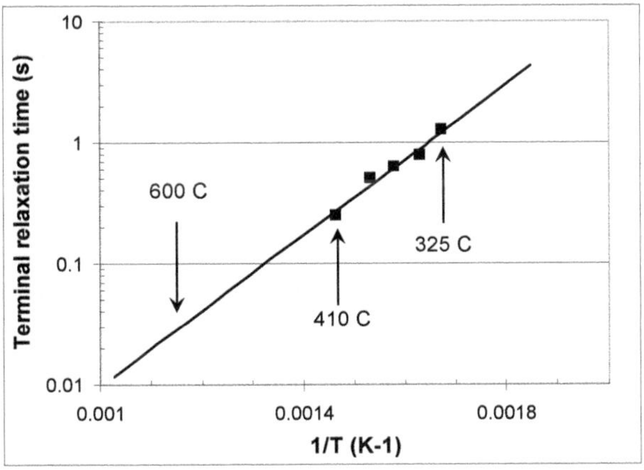

Figure 8.9. *Variation of Vitrex PEEK G450 relaxation time*

8.4.4. *Modeling of thermal aging*

According to the literature [DAY 89, COL 93], crosslinking apparently results from radical processes and predominates during PEEK thermal aging under nitrogen. It is expected to disadvantage welding because it increases the melt viscosity. To study crosslinking kinetics, real viscosity η' was measured in the 360–460°C temperature range at low frequency in order to remain in Newtonian regime. Very small variations were observed, after 6 h at 360°C, whereas the viscosity increases by almost two orders of magnitude after about 2 h at 460°C.

Far from the gel point at low conversion, the number of crosslinking events per mass unit Rc varies from initial time (subcript 0) as:

$$R_c = \frac{1}{M_{n_0}} - \frac{1}{M_n} = \frac{I_{P_0}}{M_{w_0}} - \frac{I_P}{M_w} \quad [8.20]$$

where M_n and M_w are, respectively, the number and weight average of molar mass and IP is the polydispersity index.

It will be assumed that $I_P \sim I_{P_0}$ and viscosity still obeys the power law established for linear chains:

$$\eta' = KM_w^{3.4} \quad [8.21]$$

where K is a temperature-dependent material property. More sophisticated relationships taking into account the specific influence of branched architecture in crosslinked polymers on their viscosity are available, but it seems the above equations are sufficient to predict the trend of viscosity variations. Thus, the crosslink concentration Rc can be given by:

$$R_c = \frac{I_{P_0}}{M_{w_0}}\left(1-\left(\frac{\eta_0'}{\eta'}\right)^{1/3.4}\right) \qquad [8.22]$$

R_c was plotted against exposure time in Figure 8.10(a), it varies quasi-linearly with time, even at 460°C, in the first hour of exposure.

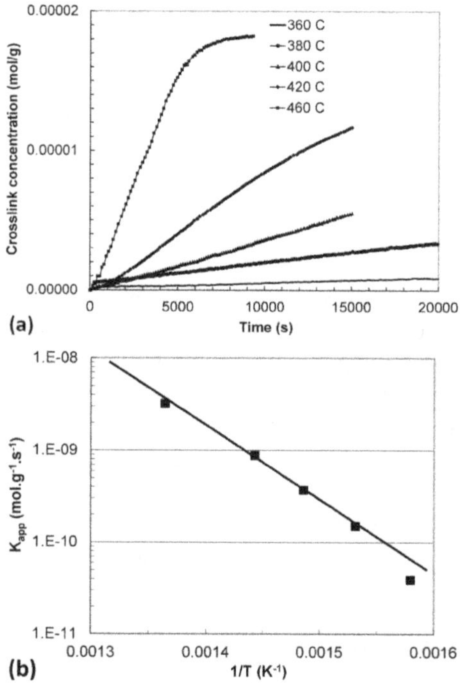

Figure 8.10. *a) Evolution of crosslink concentration R_c with time for different temperatures, b) variation of crosslink kinetic rate with temperature*

The slope of the linear domain gives the crosslinking kinetic rate K_{app}, thus the crosslinking kinetics can be approximated by a zero-order process:

$$R_c = K_{app} t \qquad [8.23]$$

where it appears that the kinetic rate obeys to Arrhenius law (Figure 8.10(b)):

$$K_{app} = A\exp\left(-\frac{E_a}{RT}\right) \qquad [8.24]$$

with $A = 3384$ mol.g^{-1}.s^{-1} and $E_a = 168$ kJ.mol^{-1}.

As for diffusion in an anisotherm process, it is possible to define the crosslink concentration R_c representing the extent of thermal aging in a given point:

$$R_c = A\int_{t_{melting\ \&\ contact}}^{t_{crystallisation}} \exp\left(-\frac{E_a}{R(T(t))}\right)dt \qquad [8.25]$$

This calculation was implemented in the thermal modeling of the process (next section), thus it was possible to plot iso-aging curves a process window. Peel tests were done on composite tapes consolidated and aged in autoclave for different temperatures. It was found that thermal aging begins to hinder adhesion for $R_c = 10^{-7}$ mol.g^{-1}.

8.4.5. *Thermal modeling of the process*

As thermal conductivity is 10 times higher in the fiber direction than in the transverse direction, a time-dependent 2D model has been built to predict the evolution of temperature during the process. The 2D model written in Matlab® uses a finite volume formulation with an explicit scheme. The heat transfer from hot nitrogen to the material is characterized by a heat transfer coefficient h_{torche} over a length l_{torche} before the roller applies the pressure (Figure 8.11). The interface is considered inexistent in the welded zone so that thermal exchanges are only due to conduction. The roller imposes a local surface heat flow characterized by its heat transfer coefficient h_{roller} and the contact length l_{roller}.

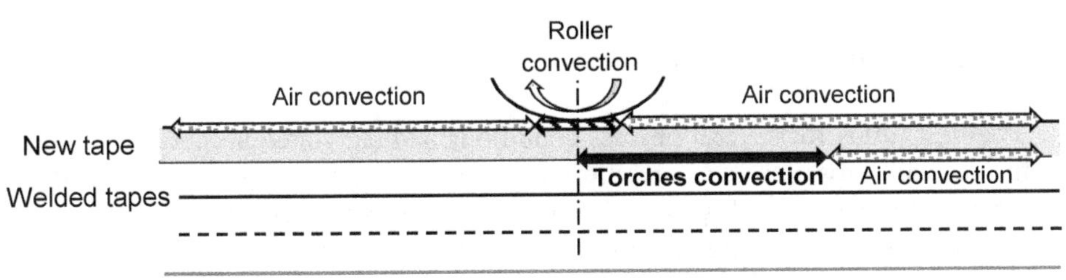

Figure 8.11. *Principle of thermal modeling*

The global heat equation contains two source terms relative to crystallization and melting. Both source terms were identified from DSC characterizations and crystallization kinetic has been implemented to define more precisely the end of macromolecular diffusion process [NIC 05].

Small thermocouples inserted between the plies have allowed both model calibration by adjusting the different heat transfer coefficients and temperature evolution validation at temperature measurement locations.

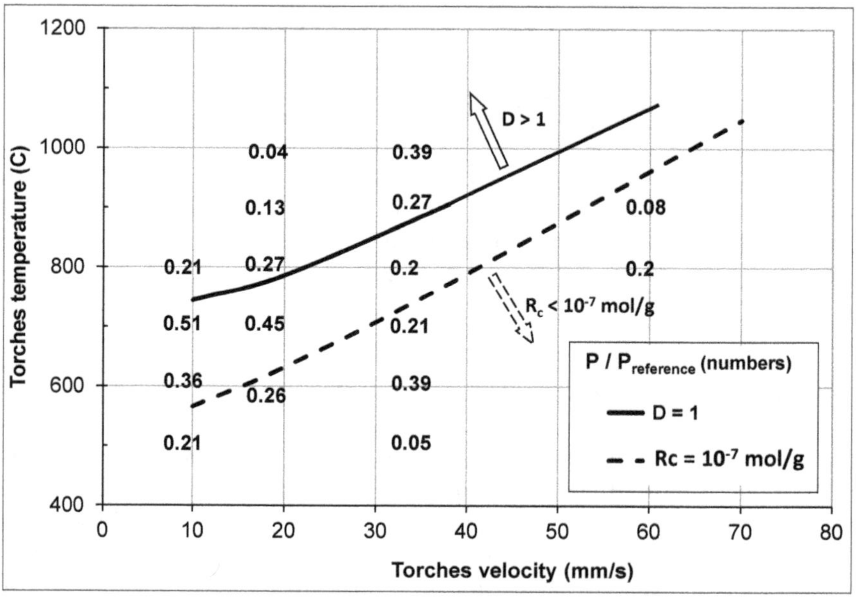

Figure 8.12. *Velocity-temperature process diagram. The numbers are the ratio $P/P_{reference}$ between process peel test values and peel test reference values obtained for a tape consolidated in autoclave. Arrows show valid domains for diffusion and crosslink concentration criteria*

8.4.6. *Weldability prediction*

Peel test ratios for several process conditions and simulated acceptable limits of diffusion and crosslink concentration criteria, respectively, D = 1 and R_c = 10^{-7} mol/g are reported in Figure 8.12. According to this process diagram, there is no intersection between acceptable macromolecular diffusion domain and crosslink concentration domain. In other words, there is no way to obtain a satisfying weldability window for this process. This is confirmed by the results of peel tests: in the best case, (Torches = 700 C, Velocity = 10 mm.s^{-1}), the peel ratio is twice smaller than for autoclave consolidated samples.

The simulated results obtained by our model explain the measured mechanical properties over a large domain of processing conditions. This type of modeling could be used advantageously to optimize or design other variants of this process, in with another heating system such as laser heating.

8.5. Application to ultrasonic welding

8.5.1. *Process description and time scale separation*

Among the different available welding techniques, ultrasonic welding is of particular interest because of its energy efficiency and rapidity. Furthermore, it does not require the presence of a foreign metallic grid like in resistance or electrofusion welding. In this process, in order to achieve the local heating, triangular bulges, called energy directors, are molded on one of the plates to be welded (Figure 8.13). The assembly is then placed under a tool called sonotrode, which applies a constant load and a harmonic ultrasonic compression simultaneously. The triangular shape is obviously the weakest part of the structure and is thus designed to concentrate the deformation. This localized strain combined with the high loading frequency induces heating because of viscous dissipation [BEN 86, NON 96, TOL 83]. The director then melts and flows at the interface to allow welding, through the formerly described physical mechanisms of intimate contact and macromolecular diffusion.

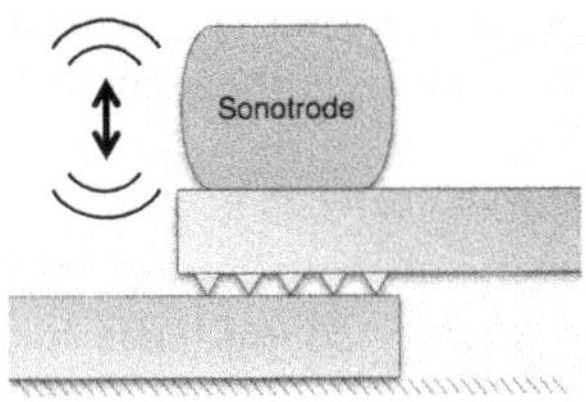

a)

b)

Figure 8.13. *Principle of the ultrasonic welding process with molded "energy directors" a) and picture of a stiffener for aeronautic applications b)*

Initial results reveal a good mechanical quality of the welding [SOC 07]. In particular, the advance of the sonotrode enables air removal along the director and avoids the trapping of bubbles. Moreover, it would allow us to assemble large parts using continuous weld lines, where the "static" process (without sonotrode motion) limits the welding area to the size of the sonotrode. This opens possibilities for this process to be used at an industrial level to assemble large parts keeping an excellent weld quality.

Nevertheless, this technique based on energy directors was not so successful due to the costs of machining and additional surface molding. More recently, an energy director-free solution, combining dry friction and viscoelastic heating, was proposed by Villegas [VIL 13] and modeled by Levy [LEV 14]. It was proved to be efficient and should be also extendable to continuous welding.

8.5.2. *Process modeling: necessity of a time homogenization framework*

The main phenomena occurring in this process, initially outlined by Benatar and Gutowski [YAR 98], can be summarized as:

1) mechanics and vibration of the parts;

2) viscoelastic heating;

3) heat transfer;

4) flow and wetting;

5) intermolecular diffusion.

Like in every welding process, the modeling of the ultrasonic technique is a complex multi-physical problem implying heat transfer, phase change, thermo-physical evolutions, adhesion, fluid flow, etc. However, a unique feature of this process is the high frequency of the loading (about 20 kHz) that makes any direct calculation unaffordable, especially in regard of the nonlinear nature of the multi-physical problem.

To circumvent this problem, a time-homogenization technique was proposed by Levy *et al.* [LEV 11] that provides a mathematically sound basis for such type of problems, to the contrary of more empirical "cycle jump" methods. It takes advantage of the good time scales separation between the welding characteristic time t_m~1 s and the period of the ultrasonic vibration t_μ=50 μs, and first introduces the scale separation factor

$$\xi = \frac{t_\mu}{t_m} = 10^{-5} . \qquad [8.26]$$

Such a good separation is rarely encountered in similar spatial homogenization techniques [BEN 78, BAK 89, MOR 95] and simply means that problems at short times can probably be well uncoupled from long time ones. All the fields

φ (temperature, displacement and velocity) are then rewritten as double-scale asymptotic expansions in powers of ξ:

$$\varphi(\boldsymbol{x},t) = \varphi_0(\boldsymbol{x},t,\tau) + \xi\varphi_1(\boldsymbol{x},t,\tau) + \xi^2\varphi_2(\boldsymbol{x},t,\tau) + \cdots, \qquad [8.27]$$

where $\tau = t/\xi$ denotes the "fast" time scale. These expansions are then introduced in the energy and momentum balances and the equations at the different orders of ξ can be identified.

On the basis of a simple Maxwell fluid rheological model for the energy directors, Levy *et al.* [LEV 11] showed that the thermomechanical problem with a vibrating imposed load can be separated into three coupled problems: (1) a thermal problem at long time scale, (2) a mechanical problem at long time scales that describes the squeezing of the energy directors and (3) a mechanical problem at short time scales. The latter happens to be an elastic problem; it results from the very fast loading of the polymer at high frequency, which in this case behaves like a solid. The first two problems are rather classical, except that the heat source of the thermal problem is directly induced by the viscoelastic dissipation resulting from problem (3). The interest of this approach is that the short times can be solved separately, thus saving a huge amount of time. For example, in the case of a large strain time-dependent problem, the elastic problem for the self-heating can be solved once only every "macro-temporal" time step. It varies only due to the shape change of the domain. The partition into three problems can furthermore be shown to be valid over a wide range of temperature.

8.5.3. *Numerical multi-physical model*

Retaining the principle of the theoretical analysis provided by the time homogenization framework, Levy *et al.* [LEV 11] proposed the following set of balance and constitutive equations, dedicated to the simulation of self-heating and squeezing of energy directors made of PEEK:

– long times mechanical problem: Stokes problem with a power-law fluid, solved in terms of the velocity $\mathbf{v}$; determination of the shape evolution of the energy director.

$$\nabla \cdot \boldsymbol{\sigma}^v - \nabla \mathrm{p} = 0 \text{ with } \nabla \cdot \mathbf{v} = 0 \text{ and } \boldsymbol{\sigma}^v = 2\eta(\dot{\gamma})^{n-1}\boldsymbol{D}, \qquad [8.28]$$

where $\boldsymbol{\sigma}^v$ denotes the viscous stress tensor, $\boldsymbol{D}$ denotes the strain rate tensor, p denotes the hydrostatic pressure, $\dot{\gamma}$ denotes the equivalent shear rate, η denotes the consistency and n denotes the power law index;

– short times mechanical problem: elastostatics; determination of vibrations amplitude.

$$\nabla \cdot \boldsymbol{\sigma}^e = 0, \text{ with } \boldsymbol{\sigma}^e = \lambda \text{trace}(\epsilon^e) + 2\mu\epsilon^e, \qquad [8.29]$$

where $\boldsymbol{\sigma}^e$ denotes the elastic stress tensor, ϵ^e denotes the strain tensor and (λ, μ) denotes the Lamé elasticity coefficients;

– long times thermal problem: heat transfer problem with a source term Qaccounting for the heat produced by self-heating.

$$\rho C\left(\frac{\partial T}{\partial t} + \mathbf{v} \cdot \boldsymbol{\nabla} T\right) = \boldsymbol{\nabla} \cdot (k \boldsymbol{\nabla} T) + Q, \text{ with } Q = \frac{1}{2} E'' \omega \epsilon^e : \epsilon^e, \qquad [8.30]$$

where ρ denotes the density, C denotes the heat capacity, k denotes the conductivity, T denotes the temperature field, E'', denotes the loss modulus and $\omega = 2\pi f_{sono}$ denotes the sonotrode pulsation.

Like in most welding processes, obtaining the material data of such multi-physical models is a key problem in order to achieve a realistic simulation tool. As the temperature ranges from room temperature to melting temperature, the thermo-dependence of all parameters has to be thoroughly determined experimentally. Particularly difficult is the determination of the loss modulus E'' at 20 kHz which is the usual ultrasonic frequency. No conventional apparatus is able to perform such viscoelastic characterization but a time-temperature equivalence is possible, in order to extrapolate results from a classical test at 100 Hz [LEV 14].

The above-described coupled problem implies an original coupled numerical problem. As explained in [LEV 11], the first issue is to handle the large strain of the energy directors. This can be done efficiently by the choice of a Eulerian description (fluid-type approach), coupled to a level-set method to describe the evolution of the interface of the flowing energy director. Unlike the two mechanical problems, the thermal problem includes a transient term and an advection term. To simplify the resolution, Levy *et al.* [LEV 11, LEV 14] adopted an operator splitting technique, which provides an efficient way to solve separately the convective part. Following such an approach, the multi-physical problem formed by equations [8.28]–[8.30] can be solved on the domain at time t, either with an iterative or a monolithic approach.

As illustrated in Figure 8.14, the simulation code developed by Levy *et al.* [LEV 11] is able to produce acceptable physical results compared to the experimental observations.

a)

b)

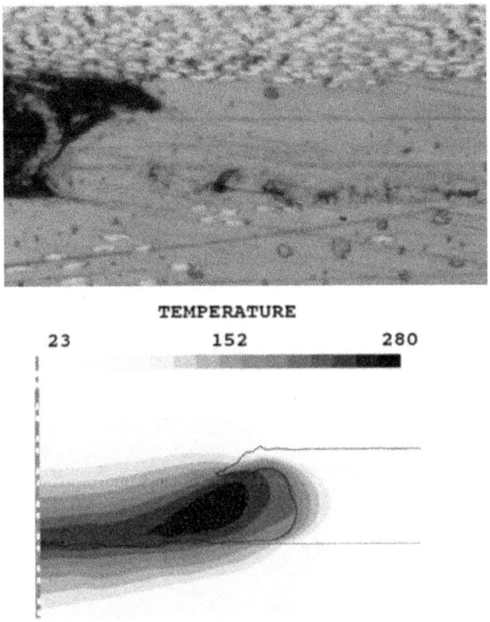

Figure 8.14. *Comparison between a stopped experiment and the reference simulation – temperature field after 0.12 s [LEV 11]. For a color version of the figure, see www.iste.co.uk/boyard/heat.zip*

8.5.4. *Ultrasonic welding with energy directors: process analysis and optimization*

Once the numerical tool is validated, it is possible to use it to better understand the effects of the numerous process and physical parameters: vibration frequency, holding force, welding duration and geometry of the directors.

Figure 8.15 first focuses on the maximum temperature reached in the initial heating stage. Before the glass transition temperature T_g, the value of the holding force has almost no effect because the shape remains the initial one. After T_g, a high holding force is observed to lower the maximum temperature. In fact, it promotes a fast squeezing, but subsequently also rapidly reduces the strain concentration effect. In order to reach the melting temperature as fast as possible, a rather low holding force would be required, contrary to the first intuition. Figure 8.15(b) illustrates the high importance of the "tip effect". The strain concentration and subsequent self-heating are directly linked to the sharpness of the energy director. In the limit of what machining would allow, a very tip radius is required.

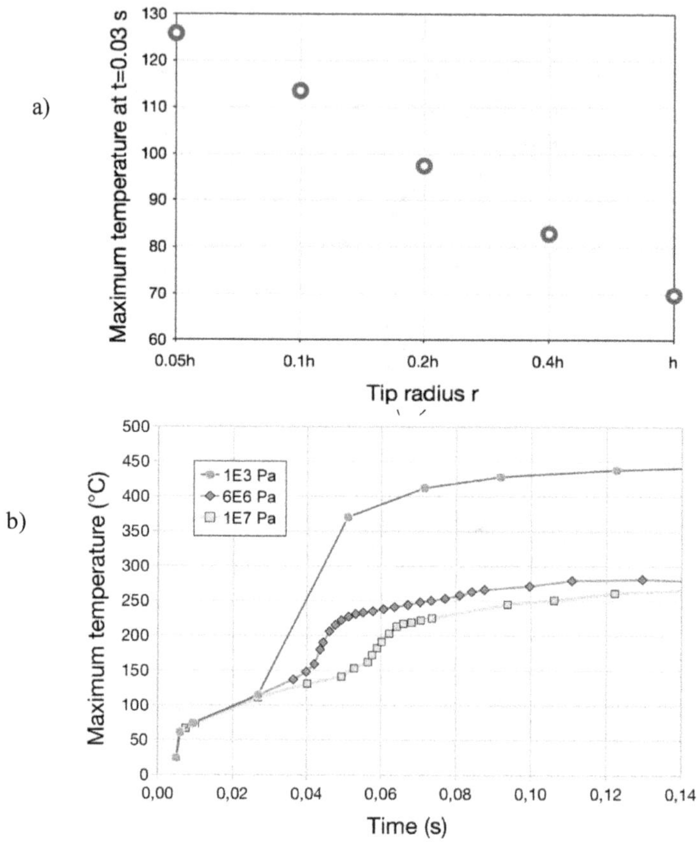

Figure 8.15. *Process analysis and optimization: a) effect of the holding force on the maximum temperature, and b) effect of the energy director's tip radius [LEV 14]. For a color version of the figure, see www.iste.co.uk/boyard/heat.zip*

The effects of triangular directors sharpness are illustrated in Figure 8.16. In the heating stage (Figure 8.16(a)), it is shown that a sharper director promotes a more localized heat generation and temperature distribution. The consequence on the flow stage is then shown in Figure 8.16(b). Comparing the flow fronts at t = 0.21 s, we can check that the sharp director induces a faster flow, and therefore a shorter welding time. Nevertheless, the analysis of flow fronts during squeezing also suggests a higher risk of porosity entrapment. Optimal shapes of these energy directors should therefore combine the antagonist features observed with triangular shapes. They should have a more complex cross-section, made of a sharp part, close to the tip, progressively moving into a larger part near the basis.

a)

b)

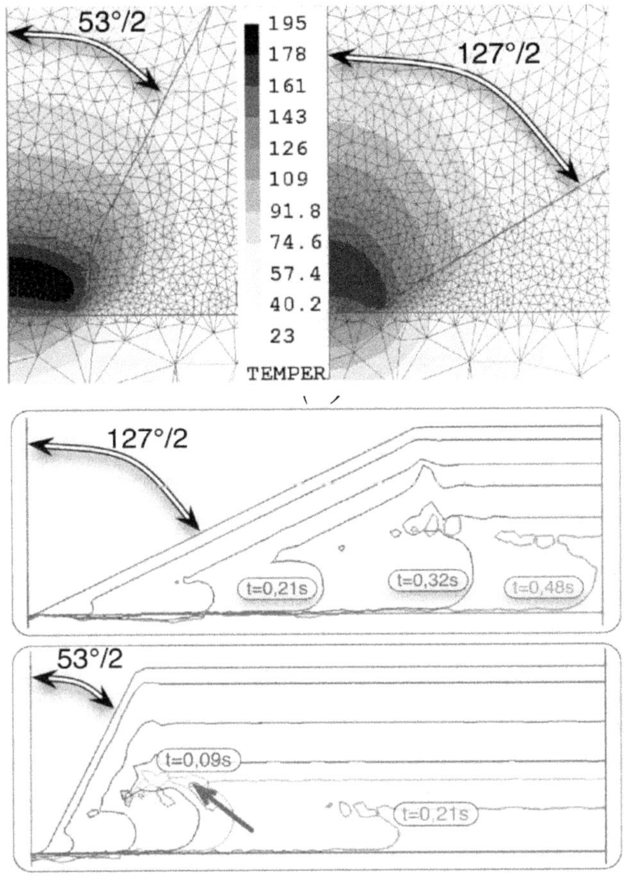

Figure 8.16. *Effect of the geometry of energy directors: a) temperature distribution and b) flow front at different instants [LEV 14]. For a color version of the figure, see www.iste.co.uk/boyard/heat.zip*

Finally, the classical healing models can also be introduced in such complex simulation tools and provide information about the expectable quality of adhesion. Equation like [8.19] is an ODE that can be easily solved on the flow front of the energy director, as soon as it comes in contact with the lower substrate. Provided a representative law for the characteristic time $\lambda(T)$ and a good calculation of the temperature field, a diffusion criterion or a healing degree can be estimated, as shown in Figure 8.17.

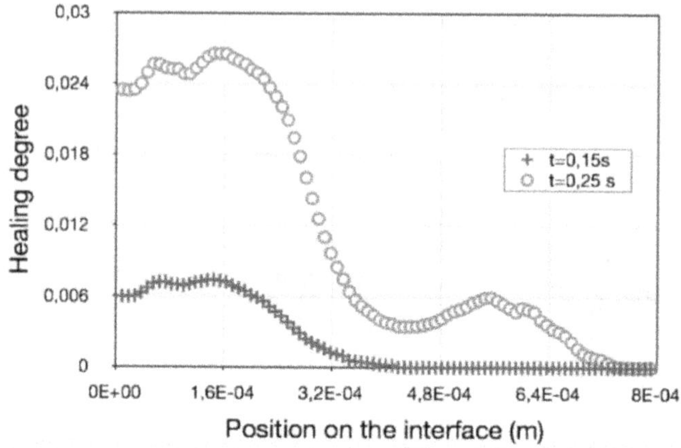

Figure 8.17. *Predicted healing degree for the US welding of PEEK triangular energy directors [LEV 14]*

8.6. Conclusion

Optimizing healing of welded thermoplastic material or matrix requires the understanding of physical mechanisms involved in welding processes, which are mainly surface rearrangement and surface approach, wetting of surfaces, macromolecular diffusion and even cocrystallization at interface of semi-crystalline polymers. As it is quite difficult to separate wetting phenomenon from surface rearrangement, as it seems that time scale of wetting phenomenon is one order of magnitude less than surface rearrangement, most researchers have modeled these two stages into one stage to obtain the intimate contact which is necessary to allow macromolecular diffusion. For not fully welded semi-crystalline polymers, co-crystallization at interface surely increases the strength of weld lines, but in the state of current scientific knowledge, it is not possible to quantify the role of partial diffusion and co-crystallization in the strength of weld lines. Therefore, it is reasonable for safety reasons of an industrial process to consider that diffusion should be complete before crystallization.

In the last decade, the growing use of thermoplastic matrices inside composite materials has reinforced the interest of the community for the understanding and modeling of the two essential physical phenomena: intimate contact and macromolecular diffusion. As these two physical phenomena are very temperature-dependent, precise local temperature measurements or a thermal modeling of processes is necessary to record the thermal history at interfaces in order to quantify these two physical phenomena. As it was shown in this chapter through successful technological applications, these concepts can now be integrated in the modeling of welding processes in order to predict the quality of adhesion in rather realistic situations. It is rather clear that the theory for describing the healing in that way is becoming rather mature. As thermoplastics may degrade quite quickly in the melted state at high temperature, i.e. have molecular masses, which vary either by polymer chain cut or crosslinking, molecular mobility is widely modified, therefore intimate contact and diffusion may be affected. Thus, it is important to take into account this possible degradation and to model it to refine processing windows.

Regarding the intimate contact, no model takes into account neither the possible non-Newtonian rheology of the polymer nor the tri-dimensional nature of the surfaces roughness. Though some of them try to account more precisely for the statistical distribution of roughness, the squeezing mechanics remain based on a lubrication assumption, which is questionable. Another critical issue, which was not discussed in this chapter, is the experimental validation of those physical models knowing that the only way to estimate the bonding degree is mechanical testing, often through single lap shear tests or crack opening tests. No in-line measurement of D_{ic}, D_h or diffusion criterion D is available, especially at short times. This is the reason why the control of the duration of application of welding pressure and heating should be very precise, which represents an important challenge.

8.7. Acknowledgments

The authors acknowledge DGA and EADS for their financial support. We wish to thank Célia Nicodeau, Arthur Levy, Jacques Verdu, Jacques Cinquin and Francisco Chinesta for their contribution to the presented works.

8.8. Bibliography

[AGE 01] AGEORGES C., YE L., HOU M., "Advances in fusion bonding techniques for joining thermoplastic matrix composites: a review", *Composites Part A: Applied Science and Manufacturing*, vol. 32, pp. 839–857, 2001.

[ATK 89] ATKISON J.R., WARD I.M., "The joining of biaxially oriented polyethylenepipes", *Polymer Engineering and Science*, vol. 29, pp. 1638–1641, 1989.

[BAK 89] BAKHVALOV N.S., PANASENKO B., "Homogenisation: averaging processes in periodic media–mathematical problems in the mechanics of composite materials", *Mathematics and its Applications Soviet Series*, vol. 36, Kluwer, Dordrecht, 1989.

[BEN 78] BENSOUSSAN A., LIONS J.-L., PAPANICOLAOU G., *Asymptotic Analysis for Periodic Structures*, North-Holland, Amsterdam, 1978.

[BEN 86] BENATAR A., GUTOWSKI T.G., "Methods for fusion bonding thermoplastic composites", *SAMPE Q*, vol. 18, pp. 35–42, 1986.

[BUT 98] BUTLER C.A., MCCULLOUGH R.L., PITCHUMANI R. *et al.*, "An analysis of mechanisms governing fusion bonding of thermoplastic composites", *Journal of Thermoplastic Composite Materials*, vol. 11, pp. 338–363, 1998.

[CHE 10] CHEN Y., FU P., ZHANG C. *et al.*, "Numerical simulation of laminar heat transfer in microchannels with rough surfaces characterized by fractal Cantor structures", *International Journal of Heat and Fluid Flow*, vol. 31, pp. 622–629, 2010.

[COG 89] COGSWELL F.N., MEAKIN P.J., SMILEY A.J. *et al.*, "Thermoplastic interlayer bonding for aromatic polymer composites", *34th International SAMPE Symposium*, pp. 2315–2325, 1989.

[COI 05] COIFFIER-COLAS C., SIBOIS H., LEFEBURE P., "Future for high performance thermoplastic composites and aircraft helicopters", *Proceedings of JEC Composites*, Paris, pp. 14–19, 2005.

[COL 93] COLE K.G., CASELLA I.G., "Fourier transform infra-red spectroscopic study of thermal degradation in poly(ether ether ketone)-carbon composites", *Polymer*, vol. 34, pp. 740–745, 1993.

[DAR 85] DARA P.H., LOOS A.C., Thermoplastic matrix composite processing model, Virginia Polytechnic Institute Report, CCMS-85-10, Hampton, VA, 1985.

[DAY 89] DAY M., COONEY J.D., WILES D.M., "Kinetic study of the thermal decomposition of poly(aryl-ether-ether-ketone) (PEEK) in nitrogen", *Polymer Engineering and Science*, vol. 29, pp. 19–22, 1989.

[FAY 03] FAYOLLE B., AUDOUIN L., VERDU J., "Radiation induced embrittlement of PTFE", *Polymer*, vol. 44, pp. 2773–2780, 2003.

[FER 80] FERRY J.D., *Viscoelastic Properties of Polymers*, 3rd ed., Wiley, New York, 1980.

[FET 99] FETTERS L.J., LOHSE D.J., MILNER S.T. *et al.*, "Packing length influence in linear polymer melts on the entanglement, critical, and reptation molecular weights", *Macromolecules*, vol. 32, pp. 6847–6851, 1999.

[FET 07] FETTERS L.J., LOHSE D.J., COLBY R.H., "Chain dimensions and entanglement spacings", in MARK J.E. (ed.), *Physical Properties of Polymers Handbook–Second Edition*, Springer, New York, 2007.

[FLO 49] FLORY P.J., "The configuration of real polymer chains", *Journal of Chemical Physics*, vol. 17, pp. 303–310, 1949.

[GEN 71] DE GENNES P.G., "Reptation of a polymer chain in the presence of fixed obstacles", *Journal of Chemical Physics*, vol. 55, pp. 572–579, 1971.

[GEN 79] DE GENNES P.G., *Scaling Concepts in Polymer Physics*, Cornell University Press, New York, 1979.

[GEN 97] GENT A., KIM E.G., YE P., "Autohesion of crosslinked polyethylene", *Journal of Polymer Science Part B: Polymer Physics*, vol. 35, pp. 615–622, 1997.

[GRA 74] GRAESSLEY W.W., "The entanglement concept in polymer melt rheology", *Advances in Polymer Science*, vol. 16, pp. 1–179, 1974.

[GRA 80] GRAESSLEY W.W., "Some phenomenological consequences of the Doi–Edwards theory of viscoelasticity", *Journal of Polymer Science. Phys. Ed.*, vol. 18, pp. 27–34, 1980.

[GRI 90] GRIMM R.A., "Fusion welding techniques for plastics", *Welding Journal*, vol. 69, pp. 23–28, 1990.

[GRI 95] GRIMM R.A., "Welding processes for plastics", *Advanced Material Processes*, vol. 147, pp. 27–30, 1995.

[GRU 12] GRUBER M.B., LOCKWOOD I.Z., DOLAN T.L. *et al.*, "Thermoplastic in situ placement requires better impregnated tapes and tows", *Proceedings of the SAMPE Conference*, Baltimore, USA, vol. 57, pp. 2164–2178, 2012.

[KAU 89] KAUSCH H.H., TIRRELL M., "Polymer interdiffusion", *Annual Review of Materials Science*, vol. 19, pp. 341–377, 1989.

[KEN 92] KENNEY E., "Designing plastic parts for ultrasonic assembly", *Machine Design*, vol. 21, pp. 65–68, 1992.

[LAM 05] LAMETHE J.-F., BEAUCHENE P., LEGER L., "Polymer dynamics applied to PEEK matrix composite welding", *Aerospace Science and Technology*, vol. 9, no. 3, pp. 233–240, 2005.

[LEE 84] LEE W.I., SPRINGER G.S., "Interaction of electromagnetic radiation with organic matrix composites", *Journal of Composite Materials*, vol. 18, pp. 357–386, 1984.

[LEE 87] LEE W.I., SPRINGER G.S., "A model of the manufacturing process of thermoplastic matrix composites", *Journal of Composite Materials*, vol. 21, pp. 1017–1055, 1987.

[LEV 11a] LEVY A., LE CORRE S., CHEVAUGEON N. *et al.*, "A level set based approach for the finite element simulation of a forming process involving multiphysics coupling: ultrasonic welding of thermoplastic composites", *European Journal of Mechanics*, vol. 9, pp. 501–509, 2011.

[LEV 11b] LEVY A., LE CORRE S., POITOUA. *et al.*, "Ultrasonic welding of thermoplastic composites, modeling of the process using time homogenization", *Int J. Multiscale Comput. Engrg.*, vol. 9, pp. 53–72, 2011.

[LEV 14a] LEVY A., HEIDER D., TIERNEY J. *et al.*, "Inter-layer thermal contact resistance evolution with the degree of intimate contact in the processing of thermoplastic composite laminates", *Journal of Composite Materials*, vol. 48, pp. 491–503, 2014.

[LEV 14b] LEVY A., LE CORRE S., POITOU A., "Ultrasonic welding of thermoplastic composites: a numerical analysis at the mesoscopic scale relating processing parameters, flow of polymer and quality of adhesion", *International Journal of Material Forming*, vol. 7, pp. 39–51, 2014.

[LEV 14c] LEVY A., LE CORRE S., FERNANDEZ-VILLEGAS I., "Modelling of the heating phenomena in ultrasonic welding of thermoplastic composites with flat energy directors", *Journal Materials Processing Technology*, vol. 214, pp. 1361–1371, 2014.

[MAG 89] MAGUIRE D.M., "Joining thermoplastic composites", *SAMPE J.*, vol. 25, pp. 11–14, 1989.

[MAN 92] MANTELL S.C., SPRINGER G.S., "Manufacturing process models for thermoplastic composites", *Journal of Composite Materials*, vol. 26, pp. 2348–2377, 1992.

[MOR 95] MOREAU G., CAILLERIE D., "Continuum modeling of lattice structures in large displacement framework", *Polymer Engineering and Science*, vol. 20, pp. 53–70, 1995.

[NIC 05a] NICODEAU C., Modélisation du soudage en continu des composites à matrice thermoplastique, PhD Thesis, Ecole Nationale Supérieure d'Arts et Métiers de Paris, available at: http://tel.archives-ouvertes.fr/ pastel-00001506, 2005.

[NIC 05b] NICODEAU C., RÉGNIER G., VERDU J. *et al.*, "Welding of thermoplastic matrix composites: prediction of macromolecules diffusion at the interface", *Proceedings of the 8th ESAFORM Conference*, Cluj Napoca, Roumania, pp. 897–900, 2005.

[NON 96] NONHOF C., LUITEN G., "Estimates for process conditions during the ultrasonic welding of thermoplastics", *Polym. Eng. Sci.*, vol. 36, pp. 1177–1183, 1996.

[REG 07] RÉGNIER G., NICODEAU C., VERDU J. *et al.*, "A multi-physic and multi-scale approach to model the continuous welding of thermoplastic matrix composites", *Proceedings of ICCM16*, Kyoto, Japan, pp. 1–6, 2007.

[ROU 71] ROUSE P.E., "A theory of the linear viscoelastic properties of dilute solutions of coiling polymers", *Journal of Chemical Physics*, vol. 55, pp. 572–579, 1971.

[RUB 03] RUBINSTEIN M., COLBY R.H., *Polymer Physics*, Oxford University Press, Oxford, 2003.

[SAI 89] SAINT-ROYRE D., GUEUGNAUT D., REVERET D., "Test methodology for the determination of optimum fusion welding conditions of polyethylene", *Journal of Applied Polymer Science*, vol. 38, pp. 147–162, 1989.

[SCH 09] SCHELL J.S.U., GUILLEMINOT J., BINETRUY C. *et al.*, "Computational and experimental analysis of fusion bonding in thermoplastic composites: influence of process parameters", *Journal of Materials Processing Technology*, vol. 209, pp. 5211–5219, 2009.

[SOC 07] SOCCARD E., Ultrasonic assembly method, Patent no. WO 2007/003626 A1 WO 2007/003626 A1, EADS CCR, 2007.

[SON 97] SONMEZ F., HAHN H., "Analysis of the on-line consolidation process in thermoplastic composite tape placement", *Journal of Thermoplastic Composite Materials*, vol. 10, p. 543, 1997.

[TAY 88] TAYLOR N.S., WATSON M.N., "Welding techniques for plastics and composites", *Joining & Materials*, vol. 1, pp. 70–73, 1988.

[TIE 06] TIERNEY J., GILLESPIE J.W., "Modeling of in situ strength development for the thermoplastic composite tow placement process", *Journal of Composite Materials*, vol. 40, pp. 1487–1506, 2006.

[TOL 83] TOLUNAY M.N., DAWSON P.R., WANG K.K., "Heating and bonding mechanisms in ultrasonic welding of thermoplastics", *Polymer Engineering and Science*, vol. 23, pp. 726–733, 1983.

[TSA 09] TSAO C.-W., DEVOE D., "Bonding of thermoplastic polymer microfluidics", *Microfluidics and Nanofluidics*, vol. 6, pp. 1–16, 2009.

[VAN 07] VAN RUYMBEKE E., LIU C.-Y., BAILLY C., "Quantitative tube model predictions for the linear viscoelasticity of linear", *Rheology Reviews*, pp. 53–134, 2007.

[VIL 13] VILLEGAS I.F., "In situ monitoring of ultrasonic welding of thermoplastic composites through power and displacement data", *Journal of Thermoplastic Composite Materials*, vol. 24, pp. 1007–1024, 2013.

[VIN 89] VINSON J.R., "Adhesive bonding of polymer composites", *Polymer Engineering and Science*, vol. 29, pp. 1325–1331, 1989.

[WAR 96] WARREN T.L., MAJUMDAR A., KRAJCINOVIC D., "A fractal model for the rigid-perfectly plastic contact of rough surfaces", *Journal of Applied Mechanics*, vol. 63, pp. 47–54, 1996.

[WIL 90] WILLIAMS G., GREEN S., MCAFEE J. *et al.* "Induction welding of thermoplastic composites", *IMechE*, vol. C400/034, pp. 133–136, 1990.

[WOO 81] WOOL R.P., O'CONNOR K.M., "A theory of crack healing in polymers", *Journal of Applied Physics,* vol. 52, pp. 5953–5963, 1981.

[WOO 83] WOOL R.P., O'CONNOR K.M., "A theory of healing at apolymer interface", *Macromolecules*, vol. 16, pp. 1115–1120, 1983.

[WOO 89] WOOL R.P., YUAN B.L., MCGAREL O.J., "Welding of polymer interfaces", *Polymer Engineering and Science*, vol. 29, pp. 1340–1367, 1989.

[XUE 00] XUE Y.-Q., TERVOORT T.A., RASTOGI S. *et al.*, "Welding behavior of semicrystalline polymers. 2. Effect of cocrystallization on autohesion", *Macromolecules*, vol. 33, pp. 7084–7087, 2000.

[YAN 01a] YANG F., PITCHUMANI R., "A fractal Cantor set based description of interlaminar contact evolution during thermoplastic composites processing", *Journal of Materials Science*, vol. 36, pp. 4661–4671, 2001.

[YAN 01b] YANG F., PITCHUMANI R., "Fractal description of interlaminar contact development during thermoplastic composites processing", *Journal of Reinforced Plastics and Composites*, vol. 20, pp. 536–546, 2001.

[YAN 02a] YANG F., PITCHUMANI R., "Healing of thermoplastic polymers at an interface under nonisothermal conditions", *Macromolecules*, vol. 35, pp. 3213–3224, 2002.

[YAN 02b] YANG F., PITCHUMANI R., "Interlaminar contact development during thermoplastic fusion bonding", *Polymer Engineering and Science*, vol. 42, pp. 424–438, 2002.

[YAN 03] YANG F., PITCHUMANI R., "Nonisothermal healing and interlaminar bond strength evolution during thermoplastic matrix composites processing", *Polymer Composites,* vol. 24, pp. 263–278, 2003.

[YAR 98] YARLAGADDA P.K., CHAI T.C., "An investigation into welding of engineering thermoplastics using focused microwave energy", *Journal of Materials Processing Technology*, vol. 74, pp. 199–212, 1998.

[YUA 90] YUAN B.L., WOOL R.P., "Strength development at incompatible semicrystalline polymer interfaces", *Polymer Engineering Science*, vol. 30, pp. 1454–1464, 1990.

[ZAN 01] ZANETTO J.E., PLUMMER C.J.G., MANSON J.A., "Fusion bonding of polyamide 12", *Polymer Engineering and Science*, vol. 41, pp. 890–897, 2001.

9

Multiphysics for Simulation of Forming Processes

This chapter is dedicated to the simulation of composite forming processes. It concerns numerical methods for molding, at the different material scales, involving multiphysics analysis. From the different possible numerical techniques, we focus on a stabilized finite element method for the resolution of the continuum equations (mass, momentum and energy conservation), in a Eulerian framework, with a level-set approach for interface displacement treatment. Internal variables, such as the reticulation rate or fiber fraction and orientation may be computed using models defined through ordinary differential equations. Examples at the process scale, as well as at the composite inner structure scales illustrate the capabilities of such numerical techniques.

9.1. Introduction

In recent years, the use of composite materials has been widely diffused and continues to evolve in the aeronautics, aerospace and automotive industries. Most of the composites used are composed of an organic matrix and fiber reinforcements, ensuring a certain level of physical properties, designed taking into account the material's anisotropy, with a large weight reduction when compared to traditionally used metallic alloys. Thermoplastic or thermoset may be used as resins, whereas we may have glass, carbon or even natural-based fibers.

To produce composite parts, processes may be more generally grouped into three categories [ADV 09, ADV 10]. The first concerns *short fiber reinforced matrices*, involving fiber transport with resin flow, as a non-Newtonian suspension being *injected* into a closed mold or through a die. The second involves highly viscous thermoplastics or thermoset resins containing *continuous or long discontinuous fibers*, as well as high fiber content rate. In this case, the reinforcement and the resin

Chapter written by Luisa SILVA, Patrice LAURE, Thierry COUPEZ and Hugues DIGONNET.

are heated and deform together under an applied (compression) stress to form the required composite shape like, for example, in *compression molding* of sheet molding compounds. Finally, the third category concerns porous media processes, when the reinforcement is composed of continuous fiber networks or textiles placed in a (closed or open) mold into which a low viscosity resin is injected or placed, heated and compressed, in such a way that it impregnates the fibrous architecture.

Complex multiphysics phenomena, strongly coupled, arise in these different processes: short fibers orient, interact and may aggregate; preforms are compressed and may locally be subject to large deformations; the fluid flows in a medium that has very high anisotropic features, undergoing simultaneously thermal and chemical solicitations. Control of the process, in all cases, remains difficult, and internal properties such as local fiber fraction and orientation, or residual stresses, strongly vary as a function of process conditions. Numerical techniques may bring answers when comprehension of what happens during the process is difficult, and may provide quantitative values for the influence of the different process parameters, allowing its improvement, enhancing its productivity and reducing the development time.

9.2. Multiscale, multiphysics and multidomain modeling

Modeling of composite forming (and in particular when the resin is at the liquid state) is thus very complex and may include simulating the preforming and the flow processing stages. Preforming includes draping, braiding or fiber placement and has also been the object of the development of specific numerical applications, like in [KHA 10, ALL 11], not illustrated in this chapter. In recent decades, there has also been huge progress on simulation of the flow step, with different codes simulating the process, at different scales.

To develop numerical models, we need to first decide the scale of description of the physical mechanisms to study: microscopic, mesoscopic or macroscopic (Figure 9.1). At the microscopic scale, we distinguish each fiber and resin flow is computed. This may induce fiber motions, important, for example, in the comprehension of the rheological behavior of suspensions. For continuous fibers assembled in tows, the mesoscopic scale includes not only the study of the resin flow between the tows, but also impregnating them. Most often, the tow is considered as a homogeneous porous media. This dual scale structure is important, since we need to ensure that the resin fully saturates the inter-tow spaces, as well as intra-tow spaces, to prevent the appearance of macro- or microvoids. At the macroscopic scale, a homogeneous equivalent media being injected or compressed is considered, and simulation occurs for the forming of the whole part. Even if the latter is the one of interest for industrial and process development purposes, material parameters and models may be very complex, with data difficult to obtain experimentally. Thus, it is useful to couple simulation at all scales: at lower levels, we get closer to what experiences the material and less inner or homogenized properties are necessary.

At all scales, the same general continuum mechanics approach is used and detailed in the following. Mass, momentum and conservation equations are solved and constitutive laws and internal variable evolution depend on the scale and on the observed phenomena. Flow equations and classical thermo-rheological-kinetic coupling are illustrated in the next section.

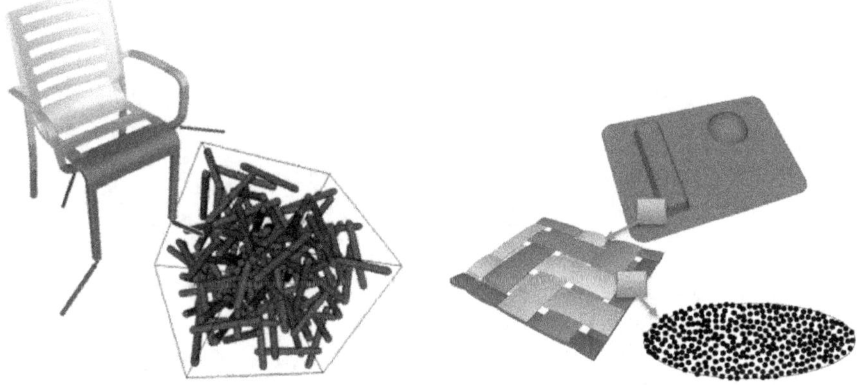

a) b)

Figure 9.1. *Spatial scales for composites: a) fiber suspensions, with a macroscopic or part scale and a microscopic scale, composed of resin and chopped fibers; b) continuously reinforced composites, with a macroscopic scale, a mesoscopic (tow) scale, and a microscopic (fiber) scale. For a color version of the figure, see www.iste.co.uk/boyard/heat.zip*

9.2.1. *Flow equations*

Let us consider the flow of a fluid around a bundle of rigid fibers (chopped or continuous), thus placing ourselves at the microscopic scale. The injected resin may be considered as a Newtonian or non-Newtonian incompressible fluid and the following Navier–Stokes equations, coupled to an appropriate set of boundary conditions, describe the fluid flow:

$$\left\{\begin{aligned} \rho_f\left(\frac{\partial \vec{v}}{\partial t} + \vec{v}\cdot\nabla\vec{v}\right) - \nabla\cdot\underline{\sigma} = \rho\vec{g} \\ \nabla\cdot\vec{v} = 0 \\ \underline{\sigma} = 2\,\eta_f\underline{\varepsilon}(\vec{v}) - p\underline{I} \\ \underline{\varepsilon}(\vec{v}) = \frac{1}{2}\left[\nabla\vec{v} + {}^t\nabla\vec{v}\right] \end{aligned}\right. \qquad [9.1]$$

with $\vec{v}$ the resin velocity, p the pressure, ρ the resin density, η_f the dynamic viscosity of the resin, $\underline{\sigma}$ the stress tensor, $\underline{\varepsilon}(\vec{v})$ the strain rate tensor and $\underline{I}$ the identity tensor. In most cases, injection pressures and velocities imply that inertia and gravity terms may be neglected, leading to the Stokes equations. This set of equations then needs to be coupled to a solid behavior model to take into account fiber motion or displacement. Figure 9.2 shows an example of flow computation of a Newtonian resin between a network of rigid static fibers, to illustrate the type of expected results, at the fiber scale.

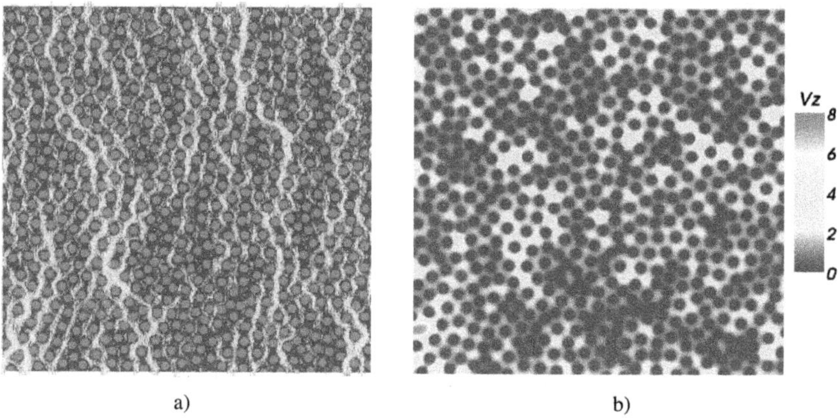

Figure 9.2. *Illustration of resin flow computation across an array of unidirectional fibers: a) flow path and b) velocity norm distribution [HIT 11]. For a color version of the figure, see www.iste.co.uk/boyard/heat.zip*

To establish conservation equations at the mesoscopic or macroscopic scales, averages or semi-analytic homogenization may be used, leading to two general cases.

In the first case, fibers are chopped, remain rather short and their concentration when impregnated by the fluid is dilute or semi-dilute. Then, the overall media being injected in a part (at the macroscopic scale) can be considered as a homogeneous fluid, with a stress tensor that may be decomposed in a fluid f and a fiber s (anisotropic) contributions $\underline{\sigma} = \underline{\sigma}_f + \underline{\sigma}_s$. where $\underline{\sigma}_f$ is defined as in equation [9.1]. The latter means that we need to follow, during part forming, the local evolution of fiber orientation and fraction. More details will be given through an illustration in section 9.3.4.1.

In the second case, fibers are gathered in yarns composing a textile reinforcement or chopped but assembled in a dense mat, and the pore scale is very small when compared with the flow length scale. In this case, Darcy's law [DAR 56] is the most used to model flow through this porous media, and is established by performing a volume average of the Navier–Stokes equations describing the flow at the microscopic scale, supposing that the solid skeleton composing the reinforcement is static and non-deformable, and that the porous media is saturated. It is given by:

$$\eta_f \underline{K}^{-1}\vec{q} + \nabla p = 0 \qquad [9.2]$$

Here, $\vec{q}$ is the discharge rate, such that $\vec{q} = \phi\vec{v}$, where ϕ is the fluid volume fraction, or porosity. The permeability, $\underline{K}$, is a tensor for an anisotropic porous media. In the case of unidirectional fiber packing, if the z axis of the coordinate system is taken in the same direction as the fiber axis, the permeability tensor can then be written as follows:

$$\underline{K} = \begin{pmatrix} K_{\perp_1} & 0 & 0 \\ 0 & K_{\perp_2} & 0 \\ 0 & 0 & K_{\parallel} \end{pmatrix} \qquad [9.3]$$

with $K_{\perp_1}$ and $K_{\perp_2}$ the transverse permeability of the packing and $K_{\parallel}$ the permeability along the fiber axis.

Several articles use Brinkman's equations [BRI 49] instead of Darcy's law to describe this type of flow, by generalizing Navier–Stokes equations supposing the flow of a Newtonian fluid through a swarm of fixed particles and writing:

$$\eta_f \underline{K}^{-1}\vec{q} - \nabla \cdot [2\eta_e \underline{\varepsilon}(\vec{v})] + \nabla p = 0 \qquad [9.4]$$

where η_e is the effective viscosity, which may be different from η_f. One advantage of equation [9.4] is that it is an extension of the Stokes equation. Thus, numerical tools developed for the latter may be used in the Brinkman case with a small amount of implementations. However, its physical domain of validity is very restricted (small fiber concentrations, when $\phi \to 1$, corresponding to very large permeabilities), and theoretical analysis shows that it yields to $\eta_e = \eta_f$ [AUR 09]. Nevertheless, the Brinkman form will be preserved in the following, since it allows us to treat numerically both the Stokes and Darcy cases, by considering extreme values of $\underline{K}$ and η_e. Section 9.3.4.2 shows an example of flow on one of these cases.

9.2.2. *Thermal-rheological-kinetical coupling*

Liquid forming processes involve more complex physical phenomena than "just" solving a fluid (and solid) mechanics problem. In fact, modeling these processes involves studying the coupling between this problem, the thermal one and, for some resins, like thermosets, the polymerization or degree of curing reaction. In fact, resin cure affects its viscosity, which modifies the flow pattern. Higher viscosities also imply more pressure and higher deformation and motion of the fiber reinforcement, which changes the permeability, influencing the resin flow. Temperature directly affects the rheology, changing the viscosity and modifying the reinforcement geometry by thermal dilatation. In addition, polymerization impacts thermal transfer, since curing is exothermal, producing heat, which also accelerates the reaction.

The temperature, governed by the energy conservation equation, depends on the heat transfer by conduction in the resin, fibers and with the mold, the viscous dissipation (which may often be neglected) and on the heat released by the chemical reaction (function of the reaction kinetics). Let us suppose that we wish to study directly the homogenized composite media. This will allow us to establish a general form of the energy equation. Thus, the heat equation in the absence of viscous dissipation, of the convection-diffusion-reaction type, is written as follows:

$$\rho c_p(\frac{\partial T}{\partial t} + \vec{v} \cdot \nabla T) - \nabla \cdot (\underline{\lambda} \nabla T) = H_\alpha f(\alpha) \qquad [9.5]$$

Here, ρ and c_p are, respectively, the composite density and specific heat, whereas $\underline{\lambda}$ is the thermal conductivity, which is, for an anisotropic media, also a tensor. All are dependent on the fiber fraction and the latter most often on other inner geometrical parameters, such as orientation. The source term corresponds to the polymerization or cure reaction and is a function of the cure enthalpy, H_α, and of the cure rate, $f(\alpha)$. The evolution of the degree of cure α is obtained by solving a simple advection equation with a source term, the cure rate:

$$\frac{\partial \alpha}{\partial t} + \vec{v} \cdot \nabla \alpha = f(\alpha) \qquad [9.6]$$

Here, α is the degree of cure and $f(\alpha)$ is a function of the cure kinetics model chosen. One of the most general and widely used is the Kamal and Sourour model [SOU 76], which has the form:

$$f(\alpha) = \left[k_1 \exp(\frac{E_1}{RT}) + k_2 \exp(\frac{E_2}{RT})\alpha^m\right](1-\alpha)^{1-n} \qquad [9.7]$$

and where k_i, m, n are the material parameters, whereas E_i are activation energies. The degree of cure and the temperature affect the viscosity, in opposite senses. More generally, $\eta = \eta_0 g(\vec{v}) \cdot g(T) \cdot g(\alpha)$. Dependence on $\vec{v}$ through the strain rate tensor is taken into account through purely viscous models for classical non-Newtonian resins, and the influence of temperature may be governed by Arrhenius law [AGA 96]:

$$g(T) = \exp\left[\frac{E_a}{R}(\frac{1}{T} - \frac{1}{T_{\text{ref}}})\right] \qquad [9.8]$$

where E_a is the activation energy, R is the universal gas constant and T_{ref} is the reference temperature. Finally, the degree of cure implies a modification of the viscosity according to, for example, a percolation model, which is widely used for epoxy resins [SER 90]:

$$g(\alpha) = (1 - \frac{\alpha}{\alpha_{\text{gel}}})^p \qquad [9.9]$$

where α_{gel} is the extent of the reaction at the percolation threshold (the gel point) and p is the percolation exponent, both material properties to be identified experimentally. The reference temperature T_{ref} is usually the glass transition temperature T_g, influenced by the cure kinetics (when vitrification occurs) and that may be modeled using the DiBenedetto model [DIB 87]:

$$T_g = T_g^0 + \frac{(T_g^1 - T_g^0)\kappa\alpha}{1 - (1 - \kappa)\alpha} \qquad [9.10]$$

where T_g^0 and T_g^1 are the glass transition temperature values of the uncured and fully cured resin and κ is an adjustable structure-dependent parameter. To illustrate what type of computations use these features, we consider the overmolding of an electrical part with an epoxy resin, Figure 9.3 [FOU 08].

9.2.3. *Orientation and structure development during processing*

Short or long fiber resin reinforced polymers are mostly applied in injection or compression molding. In particular, injection molding induces an oriented layered structure for this type of composite, where orientation depends on the flow pattern. For example, in the vicinity of the injection gate, the material's flow front adopts a radial flow extension which is divergent and we get a transverse orientation compared to the flow direction. Far from the gate, the shear rate is weaker and fibers preserve a state of orientation close to the initial one. Solidification occurring near the wall contributes to the formation of a frozen-in oriented layer. Near this layer, the shear rate is very important and fibers orient in the flow direction. Beyond the ones referred to, other parameters influence the orientation state such as the injection speed and the holding pressure [LAU 11, MIL 08, MIL 12].

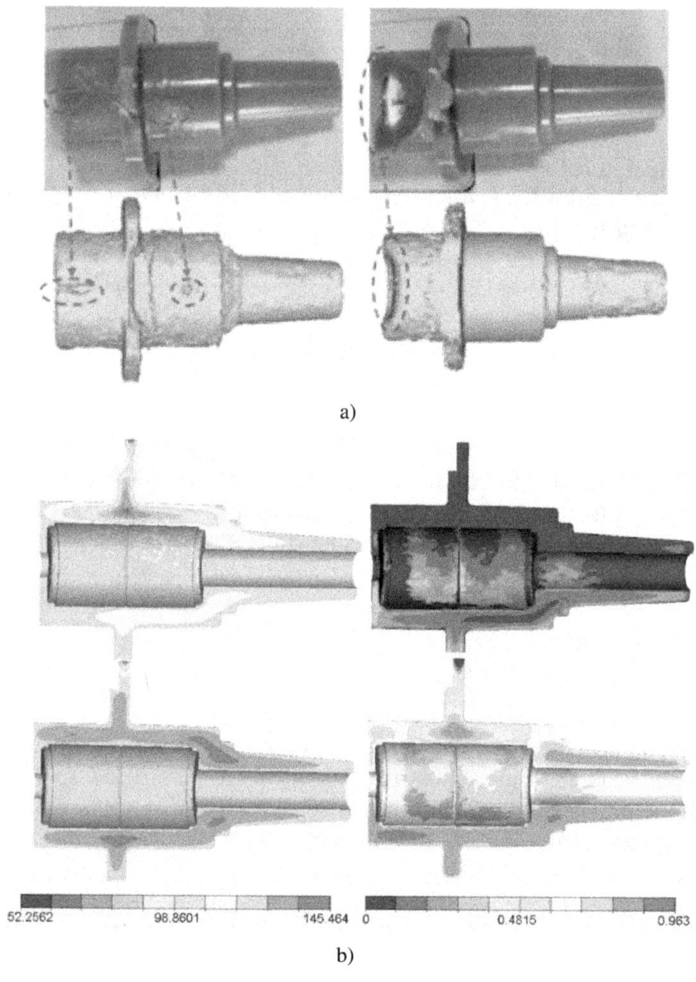

Figure 9.3. *Numerical results obtained in overmolding of an electrical part with an epoxy resin: a) comparison of experimental end of fill positions and numerical predictions for two different filling configurations; b) temperature (left) and degree of curing (right) at the end of filling (above) and at the end of the post-filling (below) stages [FOU 08]. For a color version of the figure, see www.iste.co.uk/boyard/heat.zip*

At the macroscopic scale, the most general descriptor of an oriented state is the probability distribution function of the orientation, $\psi(\vec{p}, t)$, which represents the probability to find, at time t, a fiber with axis oriented in the direction of the unit vector $\vec{p}$. Computation of ψ can be performed using the Fokker and Plank equation [PUA 11], but this quantity is difficult to handle numerically, and thus not commonly used. Other authors [HAN 62] introduced the second-order $\underline{a}_2$ (and more) orientation tensors, expressed as a function of ψ and $\vec{p}$, as a quantitative measure of the orientation state, such that:

$$\underline{a}_2 = \int_{\vec{p}} \psi(\vec{p},t)\vec{p}\otimes\vec{p} \qquad [9.11]$$

$$\underline{a}_4 = \int_{\vec{p}} \psi(\vec{p},t)\vec{p}\otimes\vec{p}\otimes\vec{p}\otimes\vec{p} \qquad [9.12]$$

$\underline{a}_2$ is symmetrical and positive definite, and its trace is equal to 1. For a random orientation, its diagonal terms are equal to 1/3 and, for a random planar orientation, to 1/2. For a perfect orientation in the direction i, $a_{ii} = 1$, and the other values are zero.

Lipscomb [LIP 84] combined Jeffery [JEF 22] and Fokker and Plank equations to represent the evolution of the second-order orientation tensor by a convection-reaction equation, containing also a fourth-order tensor, which can be expressed as a function of the second-order one using a closure approximation. Folgar and Tucker [FOL 84] extended the previous model to take into account the interaction between fibers. They obtained:

$$\frac{\partial \underline{a}_2}{\partial t} + \vec{v}\cdot\nabla\underline{a}_2 = \underline{\omega}\underline{a}_2 - \underline{a}_2\underline{\omega} + \beta(\underline{\varepsilon}\underline{a}_2 + \underline{a}_2\underline{\varepsilon} - 2\underline{\varepsilon} : \underline{a}_4) + 2C_i|\underline{\varepsilon}|(\underline{I} - 3\underline{a}_2) \qquad [9.13]$$

where $\underline{\omega}$ is the vorticity tensor, β is a function of the fiber's aspect ratio, $|\underline{\varepsilon}|$ is the equivalent strain rate and C_i is an empirical constant, usually called the interaction coefficient. Experimental data have shown that, for concentrated suspensions, kinetics of orientation is much slower than that predicted by Folgar and Tucker's model. More recently, other authors have introduced modified versions such as the reduced strain closure model [WAN 08], the anisotropic rotary diffusion model [PHE 09] and a model taking into account fiber–fiber interactions [FER 09]. Most theories have derived the following expression for the stress tensor [TUC 91]:

$$\underline{\sigma} = -pI + 2\eta_f\left[\underline{\varepsilon} + N_p\underline{\varepsilon} : \underline{a}_4 + N_s(\underline{\varepsilon}\underline{a}_2 + \underline{a}_2\underline{\varepsilon})\right] \qquad [9.14]$$

where N_p and N_s are the parameters depending on the fluid's viscosity, fiber aspect ratio, orientation and concentration. A review is given in [LAU 11].

Figures 9.4(a), (b) and (c) illustrate, for a complex geometry and at the end of the filling stage, the principal components of $\underline{a}_2$ [MIL 08, MIL 12]. It allows a precise knowledge of the fiber orientation distribution everywhere in the part and, starting from this orientation distribution, we may deduce the thermo-elastic properties of the part.

As a first constitutive law, we may use a linear anisotropic thermo-elastic relation (supposing that each point of the part cools from T to T_0):

$$\underline{\sigma}(T_0) = \underline{C}(\underline{a}_2, T_0) \left[\underline{\varepsilon}(T_0) - \underline{\alpha}(\underline{a}_2, T_0)\Delta T\right] \qquad [9.15]$$

Components of both the stiffness tensor $\underline{C}$ and the dilatation tensor $\underline{\alpha}$ can be expressed, according to [ADV 87], as a function of the components of $\underline{a}_2$ and $\underline{a}_4$:

$$\begin{aligned} C_{ijkl} &= b_1 a_{ijkl} + b2(a_{ij}\delta_{kl} + a_{kl}\delta_{ij}) + b_3(a_{ik}\delta_{jl} + a_{il}\delta_{jk} \\ &\quad + a_{jk}\delta_{il} + a_{jl}\delta_{ik}) + b_4\delta_{ik}\delta_{kl} + b_5(\delta_{ik}\delta_{jl} + \delta_{il}\delta_{jk}) \end{aligned} \qquad [9.16]$$

$$\alpha_{ij} = P_1 a_{ij} + P_2\delta_{ij} \qquad [9.17]$$

The different constants b_i are identified for a unidirectional composite using, for example, Mori–Tanaka homogenization theory [MOR 73], the stiffness tensor of the composite $\underline{C}$ being expressed as a function of the stiffness tensors of the matrix, $\underline{C}_m$, and of the fiber $\underline{C}_f$. In the same way, parameters P_1 and P_2 can be expressed as a function of the longitudinal and transverse thermal dilatation coefficients, α_1 and α_2.

Figure 9.4 shows, for the previous example, the distribution of the elastic modulus in the different directions, computed from the stiffness tensor, and using the described semi-analytic methodology [MIL 08, MIL 12].

9.3. Advanced numerical techniques and macroscale simulations

For multiphase and multiphysics flows such as those encountered in the referred to processes, either with short or textile reinforcements, different numerical techniques have been proposed to solve the conservation equations, coupled to the constitutive ones. The most popular ones are the finite difference method (FDM) [TRO 92], the finite element method (FEM) [BRU 90, TRO 06, SIM 07, SIL 09, LAU 11], the boundary element method (BEM) UM 91, YOO 96, SOU 06, GAN 13], the smooth particle hydrodynamics (SPH) method [COM 05], as well as reduced order methods [GHN 11, ABI 14]. A review of the literature for injection molding (mostly used for short fiber composites) may be found in [SIL 09] and for liquid composite molding of continuous reinforcements in [PAR 15]. Even if computational cost may be greater, finite elements remain the most popular solution due to its capability to handle complex geometries and are the ones used and described in this section.

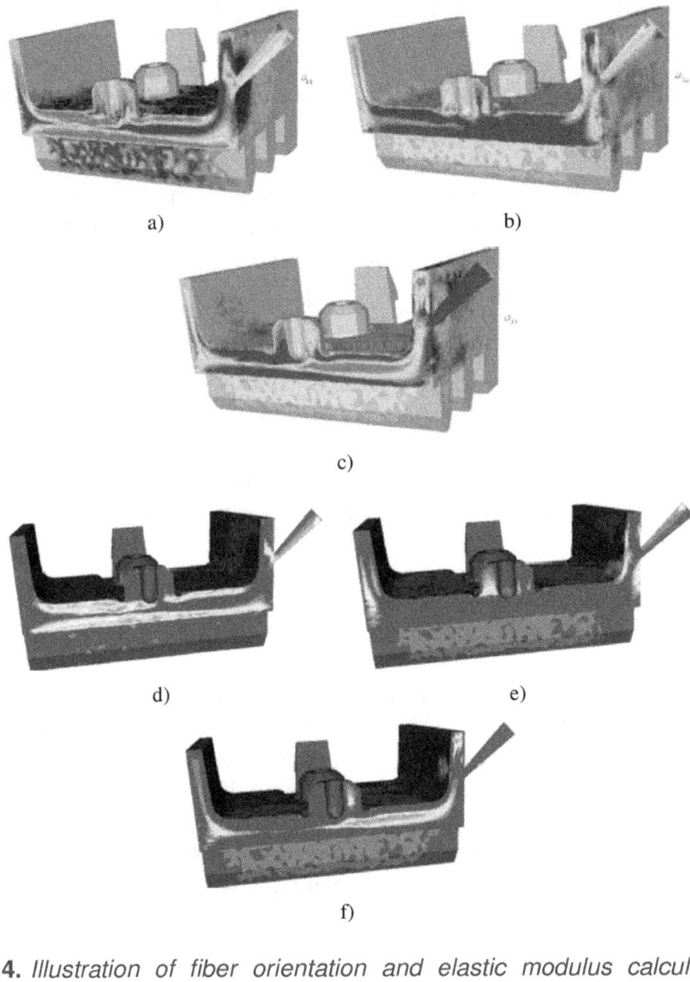

a) b) c) d) e) f)

Figure 9.4. *Illustration of fiber orientation and elastic modulus calculation on a prototype part (courtesy of Schneider Electric), where the material used was a glass fiber reinforced polycarbonate, with 20% fiber content, showing the results on half of the geometry. Distribution of the different diagonal components of $\underline{a}_2$ at the end of the filling stage: a) a_{11}, flow direction, with the injection gate represented on the right; b) a_{22}, height direction; c) a_{33}, width direction; d) E_{11}; e) E_{22} and f) E_{33} [MIL 12]. For a color version of the figure, see www.iste.co.uk/boyard/heat.zip*

In most cases, the resin does not completely fill the cavity at the beginning of the process and it is necessary to simulate an impregnation or injection step, in particular to determine void formation and vent placement. This mold filling simulation is a typical problem of a moving boundary (fluid–air interface) with the computed velocity field. To solve it, we may find in the literature two main approaches: the computation on a moving domain or a Eulerian framework, where there is one computational domain composed of all the phases; the displacement of the phases as well as their deformations, which is a Lagrangian computation of the velocity field, meaning generally that there is one computational domain per phase. Numerical simulation is still strongly dependent on the meshing capabilities of complex geometries and one of the drawbacks of Lagrangian approaches is that it often constrains the enforcement of the boundary or interface meshes in the volume mesh. This task becomes increasingly difficult since, due to domain deformation, remeshing is required at almost every time step.

In Eulerian methods, the whole domain contains all the phases (fluid, air, mold, etc.) and boundaries are represented in an implicit way. This means that the boundary is no longer given by a surface mesh or any explicit representation but implicitly by a scalar field whose value can be accessed anywhere in the domain. Two representative methods using such a field are the volume of fluid (VOF) [HIR 81] and the level-set [OSH 88] methods. In the following, a modified version of the latter will be detailed [COU 15].

9.3.1. *Implicit boundaries*

To solve conservation equations in a Eulerian framework, we need to define material properties everywhere in the computational domain. To detail this multiphase context and as a matter of simplification, let us consider a domain ω totally included in the larger one, Ω, $\Gamma = \delta\omega$ being the boundary of ω. η, ρ, $\underline{K}$ or $\underline{\lambda}$, among others, are heterogeneous fields, defined by their values inside and outside ω. In fact, these properties are discontinuous from one phase to the other, and may be represented in a general way as a variable field ξ within the mesh and through a function H, such that:

$$\xi = \xi_\omega H + \xi_{\Omega\backslash\omega}(1 - H) \qquad [9.18]$$

where H is a Heaviside function, usually equal to 0 outside ω and 1 inside (Figure 9.5(b)). Dealing with such discontinuity is not affordable by any standard numerical method and a smooth transition around the interface has been proposed [COU 15].

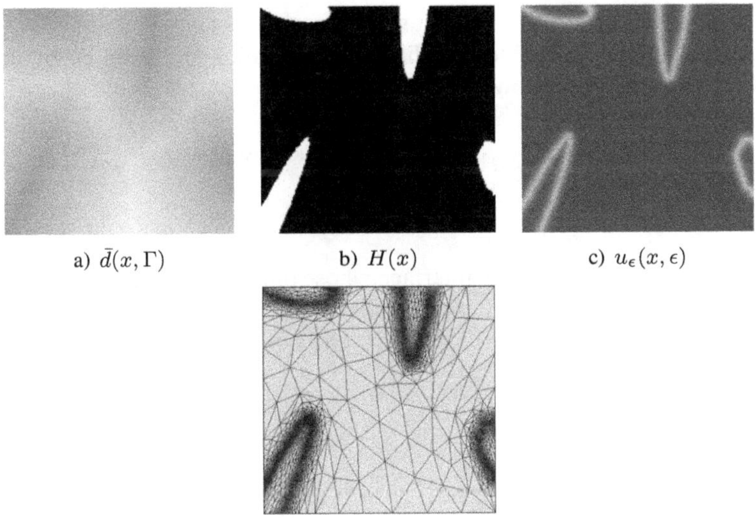

a) $\bar{d}(x, \Gamma)$ b) $H(x)$ c) $u_\epsilon(x, \epsilon)$

d) mesh

Figure 9.5. *Phase functions used to describe, in an implicit way, the different phases of implicit boundary computation. In this example, two phases are present [ZHA 15]. For a color version of the figure, see www.iste.co.uk/boyard/heat.zip*

9.3.2. *Immersed subdomains and regularization*

From the previous description, we may say that ω is *immersed* into Ω. To represent ω, an implicit function has been chosen, the immediate candidate being the signed distance function, Figure 9.5(a):

$$\bar{d}(x, \Gamma) = \begin{cases} d(x, \Gamma) \text{ if } x \in \omega \\ -d(x, \Gamma) \text{ if } x \notin \omega \end{cases} \qquad [9.19]$$

Reciprocally, the interface is completely defined by the zero value of the distance function, and the knowledge of ω is ensured by the definition of this function everywhere in Ω. From $\bar{d}$, we may calculate the Heaviside function, Figure 9.5(b), by:

$$H(\bar{d}) = \begin{cases} 1 \text{ if } \bar{d}(x, \Gamma) \geq 0 \\ 0 \text{ if } \bar{d}(x, \Gamma) < 0 \end{cases} \qquad [9.20]$$

The Heaviside function enables us to separate strictly the domains, but the counterpart is that it reintroduces the discontinuity which has been left by the use of an implicit function. The following implicit function:

$$u_\epsilon = u(d,\epsilon) = \epsilon \tanh\left(\frac{d}{\epsilon}\right) \qquad [9.21]$$

enables the introduction of a thick interface whose thickness is related to the parameter ϵ. Moreover, u_ϵ approximates d and we recover Γ from it, for whatever ϵ. Then, a smoothed Heaviside function can be introduced as well by:

$$H_\epsilon = \frac{1}{2}\left(1 + \frac{u_\epsilon}{\epsilon}\right) \qquad [9.22]$$

9.3.3. *Multiphase flow and thermokinetical numerical resolution*

Let us return to the resolution of the conservation equations, and first the flow ones, under the Brinkman form, equation [9.4]. For the case of a resin impregnating a porous media (one fluid, f, entering the cavity and air, a, venting it), material properties are controlled by the thickness parameter ϵ, such that:

$$\begin{aligned} \eta &= \eta_f H_\epsilon + \eta_a(1 - H_\epsilon) \\ \underline{K} &= \underline{K}_f H_\epsilon + \underline{K}_a(1 - H_\epsilon) \end{aligned} \qquad [9.23]$$

Brinkman's problem may be solved using a mixed finite element method [SIL 10]. The authors proposed a stabilized formulation which approximates $\vec{v}$ and p with finite dimensional spaces of continuous piecewise polynomials, such that $\vec{v}_h|_K \in P^1(K)^d$ and $p_h|_K \in P^1(K)$, $\forall K \in \mathcal{K}$. It enriches the velocity space with a space of bubbles (the finer scale), using a polynomial function of order one, with a value equal to unity at the center of the element and vanishing at its boundary. The enrichment performed stabilizes the formulation for both the limit Stokes and the Darcy cases, with a modification of bubble stabilization for both regions, to be compatible with stability and unicity of the solution conditions. The element contribution to the linear system arising can be written, in the matrix form [SIL 10]:

$$\begin{bmatrix} A_{vv} - A_{vb}(A_{bb}^{-1})^t A_{vb} & A_{vp} - A_{vb}(A_{bb}^{-1})^t A_{bp} \\ {}^t A_{vp} - {}^t A_{bp}(A_{bb}^{-1})^t A_{vb} & -{}^t A_{bp}(A_{bb}^{-1})^t A_{bp} \end{bmatrix} \begin{bmatrix} V \\ P \end{bmatrix} = \begin{bmatrix} F_v \\ P \end{bmatrix} \qquad [9.24]$$

In what concerns the solution of the energy equation, as well as the ODE representing the evolution of the different internal variables (like α or $\underline{a}_2$), these are generally equations of the convection-diffusion-reaction type. Classical techniques generate numerical instabilities, while diffusion problems may cause oscillations during the treatment of thermal shocks.

To avoid stability difficulties due to convection, some techniques have been developed that behave well on highly convective problems but are limited by their tendency to denature the solution. Techniques to avoid oscillations during the treatment of thermal shocks are well known: we must adapt the mesh size or the time step in the direction of the gradient. To be optimal on the solution of the two issues, two different stabilization techniques are widely used in the literature: the streamline upwind Petrov Galerkin (SUPG) method [BRO 82] and the residual free bubbles [BRE 98] (RFB) method.

In all cases, the proposed formulation leads to the resolution of a sparse linear system, which represents an important fraction of the overall simulation time. To solve it, the PETSc library [BRO 82] may be used, with a classical preconditioned (ILU, Block Jacobi, etc.) iterative method of the Krylov type (conjugate residual or GMRES).

9.3.4. *Composite forming simulation illustrations*

9.3.4.1. *Compression molding*

Compression molding is a high-volume, high-pressure method which is suitable for molding complex and high-performing parts. Thermosets (mainly) or thermoplastics may be compression molded in different types of reinforcements, such as unidirectional fabrics, woven textiles, random mats and chopped fibers. The use of composite materials like sheet molding compounds (SMCs), which are compressed, has increased in the last years, in particular in the automotive industry.

In the literature, several authors have modeled compression molding, but assuming an important hypothesis: a $2 - 1/2$D model; a dilute regime for fiber concentration; decoupling between cure, temperature and flow; no anisotropy effects, in particular in the thermal conductivity; perfect contact between the SMC and the mold (and between the different layers of SMC). More recently, 3D simulations involving full coupling have been shown [SIL 14a], with an illustration given in Figure 9.6. In this case, an SMC has been compressed. At the beginning, several layers were cut and disposed of in three different packs. Before curing begins to influence flow, we observe the high deformation and the flow of the samples, under heating conditions. On the right, complexity of the orientation obtained is plotted at the end of the compression.

9.3.4.2. *Resin transfer molding*

For the production of structural composites, one of the most promising routes is resin transfer molding (RTM). In this process, the reinforcement is laid in the mold cavity after which the mold closes. Once the textile has been heated, enough resin is injected from one or several gates. Consolidation occurs before the mold opens.

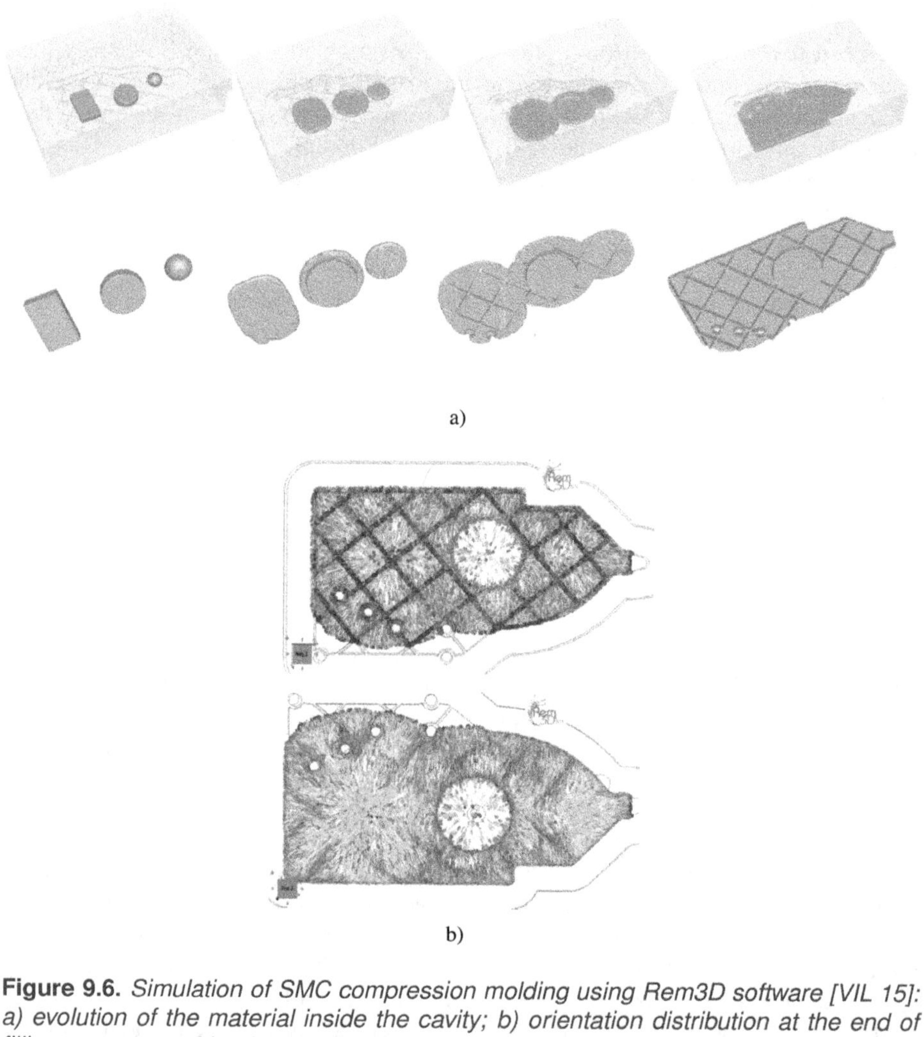

Figure 9.6. *Simulation of SMC compression molding using Rem3D software [VIL 15]: a) evolution of the material inside the cavity; b) orientation distribution at the end of filling, represented by the eigenvectors of $\underline{a}_2$ multiplied by its respective eigenvalues. For a color version of the figure, see www.iste.co.uk/boyard/heat.zip*

Several variations of this process exist. One of the most used, in particular for very large parts, is infusion: when the clamping forces required to keep the mold closed are very high, a plastic film is used as an upper mold and vacuum is imposed, so that the resin is injected. One other variation is HP-RTM, high-pressure resin transfer molding, used to produce parts with a very high-fiber volume fraction, i.e. very low permeability, enabling a reasonable injection time [BIS 13].

HP-RTM (and also compression RTM, C-RTM) simulation and experiments have been studied in [BIS 13] using the above described finite element method. The first test carried out was the injection of a resin into a small disk and a comparison between analytic and numerical solutions was made, Figures 9.7(a) and (b), showing a good agreement. As far as temperature is concerned, a strong gradient appears near the injection gate, and a strong skin flow phenomenon is developed, as seen in Figures 9.7(c) and (d). Temperature difference between a simulation made with or without thermo-rheo-kinetical coupling is also illustrated, showing its importance since, in this last case, skin flow does not occur.

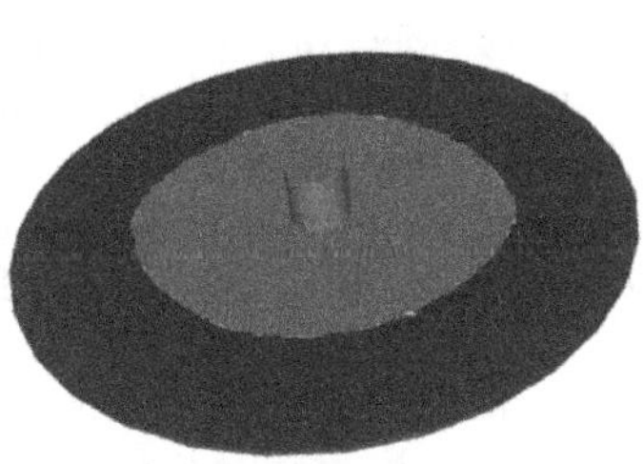

a) injection

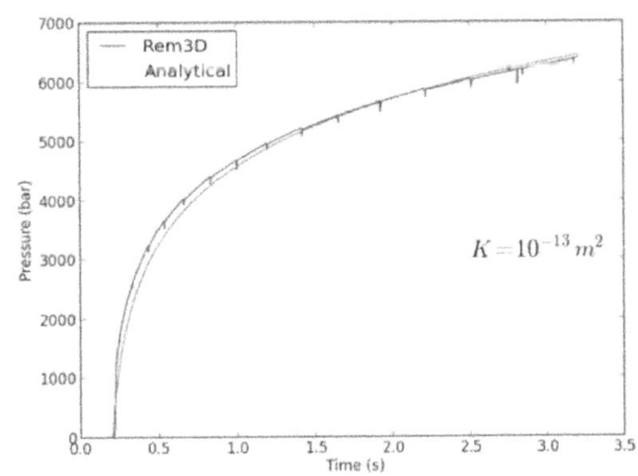

b) pressure

c) velocity norm

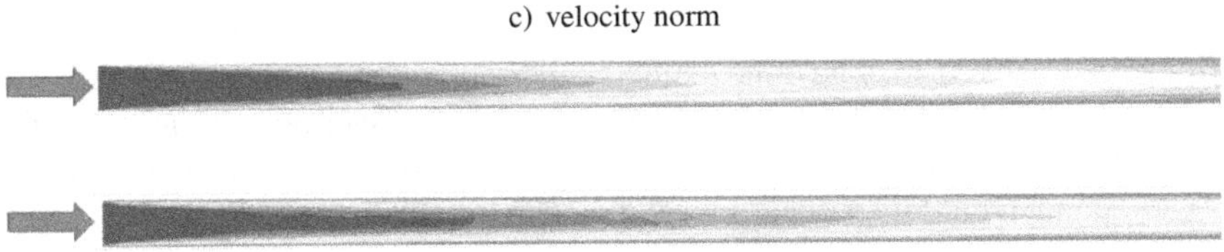

d) temperature

Figure 9.7. *Simulation of resin transfer molding using Rem3D software [VIL 15]: a) evolution of the material inside the cavity; b) pressure evolution throughout filling; c) velocity norm in a thickness cut; d) temperature in a thickness cut, with or without thermo-rheo-kinetical coupling [BIS 13]. For a color version of the figure, see www.iste.co.uk/boyard/heat.zip*

9.3.5. *Parallel mesh adaptation and high-performance computing*

Solutions obtained, for multiphase flows with very different properties, converge toward a sharp or exact interface solution when $\epsilon \to 0$. This means that the proposed methodology is based on generating a thickness ϵ small enough so that computations are accurate. Since to represent u_ϵ we also need several layers of elements within ϵ, this may only be attained using anisotropic adaptive meshing and parallel implementations.

9.3.5.1. *Automatic anisotropic adaptation*

Anisotropic mesh adaptation enables us to capture discontinuities or gradients of the solution in flow, thermal, kinetical situations [COU 11, COU 15], with a good accuracy at a very low extra number of elements. Building unstructured anisotropic meshes may be based on local mesh modifications, using a metrics field to redefine the lengths [COU 94] and a recently developed procedure, optimal in terms of remeshing, is here described.

Let $\mathcal{N}$ be a finite set of node numbers and $\mathcal{K}$ a mesh topology or a set of elements that provides the element-node connectivity. Let us consider only the mesh topologies composed of simplex elements (triangles, tetrahedra, etc.), K being a d-simplex (d is the spatial dimension) composed of $D = d + 1$ vertices. The mesh generator described adapts an initial mesh by iteratively modifying and optimizing it locally [COU 94]. This technique consists of the improvement of the quality of local cavities composed of clusters of elements by remeshing. Two principles are enforced: the *minimal volume*, which gets the conformity of the mesh, with no element overlaps; the *geometrical quality* distribution, ensuring that, if the minimal volume provides several possible cavity retriangulations, then the one chosen is such that it provides the best worst element. The geometrical quality, $Q(K)$, is evaluated for each element K and varies from 0 (worst) to 1 (best quality). It depends on $\underline{M}_K$, the metric of the element K, which is thus the information needed by the mesher to generate a new mesh.

Construction of $\underline{M}$ may be done using the interpolation error analysis on a solution field, for example, u, which may be a velocity, a phase function or other variable, or a vector containing all. Following [COU 11], let us define the edge vector as $\vec{X}^{ij} = \vec{X}^j - \vec{X}^i$, connecting nodes i and j, and the edge length distribution tensor at i as:

$$\underline{X}^i = \frac{1}{|\Gamma(i)|} \sum_{j \in \Gamma(i)} \vec{X}^{ij} \otimes \vec{X}^{ij} \qquad [9.25]$$

where $\Gamma(i)$ represents the nodes that share an edge with i. It may be established that the metrics computed at the node is: $\underline{M}^i = \frac{1}{d}(\underline{X}^i)^{-1}$.

Let us suppose that we wish to generate a new mesh that equidistributes in each edge a constant target interpolation error, e_{target}. It has been previously shown [COU 13] that this target error can be directly linked to a fixed number of nodes N (which may also be related to the available CPU power), $e_{\text{target}} = e_{\text{target}}(N)$. An *a posteriori* interpolation error analysis states that the error along an edge ij on the solution field u, e^{ij}, is $e^{ij} = \nabla u^{ij} \cdot \vec{X}^{ij}$, where $\nabla u^{ij} = \nabla u^{j} - \nabla u^{i}$, nodal gradients in i and j obtained with a gradient recovery procedure [COU 11].

The optimal stretch factor, s^{ij} that must be applied to $\mathbf{X}^{ij}$ to attain e_{target} may now be used to compute the target metric given to the mesher, which allows the new mesh (sequential) generation:

$$s^{ij} = \frac{e_{\text{target}}(N)}{e^{ij}} \quad \text{and} \quad \underline{\widetilde{M^{i}}} = \frac{1}{d}\left[\frac{1}{|\Gamma(i)|}\sum_{j\in\Gamma(i)} (s^{ij})^{2}\vec{X}^{ij}\otimes\vec{X}^{ij}\right]^{-1} \qquad [9.26]$$

An illustration of the type of result obtained in adaptation around one or several layers of fibers is illustrated in Figure 9.8, where we may observe the mesh density around the tows in (a) and (c) or the accuracy in the representation of the interface in (b).

9.3.5.2. *Parallel computing*

For complex 3D geometries and complex physical laws such as the ones concerning composites, parallel computation remains essential. It allows us to run complete simulations with a reasonable precision, inaccessible (for memory and time limits) with sequential runs. It also makes it possible to speed up the simulation run such that the global computation time will still be acceptable. In recent years, processor performance has not increased by improving the clock rate but by multiplying the number of cores in a CPU. Actual top supercomputers contain several hundred thousands to millions of cores with hundred TB to PB of memory. It is thus necessary to develop fully parallel applications that follow, at least, this multi-core CPU evolution.

Between the existing parallelization techniques, we have chosen to refer to those developed and used in the former and following examples [COU 00]. They are based on the decomposition of the computational domain into several subdomains, where each is affected to one core. In the simulations illustrated above, most of the computation time is spent on remeshing and on the resolution of several large linear systems.

As far as meshing is concerned, the chosen parallelization procedure uses the sequential mesher described before in a massive parallel context, by following an iterative procedure with two successive steps [COU 00]: independent adaptive meshing of each subdomain with blocked interfaces; constrained repartitioning and interface displacement. Even if several iterations are done, work per iteration decreases quickly. When a good quality mesh is obtained, a finite element load-balancing repartitioning step is performed. All fields defined on the initial mesh may then be mapped on the new one.

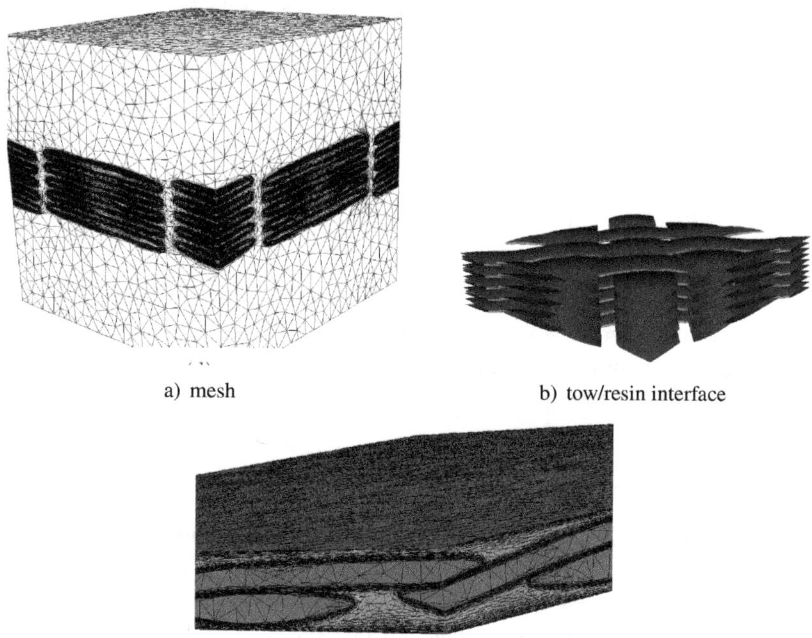

a) mesh   b) tow/resin interface

c) mesh and u_ϵ

Figure 9.8. *Adapted mesh to accurately represent, in an implicit way, the interface between tows and fluid: a) mesh; b) isozero value of u_ϵ, distance to the fiber/resin interface and c) mesh and u_ϵ representation [PUA 11]. For a color version of the figure, see www.iste.co.uk/boyard/heat.zip*

For linear system resolution, parallelization has been performed by interfacing the parallel partitioner with the PETSc library [BRO 82] and also by applying a multi-grid preconditioner. An example illustrating in more detail the capabilities of parallel computing and the performances of this method is shown in section 9.4.2. It has already allowed solving direct flow simulations on fiber samples that required meshes of around 780 million nodes and linear systems of 3.6 billion unknowns [SIL 15].

9.4. Determination of equivalent properties and microscale simulations

To obtain accurate models at the macroscopic scale, we often perform a homogenization step. Semi-analytical homogenization, such as the one presented in section 9.2.3, is a well-established and practical method to compute the effective properties of fiber reinforced materials, in particular at the solid state.

However, if we wish to later extend it to the nonlinear case (for example, by considering that the matrix is viscoelastic) or to the flow regime, these methods may not give enough accuracy. The use of micromechanical and microrheology finite element computations has thus increased in the last decade [LLO 04, SEG 05, SUN 05, TOT 10]. In these computations, properties are obtained by performing finite element simulations on 3D "Representative Elementary Volumes" (REV) of the microstructure. Generating several REVs allows us to perform a sensitivity analysis to the topology of the REV (size, number and spatial distribution of fibers or tows), as well as to the mechanical or rheological solicitations. From all the numerical techniques used to generate REVs, direct methods from 3D images will be the focus of this section.

9.4.1. *Generation of representative numerical samples*

Recently, imaging techniques, such as X-ray tomography, have been used to obtain precise geometrical and topological information on microstructures, which allowed extraordinary contributions in material science. The passage from the image to a finite element mesh allows, on one hand, the construction of a numerical representation but, on the other hand, may also reduce the size of the stored data. The first step (before building a mesh) is usually to perform the image segmentation, that is to partition the image into its multiple objects, so that it becomes easier to analyze it [SHA 01]. The result of this segmentation can be regions of the image where the pixels are similar with respect to a characteristic or property (like the density), or extracted contours. Second, a mesh is built from the segmented image by using different techniques.

Recent work [SIL 14b] proposes a novel technique to simultaneously segment and construct a finite element mesh, using directly the 3D image data, and applies it to fibrous materials. For this, the immersed image method is built, which is the interpolation of the image's pixel/voxel values (a sort of topographic distance function) on the nodes of an initial mesh. The obtained field can approximately represent the original image, depending on the discretization and on the overall number of nodes. Optimization of the number of necessary nodes can be performed through anisotropic mesh adaptation using the above described procedure, so that the obtained field accurately represents the original image. To further progress toward segmentation, the mesh is adapted not on the interpolated topographic distance function (the pixel/voxel value), but on u_ϵ. To build this function, a reinitialization method has been implemented. This procedure, coupled to mesh adaptation, allows segmentation but also allows us to build up a mesh with a smooth representation of a phase function distribution, usable in the aimed for type of application, Figure 9.9.

Figure 9.9. *Illustration of the results obtained for mesh adaptation directly from an image: a) original composite image, with* $200 \times 200 \times 220$ *voxels; b) computed distance function; c) computed regularized Heaviside function; d) final adapted mesh, with 50,000 nodes. For a color version of the figure, see www.iste.co.uk/boyard/heat.zip*

9.4.2. *Permeability of a composite*

With computed velocity and pressure on a porous REV, permeability can be determined (at the micro- or mesoscales). For semi-analytic homogenization, three different approaches are most often used: the capillary model; models arising from the hydrodynamic lubrication theory; the cell model. Hydrodynamic lubrication theory models assume the fibrous media as a regular cylinder spacing geometry and compute an analytic solution for the Stokes equations inside this REV. The most popular one for the transverse permeability of the array of cylinders is Gebart's expression [GEB 92].

In numerical homogenization, Darcy's equation, defining permeability, is directly considered. Thus, for at least three principal directions, we must compute:

$$K = \eta_f(\frac{\langle||\vec{v}||\rangle}{\langle||\nabla p||\rangle}) \qquad [9.27]$$

where $\langle||\vec{v}||\rangle$ and $\langle||\nabla p||\rangle$ are the averaged velocity and pressure gradient fields for the considered direction, only computed in the fluid domain.

At the microscopic scale, numerical results can be compared with Gebart's results, by considering a regular array of fibers, as illustrated in [PUA 11], for square and triangular packings. A pressure gradient is imposed from left to right and the normal velocity in the upper and lower planes is zero. Inside the fibers, velocity is also zero. Figure 9.10 illustrates a comparison between numerical and semi-analytic results, for different fiber volume fractions. Overall results show that the computed values are closer to lubrication model results for a low fiber volume fraction and closer to cell models for high fiber volume fraction.

However, fibers may be randomly placed in the fabric and do not generally follow a strict organized structure (even for the case where we wish to impose a certain alignment). To assess the feasibility and capability of the overall procedure, a 3D test case is presented, Figure 9.11 [SIL 14b], whose complexity implies that no statistic or equivalent organized REV reconstruction is possible.

An image (900 × 900 × 220 voxels) coming from 3D X-ray tomography [ORG 12] concerning the mat-reinforced sample is treated and a phase function equal to 1 inside the fibers (solid) and 0 in the resin (fluid) phases is built. Finite element resolution of these equations is done using the above described mixed stabilized finite element method, with a generated mesh of 5 million nodes, from the image. In terms of computational time, mesh generation, with the overall adaptation procedure, was performed in 3 h, whereas the Navier–Stokes resolution took only 10 min, on 96 CPUs. Results detailed in Figure 9.11 were obtained by imposing a pressure gradient through the sample on one direction, and imposing no-velocity in the fibers, showing that: the mesh is well adapted at the fluid–solid interface; the definition of this interface is very accurate, as illustrated by the isovalue zero of u_ε; the computed velocity field enhances well the acceleration in the interstitial spaces.

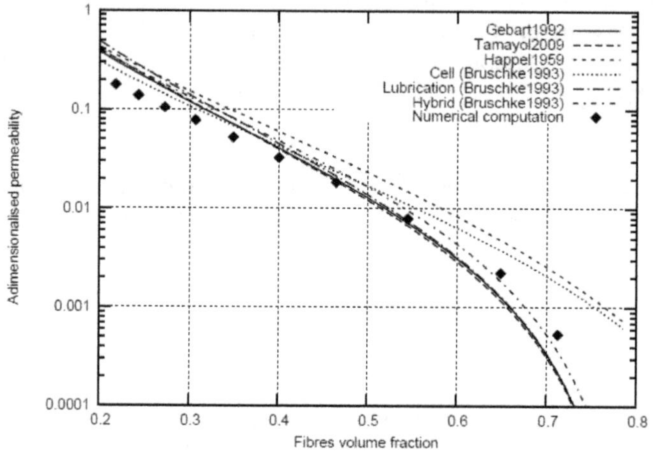

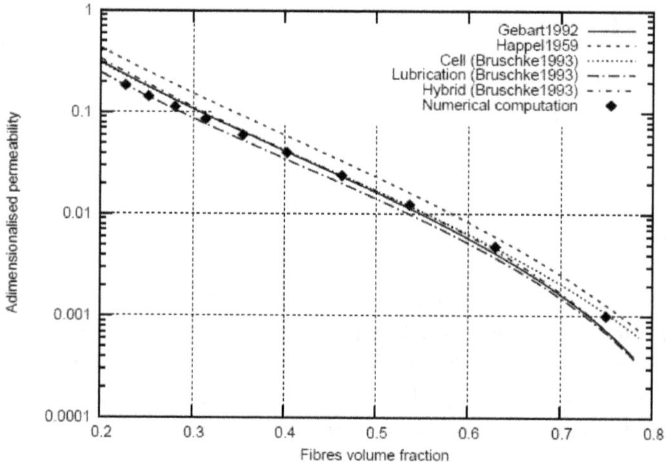

Figure 9.10. *Permeability determination for regular arrays of fibers and comparison with semi-analytic solutions for a) a square packing and b) a triangular packing, for different volume fraction of fibers [PUA 11]*

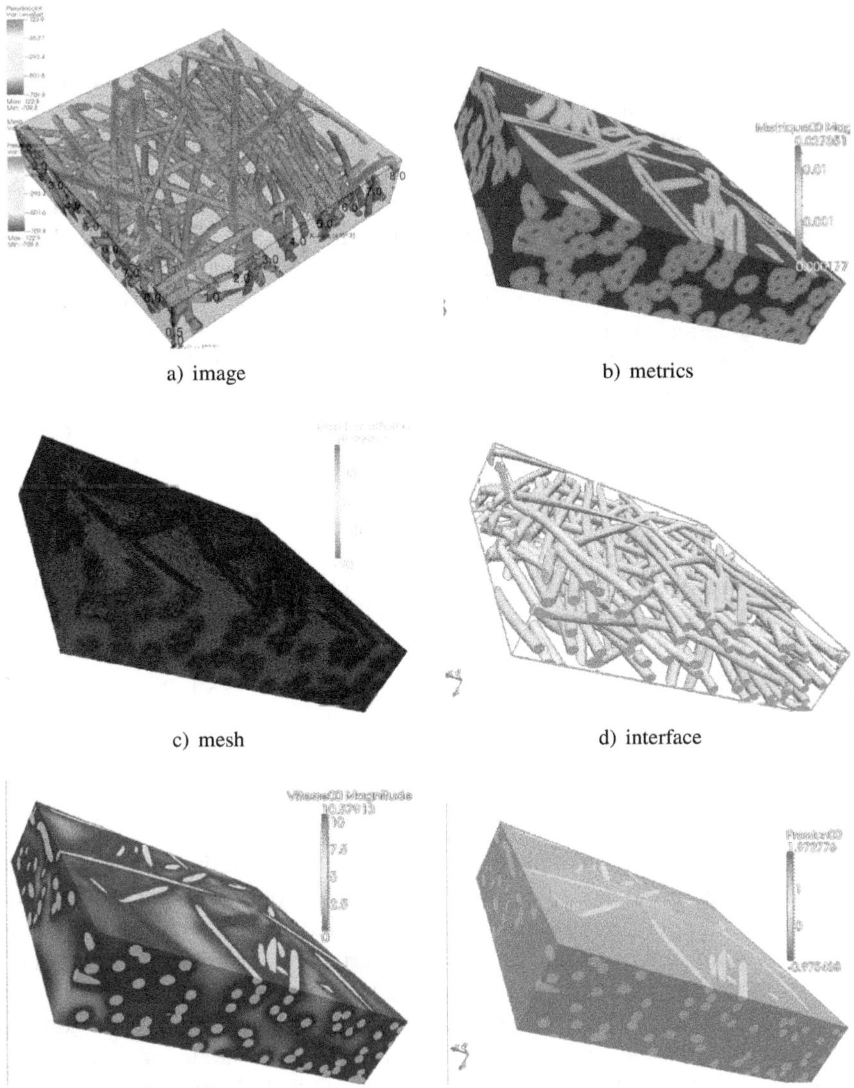

Figure 9.11. *Permeability determination for a 3D complex sample: a) original image with 178.2 millions of voxels; b) von Mises equivalent of the computed metrics field; c) final mesh of about 5 millions of nodes; d) isovalue zero of the regularized Heaviside e) computed velocity and f) pressure fields [SIL 14b]. For a color version of the figure, see www.iste.co.uk/boyard/heat.zip*

9.4.3. *Stiffness tensor determination*

As discussed previously, direct numerical thermoelastic simulations on REVs also allow the determination of the stiffness and dilatation tensors of the sample. Like before, let us consider a single mesh which contains both fibers and matrix, Figure 9.12(a). A tied contact between fiber and matrix is assumed, as well as between two independent fibers. Knowing the stiffness and the thermal dilatation for fibers and resin, equation [9.18] is used to derive the stiffness and the thermal dilatation everywhere in the REV. Several elastic thermomechanical tests are applied on this REV and numerically homogenized stiffness and thermal dilatation tensor are obtained. Numerical resolution of this heterogeneous elastic problem may be performed using a mixed finite element method [MIL 12], with forces imposed on each side of the REV. To avoid the imposition of periodic boundary conditions, homogenization is performed on a smaller included REV than the one where computations are done.

This type of computation has been performed by Miled [MIL 12], who has shown that one important issue was the optimization of the number of fibers in the REV in order to obtain a stabilized result. For unidirectional aligned fibers, Figure 9.12(b) plots the computed modulus in one of the principal directions, as a function of the number of fibers. It is compared with the Mori–Tanaka semi-analytic result showing that, for the sample represented, 22 fibers provided a convergent result, using a very fine adapted mesh.

9.5. Conclusions

In this chapter, we have detailed conservation and constitutive equations involved in multiphysics, multidomain and multiphase modeling of composites, which may be either short, long or continuously reinforced. Advanced numerical techniques involved in flow, thermo-rheo-kinetical and structure development computations were presented. They included mixed stabilized finite elements, Eulerian approaches and modified level-set techniques, automatic anisotropic mesh adaptation and parallel computing. Illustrations from different authors were presented, for forming at the macroscopic scale and property's direct numerical homogenization at the macroscopic or mesoscopic scales. Examples included short fiber reinforced thermoplastics injection molding, SMC compression molding, resin transfer molding, permeability determination at the microscopic and mesoscopic scales and stiffness tensor determination. As a conclusion, a huge step forwards has been achieved to consider complex physics in molding simulations, but other advanced phenomena need to be accounted for which will surely induce new mathematical and numerical developments.

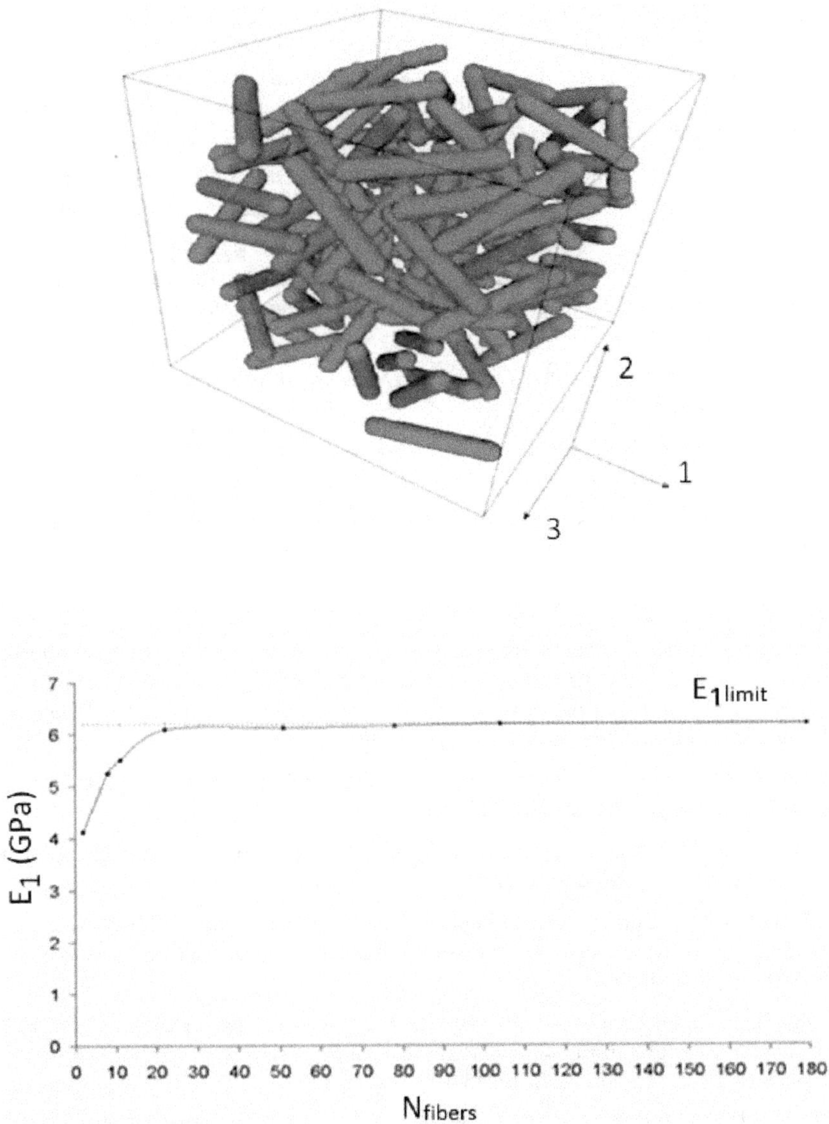

Figure 9.12. *Stiffness and dilatation tensor determination; a) typical representative elementary volume: the cube represents the computational domain (meshed) and in red, the fiber–matrix interface; b) Young's modulus as a function of the number of fibers in the REV (the Mori–Tanaka modulus is 6.21 GPa), for a given mesh size (very fine, in that case) [MIL 12]*

9.6. Bibliography

[ABI 14] ABISSET-CHAVANNE E., CHINESTA F., "Toward an optimisation of the reactive resin transfer molding process: thermo-chemico-mechanical coupled simulations", *International Journal of Material Forming*, vol. 7, no. 2, pp. 249–258, 2014.

[ADV 87] ADVANI S.G., TUCKER C.L., "The use of tensors to describe and predict fiber orientation in short fiber composites", *Journal of Rheology*, vol. 31, no. 8, pp. 751–784, 1987.

[ADV 09] ADVANI S.G., "Opportunities and challenges of multiscale modeling and simulation in polymer composite processing", *International Journal of Material Forming*, vol. 2, no. 1, pp. 39–44, 2009.

[ADV 10] ADVANI S.G., SOZER E.M., *Process Modeling in Composites Manufacturing*, 2nd ed., CRC Press, 2010.

[AGA 96] AGASSANT J.F., AVENAS P., SERGENT J.P. *et al.*, *La mise en forme des matières plastiques*, Lavoisier, 1996.

[ALL 11] ALLAOUI S., BOISSE P., CHATEL S. *et al.*, "Experimental and numerical analyses of textile reinforcement forming of a tetrahedral shape", *Composites Part A: Applied Science and Manufacturing*, vol. 42, no. 6, pp. 612 – 622, 2011.

[AMM 05] AMMAR A., CHINESTA F., "A particle strategy for solving the fokker-planck equation modelling the fiber orientation distribution in steady recirculating flows involving short fiber suspensions", in GRIEBEL M., SCHWEITZER M.A. (eds), *Meshfree Methods for Partial Differential Equations II*, Lecture Notes in Computational Science and Engineering, Springer Berlin Heidelberg, vol. 43, pp. 1–15, 2005.

[AUR 09] AURIAULT J.-A., "On the domain of validity of Brinkman's equation", *Transport in Porous Media*, vol. 79, pp. 215–223, 2009.

[BAL 15] BALAY S., ABHYANKAR S., ADAMS M.F. *et al.*, PETSc Web page, available at: http://www.mcs.anl.gov/petsc, 2015.

[BIS 13] BISSUEL A., Process modelling of HP-RTM and Compression RTM using Rem3D, Technical report, Numerical simulation of flows at microscopic scale and mesoscopic scale in the RTM process, 2013.

[BRE 98] BREZZI F., FRANCA L.P., RUSSO A., "Further considerations on residual-free bubbles for advective-diffusive equations", *Computer Methods in Applied Mechanics and Engineering*, vol. 166, nos. 1–2, pp. 25 – 33, 1998.

[BRI 49] BRINKMAN H.C., "A calculation of the viscous force exerted by a flowing fluid on a dense swarm of particles", *Applied Scientific Research*, vol. 1, no. 1, pp. 27–34, 1949.

[BRO 82] BROOKS A.N.,HUGHES T.J.R., "Streamline upwind/petrov-galerkin formulations for convection dominated flows with particular emphasis on the incompressible Navier-Stokes equations", *Computer Methods in Applied Mechanics and Engineering*, vol. 32, nos. 1–3, pp. 199 – 259, 1982.

[BRU 90] BRUSCHKE M.V., ADVANI S.G., "A finite element/control volume approach to mold filling in anisotropic porous media", *Polymer Composites*, vol. 11, no. 6, pp. 398–405, 1990.

[COM 05] COMAS-CARDONA S., GROENENBOOM P., BINETRUY C. *et al.*, "A generic mixed fe-sph method to address hydro-mechanical coupling in liquid composite moulding processes", *Composites Part A: Applied Science and Manufacturing*, vol. 36, no. 7, pp. 1004–1010, 2005.

[COU 94] COUPEZ T., "A mesh improvement method for 3D automatic remeshing", in WEATHERILL N.P., EISEMAN P.R., HAUSE J., THOMPSON J.F. (eds), *Numerical Grid Generation in Computational Fluid Dynamics and Related Fields*, Pineridge Press, pp. 615–626, 1994.

[COU 00] COUPEZ T., DIGONNET H., DUCLOUX R., "Parallel meshing and remeshing", *Applied Mathematical Modelling*, vol. 25, pp. 153–175, 2000.

[COU 11] COUPEZ T., "Metric construction by length distribution tensor and edge based error for anisotropic adaptive meshing", *Journal of Computational Physics*, vol. 230, pp. 2391–2405, 2011.

[COU 13] COUPEZ T., HACHEM E., "Solution of high-reynolds incompressible flow with stabilized finite element and adaptive anisotropic meshing", *Computer Methods in Applied Mechanics and Engineering*, vol. 267, pp. 65–85, 2013.

[COU 15] COUPEZ T., SILVA L., HACHEM E., "Implicit boundary and adaptive anisotropic meshing", *New Challenges in Grid Generation and Adaptivity for Scientific Computing*, SEMA SIMAI Springer Series, vol. 5, 2015.

[DAR 56] DARCY H., *Les fontaines publiques de la ville de Dijon*, Dalmont, Victor, 1856.

[DIB 87] DIBENEDETTO A.T., "Prediction of the glass transition temperature of polymers: a model based on the principle of corresponding states", *Journal of Polymer Science*, vol. 25, no. 9, pp. 1949–1969, 1987.

[FER 09] FÉREC J., AUSIAS G., HEUZEY M.C. *et al.*, "Modeling fiber interactions in semiconcentrated fiber suspensions", *Journal of Rheology*, vol. 53, no. 1, pp. 49–72, 2009.

[FOL 84] FOLGAR F., TUCKER C.L., "Orientation behavior of fibers in concentrated suspensions", *Journal of Reinforced Plastics and Composites*, vol. 3, no. 2, pp. 98–119, 1984.

[FOU 08] FOUDRINIER E., VENET C., SILVA L., "3d computation of reactive moulding processes", *International Journal of Material Forming*, vol. 1, no. 1, pp. 735–738, 2008.

[GAN 13] GANTOIS R., CANTAREL A., DUSSERRE G. *et al.*, "Bem-based models to simulate the resin flow at macroscale and microscale in lcm processes", *Polymer Composites*, vol. 34, no. 8, pp. 1235–1244, 2013.

[GEB 92] GEBART B.R., "Permeability of unidirectional reinforcements for RTM", *Journal of Composite Materials*, vol. 26, no. 8, pp. 1100–1133, 1992.

[GHN 11] GHNATIOS C., CHINESTA F., CUETO E. *et al.*, "Methodological approach to efficient modeling and optimization of thermal processes taking place in a die: Application to pultrusion", *Composites Part A: Applied Science and Manufacturing*, vol. 42, no. 9, pp. 1169 – 1178, 2011.

[HAN 62] HAND G.L., "A theory of anisotropic fluids", *Journal of Fluid Mechanics*, vol. 13, no. 1, pp. 33–46, 1962.

[HIR 81] HIRT C.W., NICHOLS B.D., "Volume of fluid (vof) methods for the dynamics of free boundaries", *Journal of Computational Physics*, vol. 39, no. 1, pp. 201–225, 1981.

[HIT 11] HITTI K., Direct numerical simulation of complex representative volume elements (RVEs): generation, resolution and homogenization, PhD Thesis, MINES ParisTech, 2011.

[JEF 22] JEFFERY G., "The motion of ellipsoidal particles immersed in viscous fluid", *Proceedings of the Royal Society A*, vol. 102, no. 715, pp. 161–179, 1922.

[KHA 10] KHAN M.A., MABROUKI T., VIDAL-SALLÉ E. *et al.*, "Numerical and experimental analyses of woven composite reinforcement forming using a hypoelastic behaviour. application to the double dome benchmark", *Journal of Materials Processing Technology*, vol. 210, no. 2, pp. 378–388, 2010.

[LAU 11] LAURE P., SILVA L., VINCENT M., "Modeling short fibre polymer reinforcements for composites", in BOISSE P. (ed.), *Composite Reinforcements for Optimum Performance*, Woodhead Publishing Limited, pp. 619–650, 2011.

[LIP 84] LIPSCOMB G.G., KEUNINGS R., MARRUCCI G. *et al.*, "Continuum theory for fiber suspensions", in *Advances in Rheology: Polymers: IX International Congress on Rheology*, 1984.

[LLO 04] LLORCA J., SEGURADO J., "Three-dimensional multiparticle cell simulations of deformation and damage in sphere-reinforced composites", *Materials Science and Engineering: A*, vol. 365, nos. 1–2, pp. 267 – 274, 2004.

[MIL 08] MILED H., SILVA L., AGASSANT J.F. *et al.*, "Numerical simulation of fiber orientation and resulting thermo-elastic behavior in reinforced thermo-plastics", *Mechanical Response of Composites*, volume 10 of *Computational Methods in Applied Sciences*, Springer Netherlands, pp. 293–313, 2008.

[MIL 12] MILED H., SILVA L., COUPEZ T. *et al.*, "Injection molding of fibre reinforced thermoplastics: integration of fibre orientation and mechanical properties computations", *International Polymer Processing*, vol. 27, no. 5, pp. 547–556, 2012.

[MOR 73] MORI T., TANAKA K., "Average stress in matrix and average elastic energy of materials with misfitting inclusions", *Acta Metallurgica*, vol. 21, no. 5, pp. 571 – 574, 1973.

[ORG 12] ORGÉAS L., DUMONT P.J.J., VASSAL J.P. *et al.*, "In-plane conduction of polymer composite plates reinforced with architectured networks of copper fibres", *Journal of Materials Science*, vol. 47, no. 6, pp. 2932–2942, 2012.

[OSH 88] OSHER S., SETHIAN J.A., "Fronts propagating with curvature-dependent speed: algorithms based on Hamilton-Jacobi formulations", *Journal of Computational Physics*, vol. 79, no. 1, pp. 12 – 49, 1988.

[PAR 15] PARK C.H., "15 – numerical simulation of flow processes in composites manufacturing", in BOISSE P. (ed.), *Advances in Composites Manufacturing and Process Design*, Woodhead Publishing, pp. 317–378, 2015.

[PHE 09] PHELPS J.H., TUCKER C.L. III, "An anisotropic rotary diffusion model for fiber orientation in short- and long-fiber thermoplastics", *Journal of Non-Newtonian Fluid Mechanics*, vol. 156, no. 3, pp. 165 – 176, 2009.

[PUA 11] PUAUX G., Numerical simulation of flows at microscopic scale and mesoscopic scale in the RTM process, PhD Thesis, École Nationale Supérieure des Mines de Paris, 2011.

[SEG 05] SEGURADO J., LLORCA J., "A computational micromechanics study of the effect of interface decohesion on the mechanical behavior of composites", *Acta Materialia*, vol. 53, no. 18, pp. 4931 – 4942, 2005.

[SER 90] SERRANO D., PEYRELASSE J., BONED C. *et al.*, "Application of the percolation model to gelation of an epoxy resin", *Journal of Applied Polymer Science*, vol. 39, no. 3, pp. 679–693, 1990.

[SHA 01] SHAPIRO L., STOCKMAN G., *Computer Vision*, Prentice Hall, 2001.

[SIL 09] SILVA L., AGASSANT J.F., COUPEZ T., "Three-dimensional injection molding simulation", in KAMAL M.R., ISAYEV A.I., LIU S.J. (eds), *Injection Molding, Technology and Fundamentals*, Carl Hanser Verlag, pp. 599–651, 2009.

[SIL 10] SILVA L., PUAUX G., VINCENT M. *et al.*, "A monolithic finite element approach to compute permeabilityatc microscopic scales in lcm processes", *International Journal of Material Forming*, vol. 3, no. 1, pp. 619–622, 2010.

[SIL 14a] SILVA L., VILLE L., LAURE P., "Numerical simulation of compression moulding using an adaptative monolithic finite element approach", *17th ESAFORM Conference*, 2014.

[SIL 14b] SILVA L., ZHAO J.X., DIGONNET H. *et al.*, "Flow simulations on 3d segmented images using reinitialization and anisotropic mesh adaptation", *Computational Modeling of Objects Presented in Images. Fundamentals, Methods, and Applications*, Springer International Publishing, pp. 339–350, 2014.

[SIL 15] SILVA L., COUPEZ T., DIGONNET H., "Massively parallel mesh adaptation and linear system resolution for multiphase flows", *27th International Conference on Parallel Computational Fluid Dynamics*, 2015.

[SIM 07] SIMACEK P., ADVANI S.G., "Modeling resin flow and fiber tow saturation induced by distribution media collapse in vartm", *Composites Science and Technology*, vol. 67, no. 13, pp. 2757–2769, 2007

[SOU 76] SOUROR S., KAMAL M.R., "Differential scanning calorimetry of epoxy cure: isothermal cure kinetics", *Thermochimica Acta*, vol. 14, no. 1–2, pp. 41–59, 1976.

[SOU 06] SOUKANE S., TROCHU F., "Application of the level set method to the simulation of resin transfer molding", *Composites Science and Technology*, vol. 66, no. 7, pp. 1067–1080, 2006.

[SUN 05] SUN C.J., SAFFARI P., SADEGHIPOUR K. *et al.*, "Effects of particle arrangement on stress concentrations in composites", *Materials Science and Engineering: A*, vol. 405, nos. 1–2, pp. 287 – 295, 2005.

[TOT 10] TOTRY E., MOLINA-ALDAREGUÍA J.M., GONZÁLEZ C. *et al.*, "Effect of fiber, matrix and interface properties on the in-plane shear deformation of carbon-fiber reinforced composites", *Composites Science and Technology*, vol. 70, no. 6, pp. 970 – 980, 2010.

[TRO 92] TROCHU F., GAUVIN R., "Limitations of a boundary fitted finite difference method for the simulation of the resin transfer molding process", *Journal of Reinforced Plastics and Composites*, vol. 11, pp. 772–786, 1992.

[TRO 06] TROCHU F., RUIZ E., ACHIM V. *et al.*, "Advanced numerical simulation of liquid composite molding for process analysis and optimization", *Composites Part A: Applied Science and Manufacturing*, vol. 37, no. 6, pp. 890 – 902, 2006.

[TUC 91] TUCKER C.L., "Flow regimes for fiber suspensions in narrow gaps", *Journal of Non-Newtonian Fluid Mechanics*, vol. 39, no. 3, pp. 239 – 268, 1991.

[UM 91] UM M.K., LEE W.I., "A study on the mold filling process in resin transfer molding", *Polymer Engineering and Science*, vol. 31, no. 11, pp. 765–771, 1991.

[VIL 15] VILLE L., REM3D Web page, available at: http://www.transvalor.com, 2015.

[WAN 08] WANG J., O'GARA J.F., TUCKER C.L., "An objective model for slow orientation kinetics in concentrated fiber suspensions: theory and rheological evidence", *Journal of Rheology*, vol. 52, no. 5, pp. 1179–1200, 2008.

[YOO 96] YOO Y.E., LEE W.I., "Numerical simulation of resin transfer molding process using boundary element method", *Polymer Composites*, vol. 17, no. 3, pp. 368–374, 1996.

[ZHA 15] ZHAO J.X., COUPEZ T., DECENCIÈRE E. *et al.*, "Multiphase mesh generation from 3d image by anisotropic mesh adaptation and redistancing equation", *Journal of Computational Physics*, submitted, 2015.

10

Thermal Instrumentation for the Control of Manufacturing Processes of Organic Matrix Composite Materials

Thermal characterization is useful for the control of manufacturing processes of organic matrix composite materials or every heterogeneous materials partially made up of polymers. Through transient temperature and heat flux measurements, the thermophysical properties or transfer functions related to the heat transfer in the system can be estimated in order to give information about the process or the transformations of a sample during the manufacturing. Two kinds of approaches are to be considered: the contact methods generally based on a surface instrumentation besides the process or a radiative instrumentation by considering the emission of the surface or the optical properties of the system to selected radiations.

For contact measurements, the principles of the temperature and heat flux measurement will be presented in order to design simple contact probes, with simplified assumptions about the sample or the system to be studied. Non-contact measurement methods have been recently considered through new possibilities of infrared cameras or sensors and new optical heating possibilities (lasers or multispectral sources). Instead of processing a signal coming from a monosensor, such methods need to take into account a temperature field associated with a large amount of data. Several of these processing methods and the related problems will be explained.

10.1. Introduction

The control of manufacturing processes of organic matrix composite materials (or every heterogeneous materials partially made up of polymers) is strongly related

Chapter written by Jean-Christophe BATSALE and Christophe PRADERE.

to the temperatures and thermal properties of the samples to be processed. The thermophysical properties or transfer functions related to the heat transfer in the system can be estimated in order to give information about the process or the transformations of a sample not only during the manufacturing, but also at the laboratory scale in order to anticipate the optimal design of the process or for a systematic study of a great number of samples coming from the process (screening of materials and processes).

Two kinds of approaches are to be considered: the contact methods generally considering a surface instrumentation besides the process or a radiative instrumentation by considering the emission of the surface of the wall or the transparency of the system to selected radiations.

The main difficulties of such measurements are related to the wide variety of situations especially the characteristic scales to be investigated (contact between fiber and matrix, multilayered samples or manufactory scale, etc.). However, the accessibility to the sample is often difficult and the methods need to be non-intrusive. The main principle of such methods will be to analyze the transient temperature responses to a calibrated heat flux. The data processing will then be essential. Several examples will be given (from microfluidics to large-scale non-destructive evaluations).

10.2. Methods based on contact measurement

Because the phase change of polymers, viscosity and transport properties of the liquids are strongly dependant on the local absolute temperature, it is necessary to implement temperature probes inside or beside the medium to be processed. The characterization of the process needs not only the knowledge of a temperature but also a temperature gradient or a heat flux measurement in order to access to the thermophysical properties in liquids or solids. A quick review of the classical sensors and methods for temperature and heat flux measurement is presented and related to thermal probes for *in situ* thermophysical properties measurements.

10.2.1. *Temperature sensors and fluxmeters*

10.2.1.1. *Temperature probes (thermoresistances and thermocouples)*

10.2.1.1.1. Thermoresistances, thermistors or resistance temperature detectors (RTDs)

Thermoresistances, thermistors or resistance temperature detectors (RTDs) are sensors where the thermosensitive parameter is the electrical resistance (see

[BAR 01, GAR 11, MCG 88]). Platinum was proposed by Sir William Siemens as an element for resistance temperature detector in 1871 [SIE 71]. Thermistors differ from RTDs in that the material used in a thermistor is generally a ceramic or a polymer, while RTDs use pure metals. The temperature response is also different; RTDs are useful over larger temperature ranges, while thermistors typically achieve a greater sensitivity within a limited temperature range, typically −90 to 130°C. The stability of thermistors is less reliable than the stability of RTD.

The time constant of such devices is generally about a few seconds or several tens of seconds because the resistor is associated with an insulating layer and put in contact with the substrate to be measured. Even if a modeling and very thin layers can be considered, such devices are widely used for quasi-steady measurements. However, they are sensitive and reliable to detect an absolute temperature.

By contrast, thermocouples of smaller size can be implemented in order to obtain smaller response times (about a few ms) but generally suited to measure only temperature differences reported to a reference.

10.2.1.1.2. Thermocouples

Thermocouples are based on physical phenomena called thermolectric effects. Such effects are the Seebeck, the Peltier and the Tompson effects (see [THO 12]). All these effects involve a direct conversion of temperature differences to electric voltage or inversely. The voltage created by a temperature difference on each side of a thermoelectric device allows us to set out a temperature measurement device.

The main materials to be used for thermocouple junctions are related to thermolectric power in V/°C. Tables related to the current type of thermocouples J (iron-constantan), K (chromel-alumel), T (copper-constantan), S (platinum-platimum/10%/rhodium), B (platinum/30%/rhodium-platinum/6%/rhodium), E (chromel-constantan), R (platinum/13%/rhodium-platinum) and N (nicrosil-nisil) are given on the sites of National Laboratories for Metrology [NIS 90, EUR 16].

10.2.1.1.3. Systematic errors related to the local disturbance of field temperature (inertia and position errors with contact monosensors)

A sensor is an electronic or electric device which gives a signal processed from a chain of devices (amplifiers, samplers, etc.) and transforms the ideal measurement into a noisy and biased observation. In order to take into account the measurement noise, the properties of the mean value and the knowledge of the variance of the stationary signal without any transient excitation are necessary.

A thermocouple or even a thermoresistance is a solid sensor, often with metallic parts, which is also perturbating the temperature field especially when high temperature gradients are occurring. Several precautions are taken such as putting the sensor and the connection wire along an assumed isotherm curve or plane. The temperature measurement during manufacturing processes such as molding or inside polymer flow is even more difficult. It needs the modeling and the design of specific sensors (see, for instance, Launay [LAU 13] and [LAU 14]).

With regard to the theoretical aspects of the instrumentation, Bourouga [BOU 00] proposed criteria for correct locations of thermocouples to obtain unbiased results and also to optimize the experiment for wall heat flux or temperature estimation.

Especially in the case of transient experiment, the inertia cannot be avoided. Generally, the output signal of the sensor is considered as a convolution product which takes into account the thermal contact resistance or an exchange coefficient h and an apparent heat capacity ρcL for the system such as:

$$Y(t)=\frac{K}{\rho cL}\int_0^t \exp\left(-\frac{h}{\rho cL}\tau\right)U(t-\tau)d\tau=\int_0^t H(\tau)U(t-\tau)d\tau=\int_0^t H(t-\tau)U(\tau)d\tau$$

K is an arbitrary multiplicative constant and $U(t)$ is the idealized signal without inertia and the inertial and resistance effects can be represented under an impulsional response: $H(t)$.

The implementation of the previous expression often requires discretization and several precautions related to the measurement noise. One way to avoid such difficulties is to estimate the transfer function between the measured temperature response and a surface heat flux which causes the transient gradient. A prerequisite condition is to be able to heat the system with a calibrated heat flux and to estimate from a model the impulse response or the transfer function (see [BAT 00]). Such processing needs are very similar to the approaches related to the heat flux estimations which will be evocated in section 10.2.2.1.3.

10.2.2. *Heat flux estimation*

The heat flux measurement is at least as important as the temperature measurement. Generally, the manufacturing process is not only at a high temperature, but also under temperature gradients, generally non-steady. The heat flux estimation can reveal very suitable and important information about the heat sources inside the sample to be processed (phase change enthalpies, chemical reactions or simple parietal heat losses).

The fluxmetersor heat flux sensors (HFSs) can be considered as instruments related to steady heat flux for a given situation, which deliver a global signal without additional processing. Some of these instruments will be described later on. More generally, heat flux measurement is a difficult task because the heat transfer during the manufacturing process is multidimensional and transient inside complex geometries and systems. It is often preferred to consider a heat flux "estimation" from optimized temperature observations. A section will be devoted to the description of such estimation approaches.

10.2.2.1. *Fluxmeters, heat flux sensors and direct measurement methods*

10.2.2.1.1. Heat flux sensor with gradient

The principle of this heat flux measurement (see [BAR 01], [GAR 11] and [THU 96]) consists of directly applying Fourier's law by measuring a temperature difference within the wall itself (intrinsic method) or by covering it with an additional wall (heat flux sensor (HFS)). The surface characteristics of this HFS should be close to those of the wall. The wall of the HFS can be homogeneous (the temperature difference is measured between its two main faces – normal gradient heat flux sensor) or it can be heterogeneous creating heterogeneous temperature that is measured (tangential gradient heat flux sensor). The installation of such HFS on a wall, more or less, disturbs the heat flux which crosses it. All must be done so that internal and contact thermal resistances are minimal. In these devices, the measurement of the temperature difference is performed using several thin film thermocouples or thermoresistances. These HFSs can work whatever the heat flux direction in steady state or for slowly variable temperature.

A particularly efficient HFS is based on bismuth telluride (Bi_2Te_3) sensors. Bi_2Te_3 is a semi-conductor which, when alloyed with antinomy or selenium, is an efficient thermoelectric material for refrigeration or portable power generation. The inverse use of such an electronic heat pump is offering an HFS with a very high thermoelectric power. Unfortunately, the thickness of the wall often allows a too slow characteristic time. Very active research is oriented to the miniaturization and the improvement of the thermoelectric power of such devices (see, for instance, [STA 99]).

10.2.2.1.2. Inertia heat flux sensor and heat flux sensor with electric dissipation (zero method)

Inertia heat flux sensors (see [THU 96]) only work for variable temperatures and if the heat flux is received by the wall. The HFS replaces a piece of the wall and is isolated from this one. Its surface characteristics are identical to those of the wall. The temperature increase in the HFS is proportional to the absorbed heat flux and inversely proportional to its capacity. The choice of this capacity is very important because it determines the measurement sensitivity.

The principle of this HFS with electric dissipation consists of substituting a piece of the wall at its surface with a small heating part insulated toward the wall. The electric heating output is adjusted so that the surface temperatures of the wall and the heating part are equal ($\Delta T=0$). Thus, the dissipated electric flux is equal to the heat flux which leaves the wall in its immediate vicinity. This HFS only works for heat flux leaving the wall and for steady state or slowly variable temperature.

10.2.2.1.3. Indirect heat flux measurement, transient heat flux from the inversion of a temperature measurement

A transient heat flux estimation can be obtained from temperature measurements realized within the medium and using “inverse methods”. This procedure involves solutions of “ill-posed problems”. Indeed, we cannot insure a solution, nor its uniqueness or stability. To solve such difficulties, the technique consists of replacing the ill-posed problem by a well-posed approximate problem.

One example [BAT 15] of such a situation can be often considered as a convolution problem. The deconvolution of a signal is also an operation often required in the case of processing of experimental data (see the previous thermocouple signal processing in section 10.2.1.1.3), for instance when searching the transfer function of a system or sensor, in image processing, optics, geophysics, etc. We give again the heat transfer example of some heat capacity exchanging with convective heat losses with the surrounding medium, such as:

$$C\frac{dT}{dt} = q(t) - hT \text{ with the initial condition } t = 0,\ T = 0.$$

where:

$q(t)$: calibrated heating (W) for the excitation of the system;

h: thermal exchange coefficient in WK^{-1};

C: heat capacity in JK^{-1}.

We assume here that $C = 1, T_\infty = 0$ and that the boundary surface of the body is 1.

Solving this equation by using the Laplace transform of the temperature and heat flux and inverting yields the solution in the form of the following convolution product:

$$T(t) = \int_0^t q(t-\tau)\exp(-h\tau)d\tau$$

The approach is proposed here in, such as:

1) choose some heat flux function, such as *q(t)*;

2) compute the corresponding analytical solution T(t) as the convolution product above;

3) add some random error: $\varepsilon(t)$ such as the observed temperature is: $y(t) = T(t) + \varepsilon(t)$;

4) retrieve the heat flux by inverting the product of this signal (deconvolution);

5) repeat for different values of the signal-to-noise ratio (different levels of std).

The results are depicted in Figure 10.1. For a low standard deviation of the error (std = 0.03 K), the heat flux is conveniently retrieved by the deconvolution operation. When increasing the noise level (std = 0.3 K), the drastic amplification of the errors in the deconvolution operation makes the result absolutely inaccurate. The visual effect of the noise level in the curves where the temperature outputs are drawn shows that the increase in noise between the two situations which makes the solution accurate or unavailable is not significative. It is apparent with this example that the deconvolution of an experimental signal may be an ill-posed problem, depending on the functional form of the impulse response, due to its unstable nature.

The solution is found by minimizing a norm of least square type. A heat transfer model (analytical or numerical) is required to solve the direct problem at each optimization step. These methods require significant developments [THI 77, BEC 77, ORL 11, OZI 00]. They will not be presented here. We will just underline that the solution of the inverse problem allows us to compute the temperature residuals between final and measured temperatures. These residuals are of great importance because they allow us to check the validity of the chosen heat transfer model. If no signature is observed (the residuals are purely random) the model is correct, otherwise the model should be improved.

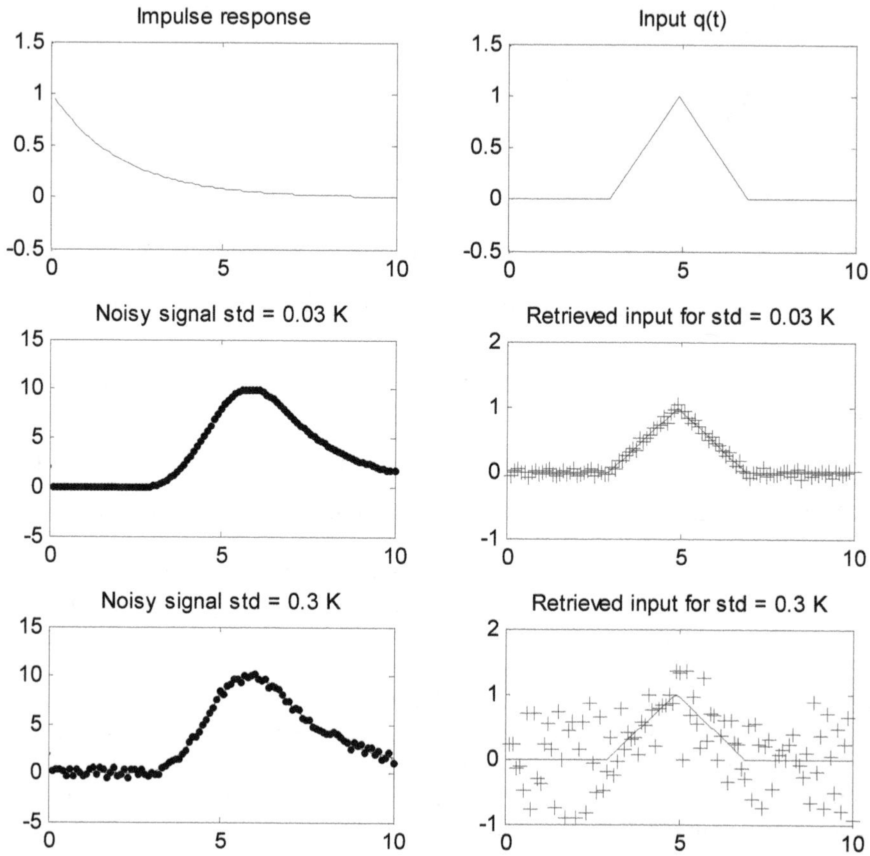

Figure 10.1. *Effect of the noise level on the deconvolution of a signal*

10.2.3. Thermal probes for thermophysical property measurement in static conditions

The association of a calibrated heat flux and the analysis of one or several temperature responses of a sample or a system is generally complementary of the previous temperature and heat flux measurements. It is classically used in order to measure thermophysical properties constitutive of a heat transfer model.

Several instruments include the heat flux generator and the temperature measurement in a single instrument called a "thermal probe". The most simple of these instruments are the "hot plane" and the "hot wire".

Thermal probes are very suitable for the control of industrial processes, because such methods are poorly intrusive and cheap. Such a principle allows very compact devices for laboratory or in *situ* measurements. The characteristic response time is well adapted to polymeric samples (some tens of seconds for classical probes and cm-sized insulating samples such as polymeric samples).

The main characteristic of these methods is to be able to perfectly master the heating energy given to the samples by the Joule effect knowledge. Such a situation is one of the most convenient in order to simultaneously estimate not only the thermal conductivity and heat capacity, but also the thermal diffusivity and thermal effusivity (see the theoretical developments of these methods in general books such as [CAR 59] and [MAI 00]).

10.2.3.1. *Hot plane*

The "hot plane" method for the estimation of the thermal effusivity is one of the simplest methods for thermophysical properties estimation. Uniform heating is applied (generally with the contact of a uniformly deposed heating resistors network) on the front face of the sample (assumed to be semi-infinite) and the transient front face temperature response is then analyzed (see [ZHA 65]). Generally, such a method needs to take into account the inertia of the heating device and the temperature probe. However, a first very suitable approximation of the temperature response is given by expression:

$$T(t) = R_c Q + \frac{Q}{\sqrt{\lambda \rho c_p}} 2\sqrt{\frac{t}{\pi}}$$

where:

R_C: thermal contact resistance between the probe and the heating device ($K\,W^{-1}m^2$);

Q: calibrated heating (Wm^{-2}) for the excitation of the system;

$\sqrt{\lambda \rho c_p}$: thermal effusivity of the considered semi-infinite sample (with λ: thermal conductivity in $W\,m^{-1}K^{-1}$, ρc_p: volumic heat capacity in $J\,m^{-3}K^{-1}$).

The success of this method is coming from the possibility to separate the probe effects (inertia and contact resistance) from the effusivity of the sample, by the separated estimation of the slope (versus $\sqrt{t}$) and the origin ordinate of the thermogram. Unfortunately, the sample if not always as thick as to be considered as "semi-infinite". However, the knowledge of the thermal effusivity can be completed by the thermal conductivity or the heat capacity. Then, the hot plane method can be combined with a "hot wire" method related to the thermal conductivity estimation.

10.2.3.2. *Hot wire*

The hot wire method is very similar from the previous hot-plane method (radial instead of plane). It consists of analyzing the temperature response to a "line source". Such a method has induced a lot of experimental applications and variants with solid sensors (platinum wire for both heating and temperature measurement, thermocouple-wire systems, etc.; see [DAV 84 and ZHA 93]). Unlike the "hot plane", this method is sensitive to the thermal conductivity instead of the thermal effusivity.

$$T(t) = \frac{q}{2\pi\lambda L} Ln(t) + k$$

where:

q: calibrated heating (W) for the excitation of the system;

λ: thermal conductivity $(W\,m^{-1}K^{-1})$;

L: length of the wire (m);

k: constant depending on the thermal diffusivity of the sample and the contact resistance and inertia between the wire and the sample.

By the same way as the previous method, it is possible to separate the probe effects (inertia and contact resistance) from the thermal conductivity of the sample, by estimating separately the slope (versus $\ln(t)$) and the origin ordinate of the thermogram.

10.2.3.3. *Hot disk, hot strip and other combined probes*

From the principle of the hot plane and hot wire, a lot of hybrid instruments have been developed. One of the first kinds of developments consists of developing a thin plane probe which can be considered as a hot plane at short time and which takes into account the 2D or 3D effects at longer time. It is then possible to estimate simultaneously the thermal effusivity and thermal conductivity.

Among the other hybrid instruments is the hot disk method [GUS 91, BOH 00]. Such a method is now developed with several dimensions for liquids or solids and for relatively high temperatures (until 700°C). Comparisons of similar methods are also available [RID 09].

A lot of variantion will consist of considering multilayered media [LAD 00, FUD 05, PER 09] and also the use of a heating probe equipped with one or several deported thermocouples [THO 10, THO 08] in order to consider non-isotropic samples and boundary effects. Other combinations of heating systems and fluxmeters allow us to estimate the thermophysical properties related to the saturation of composite samples [VIL 13].

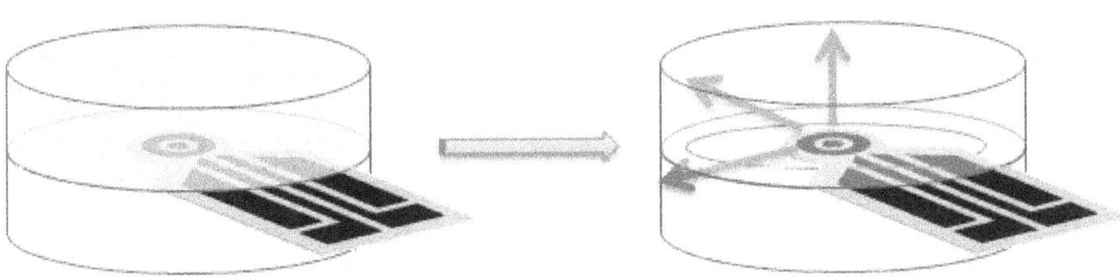

Figure 10.2. *Hot disk probe. The circular area is devoted to the simultaneous heating and temperature measurement. For a color version of the figure, see www.iste.co.uk/boyard/heat.zip*

10.3. Contactless heating

Even if the probe methods are suitable in order to allow a perfect knowledge of the heating energy, the limited number of temperature measurement allows only the characterization of homogeneous samples of sizes greater than several centimeters with simplified geometries.

More recent contactless methods (often called photothermal methods) are generally based on radiative heating and radiative temperature measurement. They allow us to avoid the inertia of the contact probes and then access to much shorter characteristic times and sizes of heterogeneous samples.

In addition, contactless radiative detectors can be assembled in focal plane arrays of detectors in order to estimate a great amount of temperatures assimilated to transient fields. Such principles offered by infrared cameras are opening the way to a lot of characterization methods: non-destructive evaluation methods for composite structures, pure screening at laboratory scale even with fluidic experiments (especially microfluidic) and, in a near future, methods for the survey of manufacture processings. From the versatility of optical devices, the range of observation scales with IR cameras is very large (about 10 μm with IR microscopes to very large scales such as factory scale).

10.3.1. *Photothermal methods with monosensors (under microscale characterization)*

Photothermal methods based on the analysis of a unique infrared detector are useful in order to estimate the properties of thin layers (thickness less than 1 μm) on thick substrate, generally used for the characterization of protective coatings [KUS 13]. The monodetector can be used at high-recording frequency, in order to operate with a high-recording rate (about 10^6 Hz) related to small diffusion length (less than 1 μm). Even if they are not suitable for the estimation of the heating energy, such methods are adapted to measure the thermal diffusivity of layered samples without using substrates with heat pulse methods or periodic heating methods [HAY 04] or to measure the resistive properties of an insulating layer on a calibrated effusive substrate [BAT 10].

In the case of composite samples, the photothermal methods can be adapted to characterize the properties of the constitutive parts such as the fibers, the wires assembled to the polymer matrix at different scales.

One example of fiber characterization (from [PRA 09]) is illustrated in Figure 10.3. It is generally composed of a short time optical heating (with a modulated laser spot) and the recording and analysis of the IR surface emission obtained with quantic detectors such as HgCdTe and InSb IR detectors.

The very small-scale thermal characterization (about several tenths of nanometers) is a very important challenge, for instance, in order to estimate the radial thermal resistance between fiber and matrix on a composite sample under stress conditions. At this moment, a lot of efforts are made in order to set out near-field thermal microscopes (see [SAC 15] and [CRE 07]).

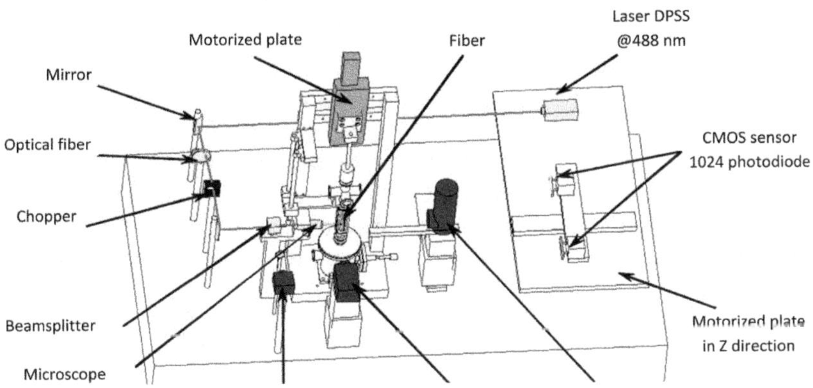

Figure 10.3. *Experiment for photothermal characterization of the longitudinal heat diffusivity of a carbon fiber at high temperature (from [PRA 09] and [PRA 06]). The diameter of the fiber is generally about a few micrometers and the longitudinal diffusion length is several millimeters. For a color version of the figure, see www.iste.co.uk/boyard/heat.zip*

10.3.2. *Thermal non-destructive evaluation: cracks or delamination detection in composite samples*

10.3.2.1. *Cracks or delamination detection in composite samples*

NDE with infrared cameras consists generally of applying a heat pulse on a non-homogeneous parallelepipedic slab and processing the temperature response $T(x, y, t)$ from one face of the sample (the front face or the rear face) (see [BAL 91]). The aim of such processing is to estimate some characteristics of the heterogeneities (structure, size, nature, position inside the sample, etc.). Instead of a 3D heterogeneous geometry, it is often convenient to consider thin samples with thin heterogeneities such as the transfer is locally 1D (versus z direction). One asymptotic expansion assuming that the heterogeneities fluctuations are small compared to the mean value of one thermophysical property of the sample yields a linear relationship, such as:

$$T(x,y,t) = q(x,y)\left(f(t,\beta_0) + \Delta\beta(x,y)\left.\frac{\partial f}{\partial \beta}\right|_{(t,\beta_0)}\right)$$

where the function $q(x, y)$ is the spatial distribution of energy coming from a flash excitation and $\Delta\beta(X, Y)$ is the spatial thermophysical property variation (diffusivity, conductivity, thickness, etc.). $q(x, y)$ and $\Delta\beta(X, Y)$ have to be specified with a finite number of parameters which will be the object of the estimation procedure. Such a procedure will then consist of processing the weighted sum of the images $T(x, y, z, t)$ where the weighting functions are the sensitivity functions $f(z,t,\beta_0)$ and $\partial f/\partial\beta|_{(z,\,t,\,\beta_0)}$. The temperature response of the front or the rear face of the sample will be recorded by a camera in order to estimate a map or a field of thermophysical properties. In the ideal case, the sample (a plane plate of small thickness L) is assumed to be thermally insulated and with a temperature field initially uniform ($T(x, y, z, t=0)=0$). If the heat transfer is supposed to be 1D, then, the temperature response related to a unique location (x,y) corresponding to a pixel, to an instantaneous thermal pulse, is given (see [PAR 61]) on the front face $(z = 0)$:

$$T(z=0,t) = \frac{Q}{\rho cL}\left(1 + 2\sum_{n=1}^{\infty} \exp\left(-\frac{n\pi at}{L^2}\right)\right) = \frac{Q}{\rho cL} f\left(\frac{at}{L^2}\right)$$

From the previous expression, the estimation problem of several parameters can be considered. Instead of the thermal diffusivity, the estimation problem of the sample thickness L, the thermal conductivity λ and the volumic heat capacity ρc can be considered. In each case, a reference approximated value of the parameter must be known (and denoted by L_0, l_0 and rc_0). In many cases, such reference values can be obtained by a previous global measurement. The following first-order asymptotic expansions can be, for example, written for the front face, at each time t_i: for example, if a thermal conductivity variation is to be estimated (see [MOU 01]):

$$T(0,t_i) \approx \frac{Q}{\rho CL} f\left(\frac{\lambda_0 t_i}{\rho CL^2}\right) + \frac{Q}{\rho CL}\frac{\Delta\lambda}{\lambda_0}\frac{\partial f}{\partial t}\bigg]_{\left(\frac{\lambda_0 t_i}{\rho CL^2}\right)}$$

It is very important to note that it is possible to replace the sensitivity functions $f(z=0, t, \beta_0)$ and $\partial f/\partial\beta|_{(zt,\,\beta_0)}$ by a linear combination of $f(z=0, t, \beta_0)$ and the time logarithmic derivative $t.\partial f/\partial\beta|_{(z=0,\,t,\,\beta_0)}$. It is then possible to implement a linear relationship for each pixel located at x,y such as:

$$T(x,y,0,t_i) \approx \beta_1(x,y)X_{\beta_1}(t_i) + \beta_2(x,y)X_{\beta_2}(t_i)$$

The functions $X_{\beta_1}(t)$ are the sensitivity functions of $T(x,y,0,t_i)$ to parameters β_j.

The parameters β_j are here defined such as: $\beta_1(x,y) = \frac{Q(x,y)}{\rho CL}$ and $\beta_2(x,y) = \frac{Q(x,y)}{\rho CL}\frac{\Delta\lambda(x,y)}{\lambda_0}$.

Several methods are based on similar decompositions. The projection of the great amount of data in a suitable truncated basis is very useful in order to handle the data.

In order to facilitate the calculation of the logarithmic derivative, Shepard [SHE 06] proposes a decomposition from a polynomial fitting, such as:

$$\mathrm{Ln}(T(x, y, z = 0, t)) = \beta_0(x, y) + \beta_1(x, y)\,\mathrm{Ln}(t) + \beta_2(x, y)\mathrm{Ln}^2(t) + \ldots$$

On the other side, Rajic [RAJ 02] or Bamford [BAM 09] proposed to consider the singular value decomposition of the global data cube (*x,y,t*), in order to reduce the amount of data and allow a statistical decomposition such as:

$$T(x, y, z = 0, t) = \sum_{k=1}^{K} U_k(x, y)\gamma_k V_k(t)$$

In fact, all the methods presented are very similar. They consist of projecting the data cube in a suitable basis of functions (in space or time, physical or not) and then to try to process the signal with a physical model. Such methods are suitable because in every pixel the whole image related to the analyzed field is of interest. Such asymptotic approaches are very convenient in the case where the thermophysical properties mapping is slightly varying versus mechanical stresses. The result shown in Figure 10.4 was used for ceramic matrix composite (CMC) characterization (see [YAG 15]).

In the next section (application to microfluidic devices), another strategy is proposed. It consists of using the physical model in order to reduce or eliminate the non-useful data (because nothing physically happens or because the sensor is providing a wrong signal).

10.3.2.2. *Photothermal determination of heat diffusivity tensor of orthotropic composite samples*

The same instruments (IR camera and radiative heating (laser or lamps)) can be used in order to estimate the thermal diffusivity tensor of a homogeneous but orthotropic composite sample.

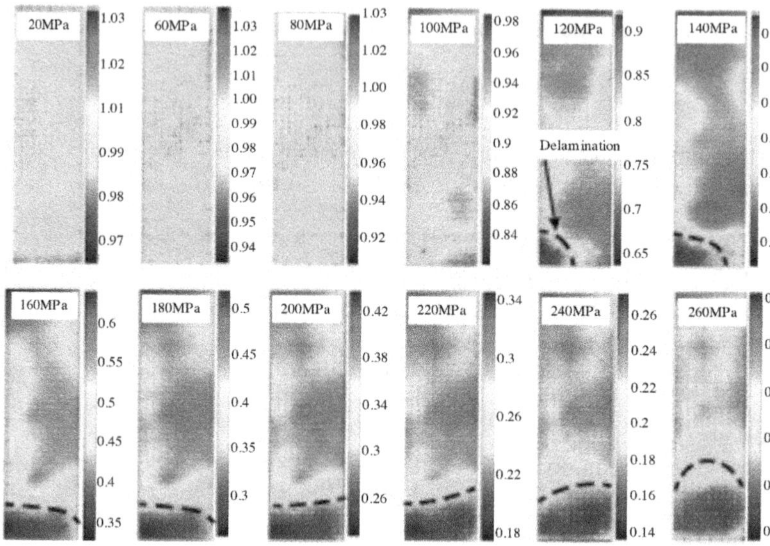

Figure 10.4. *Evolution of the transverse diffusivity mapping of a CMC sample under a tensile test. They illustrate the initiation and the propagation of the delamination (from [YAG 15]). For a color version of the figure, see www.iste.co.uk/boyard/heat.zip*

When a front face heating pulse applied on a composite sample is non-uniform, a 2D or 3D transient transfer assumption must then be considered, even with a non-isotropic sample. The general equation in the principal axes of anisotropy is then:

$$\frac{\partial 2\theta}{\partial x2}+\frac{k_y}{k_x}\frac{\partial 2\theta}{\partial y2}+\frac{k_z}{k_x}\frac{\partial 2\theta}{\partial z2}=\frac{1}{a_x}\frac{\partial\theta}{\partial t}$$

where k_x, k_y, k_z are the thermal conductivities following the principal directions x, y, z with $a_x = k_x/\rho C$: thermal diffusivity related to the direction x.

A simple situation can be reduced to a sample which is adiabatic on the lateral faces and a Fourier integral transform related to the space variables y and z and a Laplace transform related to the time variable t allows us to obtain a formal expression similar to the previous 1D case in order to allow the simultaneous estimation of parameters: a_x, k_y/k_x and k_z/k_x.

The considerations of the marginal averaging of the temperature images allowing the simultaneous estimation of the terms of thermal diffusivity tensor are, for instance, available in [PHI 95] and [KRA 04].

10.3.3. *Screening of chemical processes and microfluidic experiments*

The microfabrication techniques originally devoted to electronics have reached spectacular advances in more general domains such as chemical microreactors with microfluidic channels (lab-on-a-chip, see [JEN 01] and [MOL 04]). Such systems are very profitable because the low fluid or solid mass requirements and small volumes induce low-cost fabrication, which leads to new ways to design and intensify the experiments such as screening for catalyst testing related to chemical engineering, see [SEN 99]). Microdevices also have a lot of interesting characteristics in the field of heat transfer (high speed of the diffusive transfer, possibility of implementing localized heat sources, high surface/volume ratio and intensive local heat flux with low global energy consumption). One of the main drawbacks is related to instrumentation. Particularly, for temperature measurement, the IR thermography is a non-intrusive way for the recording and processing of a large amount of data. In the field of composite materials characterization, the inter-fibers space can be considered as a microchannel in which the polymer can flow. The main challenge is then to set out processing methods of the surface temperature field of such systems through a heat transfer model.

Instead of independent pixels, the spatial correlation of the pixel by diffusive or convective transient phenomena can be useful. This will be illustrated with the characterization of a simple transient in plane diffusion on a thin plate and then the characterization of convective effect caused by a microchannel sticked to a thin conductive plate (see [PRA 06]).

The simple in plane diffusion with source term is usually met in microelectronics. The objective can be to estimate the space and time distribution of energy given on a silicon plate from electronic chips from the observed transient temperature fields. It can also illustrate the case where a single droplet of reactant is deposited on the homogeneous conductive plate (a glass or silicon thin layer). A classical model of the temperature field with simplified assumptions is given with the following expression:

$$\frac{\partial T}{\partial t} = a\left(\frac{\partial^2 T}{\partial x^2} + \frac{\partial^2 T}{\partial y^2}\right) + Q(x,y,t)/\rho c$$

The estimation method of the source term field $Q(x, y, t)$ in $W.m^{-3}$ from a sequence of temperature images illustrated by Figures 10.5(a) and (b) consists of considering the Laplacian of the temperature field: $\Delta T = \left(\frac{\partial^2 T}{\partial x^2} + \frac{\partial^2 T}{\partial y^2}\right)$ and the time derivative $\frac{\partial T}{\partial t}$ at each time step and each pixel location, with a previous calibration for the knowledge of the thermal diffusivity a.

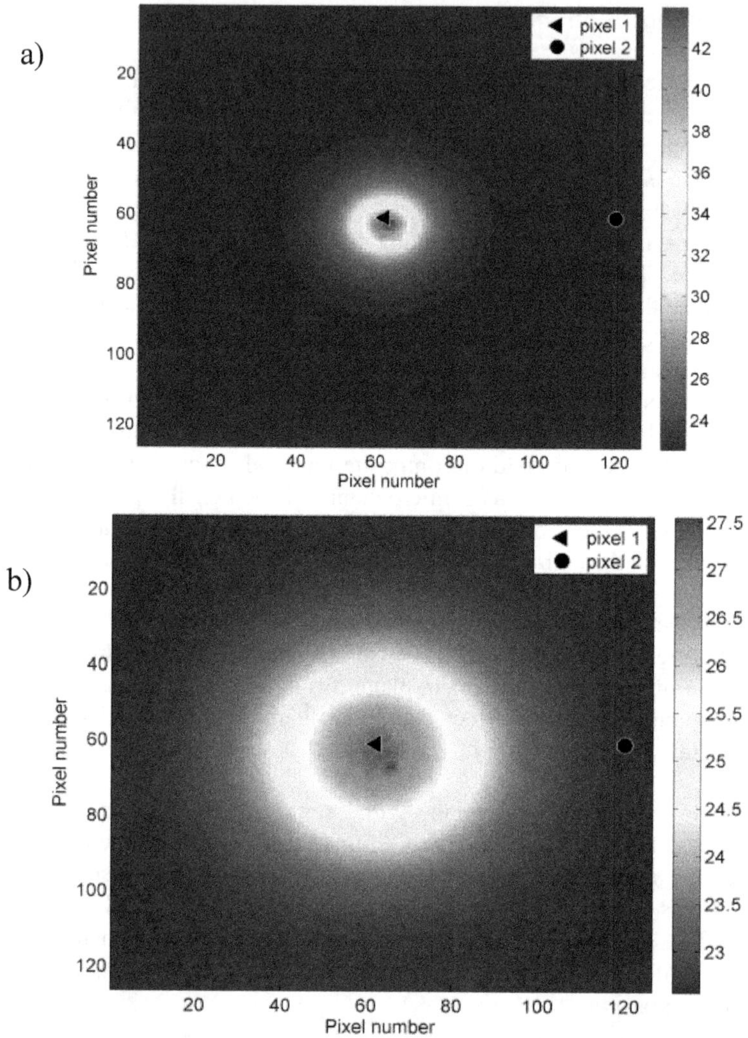

Figure 10.5. *Temperature response field obtained with a step heating made with a laser spot of 200 μm of diameter on a glass plate of 0.2 mm thick, at two different times. a) image obtained at a short time (t = 0.8 s after the beginning of the heating), b) image obtained at a longer time (t = 1.6 s). The pixel size is 10 μm. For a color version of the figure, see www.iste.co.uk/boyard/heat.zip*

One illustration of a possible processing is to study the linear relation between Laplacian and time derivative at each pixel of the image (see Figure 10.6). Such derivatives are obtained from finite differences versus space or time of the temperature information related to the pixels regularly spaced versus time and space.

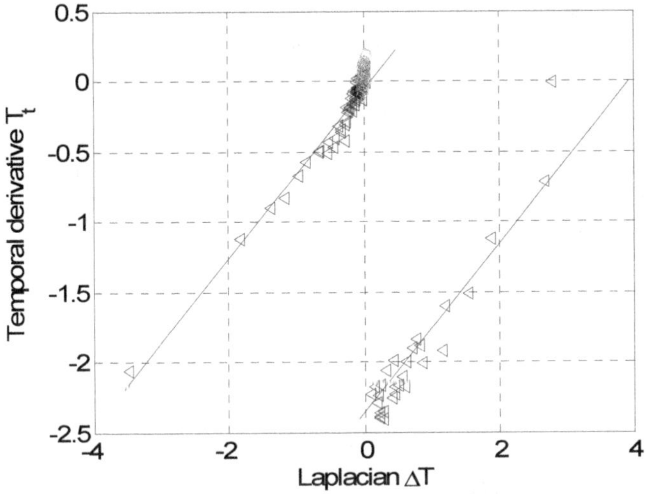

Figure 10.6. *Time derivative versus Laplacian of temperature; a) for an irrelevant pixel out of the heating zone (in gray), b) for a pixel inside the heating zone (in black)*

In Figure 10.6, the straight line with origin ordinate corresponds to pixels where a constant heating occurs. The straight line passing through the origin is corresponding to a case of pure diffusion (when the temperature field is relaxing after the heating). The source term can then be estimated versus space independently of the transport property (the thermal diffusivity: (a)). A criterion suitable to detect such zones is to consider the local correlation between the Laplacian ΔT and the time derivative here denoted by $\partial \mathrm{T} = {}^{\partial \mathrm{T}}/_{\partial \mathrm{t}}$:

$$\rho_{i,j}^{F_t} = \frac{\Sigma_{F_t} \Delta T_{i,j}^k \partial T_{i,j}^k}{\sqrt{\Sigma_{F_t} \Delta T_{i,j}^{k\ 2}} \sqrt{\Sigma_{F_t} \partial T_{i,j}^{k\ 2}}}$$

where F_t is a temporal window such as: $F_t = [k, l_t]$, with $k \in [1, N - l_t]$, k is the time step number and l_t is the width of the time window. If such a coefficient is near from 1, a diffusivity parameter is then estimable. If the correlation coefficient is near from zero, the pixel or the area is irrelevant and must be discarded.

This example is demonstrating that the great amount of thermographic data coupled with computer facilities are offering large possibilities to estimate not only one source term but also a regularly spaced distribution of sources. It offers then the possibility to design thousands of simultaneous independent experiments. It is then also possible to extend the experimental configurations to microchannels.

The idea which consists of implementing a slow in-plane transfer in a conductive plate is here used to estimate the convective transport effect due to a flow in a microchannel (of diameter from 10–100 μm) stuck to a thermally conductive plate (see Figure 10.7). The technical details of the experiment are described in [PRA 06].

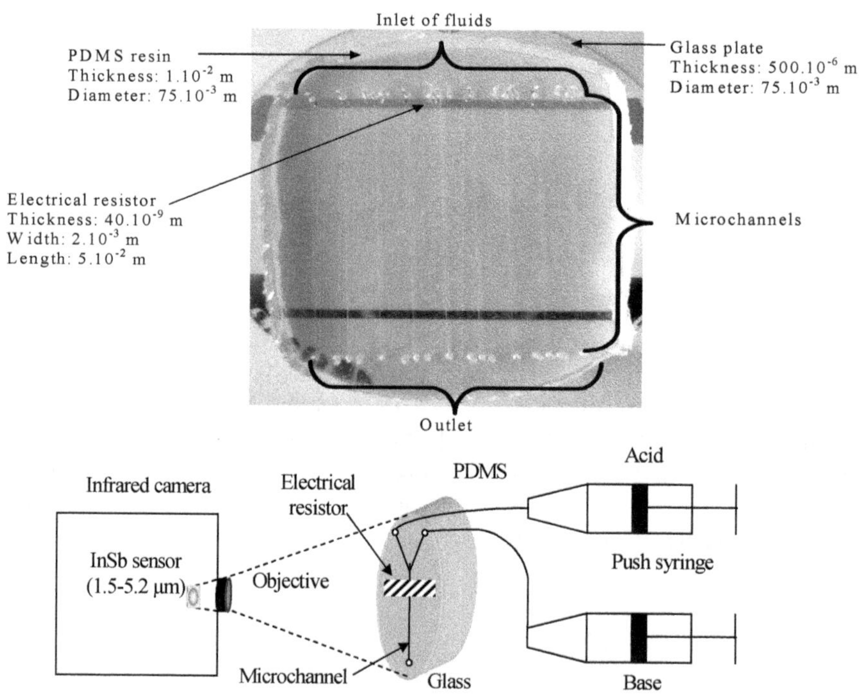

Figure 10.7. *Microfluidic chip, and principles of the microreactor experiment. For a color version of the figure, see www.iste.co.uk/boyard/heat.zip*

The first step of the study will here consist of comparing and estimating the convective effect in the channel to the 2D diffusion inside the conductive front plate.

Here, the local Laplacian of the temperature field is compared with the local temperature gradient versus the direction of the channel (*x-direction)* in order to estimate a Peclet field. In order to realize a calibration experiment, a thin heating resistor is deposited at the entrance of the channel on the conductive plate. An in-plane reference permanent temperature field $T_{i,j}^{0}$ is obtained (see Figure 10.8(a)) with stationary pure water inside the channel and the heating on the left of the plate. Then, a perturbed temperature field $T_{i,j}$ is obtained with a water flow inside the channel (see Figure 10.8(b)).

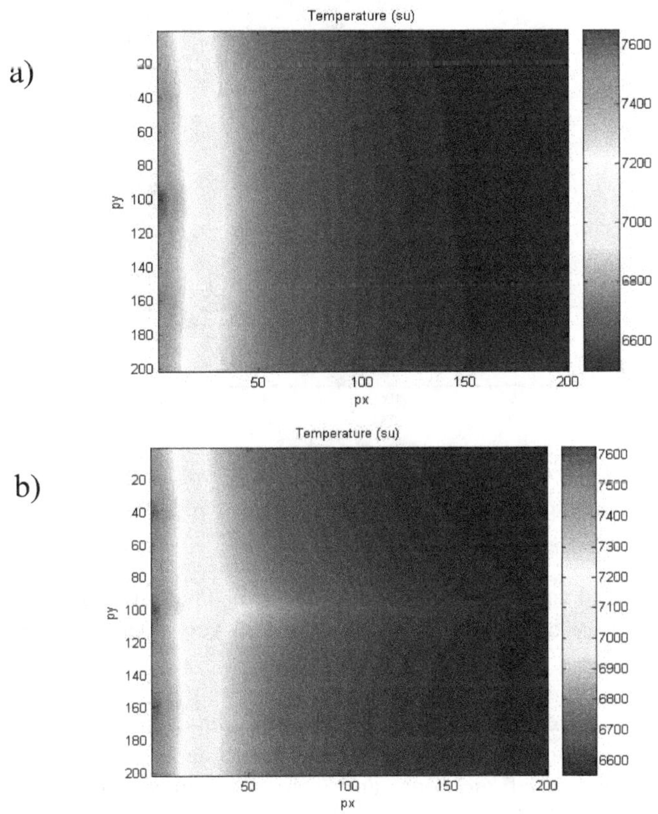

Figure 10.8. *a) Temperature field $T_{i,j}^{v}$ (without flow); b) Temperature field $T_{i,j}^{v}$ (with a microflow versus x-direction). For a color version of the figure, see www.iste.co.uk/boyard/heat.zip*

In order to estimate the non-dimensional velocity field compared to diffusion, a Peclet field $Pe_{i,j}$ is locally defined such as:

$$Pe_{i,j} = \Delta\left(T_{i,j} - T_{i,j}^{0}\right)/(\eth T_{i,j}/\eth x)$$

It is very important to remark that such estimation (see Figure 10.9) is giving a constant non-zero velocity well estimated inside the channel location and with a poor noise only in the sharp temperature gradient zone. A more detailed analysis with a correlation analysis is given in [RAV 12].

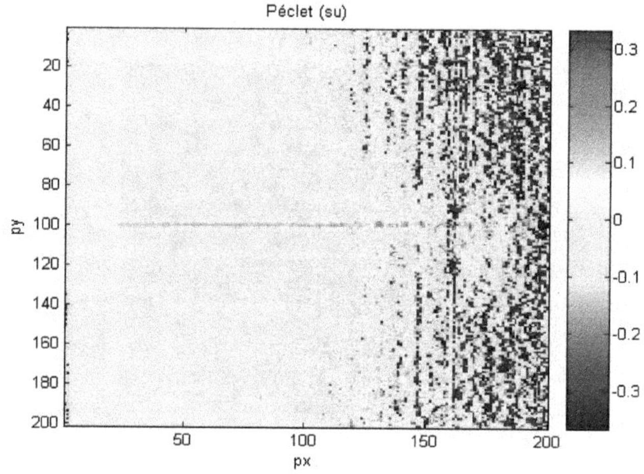

Figure 10.9. *Peclet field estimation $Pe_{i,j}$ from Figure 10.8. For a color version of the figure, see www.iste.co.uk/boyard/heat.zip*

Such a Peclet field can then be used in order to estimate the source term along the channel in the case of a specific microreaction in very near conditions of flow rate and fluids.

From these first considerations, the microfluidic applications have been largely improved. It is now possible to estimate the enthalpy of chemical reactions in a lot of channel configurations and micro or millifluidic devices (see [RAV 12, HAN 08, ROM 15]). A special attention was paid to droplet flows and the related analysis of the heat release of such complex systems (see [ROM 15]).

More recently, a complementary step was devoted to the detection of the transmission of wide-band multispectral infrared sources in such chemical devices [ROM 15]. There is now a challenge to combine heat transfer and chemical analysis in order to estimate simultaneously thermophysical and thermochemical parameters related to coupled heat and mass transport in such devices.

10.3.4. *Principles for the factory global monitoring, example of conveyor belt parameter estimation*

From the previous examples, it appears that the space and time temperature field processing with IR cameras can be applied to a lot of very different situations where the multidimensional heat diffusion, the sources terms and the convective transport can be detected and analyzed. Such infrared image processing can be implemented at the manufactory scale, for parameters estimation or only in order to trigger alarms. Generally, the simultaneous space and time derivative of the temperature field is necessary (see Figure 10.10). Such "derivation" of the signal is once again an ill-posed problem, where only several parts of the infrared images are containing a suitable information (see [BAT 15] for developments related to this subject).

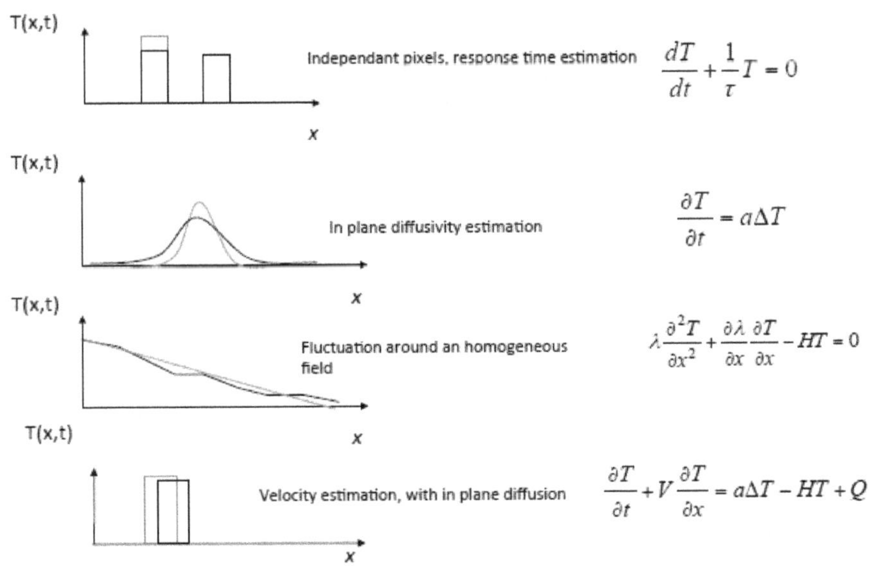

Figure 10.10. *Examples of situations (from [BAT 15]) where the parameter estimations can be deduced from a space and/or time derivation. The red and black curves are related to two different time steps. For a color version of the figure, see www.iste.co.uk/boyard/heat.zip*

One example of simultaneous estimation of velocity and diffusivity in the case of a conveyor belt characterization was previously studied in [BAM 08]. It was shown that the diffusivity estimation was difficult and correlated to the velocity. A more favorable case was devoted to a moving cylinder with periodicity conditions [BAT 07] and is illustrated in Figure 10.11. The poor sensitivity can be overcome when the velocity parameter can be estimated from a great number of observations such as in periodic phenomena (see Figure 10.12).

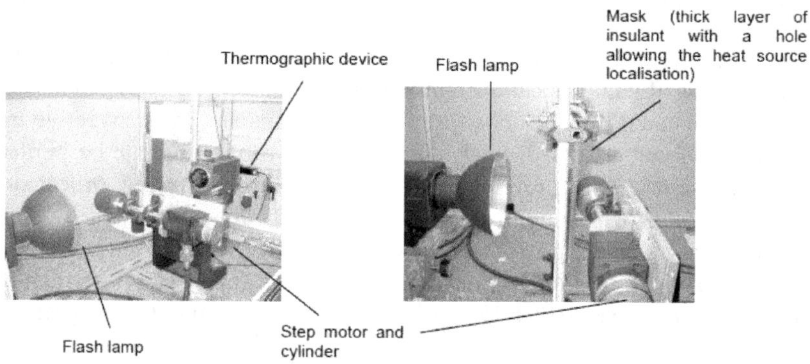

Figure 10.11. *Experimental device in order to simultaneously estimate the velocity and the thermal diffusivity of a periodically moving cylinder*

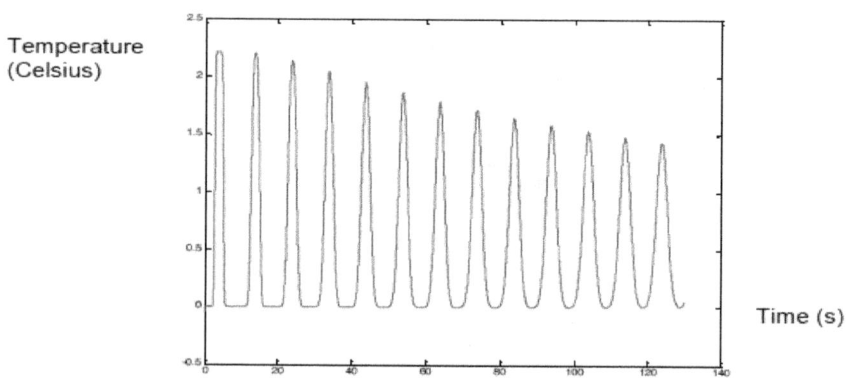

Figure 10.12. *Theoretical temperature evolution versus time at one fixed position*

From Figure 10.12, the temperature signal appears to be "quite periodical" (such a period is related to the velocity of the plate), but the pics are blurred as time grows. The estimation of the thermal diffusivity related to the state of the sample was implemented with a previous accurate estimation of the velocity. Such an estimation can be obtained with a suitable repetition of the pics and an analysis in Fourier space. The thermal diffusivity is then estimated with methods such as those presented in section 10.3.2.2.

10.3.5. *In a near future: the Big Data related to multiscale process survey*

On the one hand, the optical instruments for multiscale analysis such as IR cameras with varifocal optical systems (wide focal length range from normal to telephoto) and high-frame rate recording [CLE 10] related to geometrically controlled and periodically synchronized laser spots [GAV 15] will allow us to access to a large range of properties related to a wide range of characteristic sizes.

On the other hand, it has been recently demonstrated that IR cameras can be used as indirect bolometers for the detection of a large range of radiations, especially the terahertz radiation. The principle is to detect the mapping of thermal effects due to the radiative absorption of thin films which consequently give a thermal reemission in the thermal IR range. The polymer materials are suitably transparent to such kind of radiations [PRA 10, PRA 12].

Our team developed simple terahertz tomography systems which can be useful for the control of polymers processing in a non-metallic environment (see Figure 10.13).

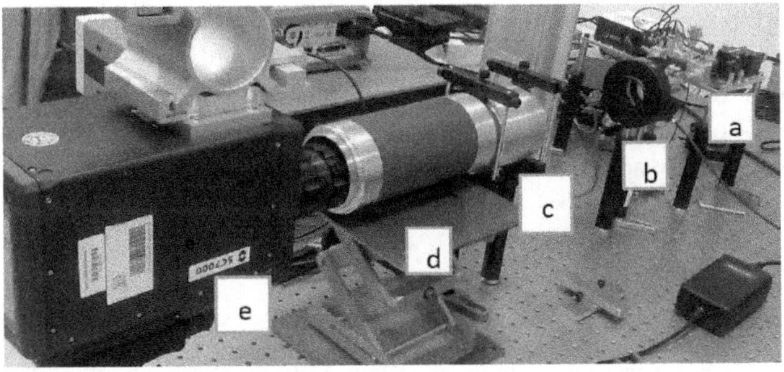

Figure 10.13. *Terahertz tomography device based on IR camera: a) Gunn diode of 110 GHz, b) mechanical shutter, c) THz objective of magnitude 1, d) THz-IR objective with tera thermal converter and e) IR camera*

We can then imagine that in the near future the data ranged in one hypercube of four-dimensional (4D) or five-dimensional (5D) data (x, y, z, t and radiation wavelength) will be available in order to control the polymer processes with multiscale analysis.

10.4. Conclusion

This chapter is an overview related to the different methods for thermal characterization which can be used for the control of manufacturing processes.

On the one hand, the contact methods (temperature sensors, heat flux measurement systems and probes for the thermophysical properties characterization) are ancient but robust and well adapted to systematic measurement of cm-sized samples (especially polymers or organic composite materials). They can be installed, for example, in contact or inside a manufacturing device in order to follow the process or the transformation of the materials.

On the other hand, the contactless methods based on radiative heating and IR temperature measurements allow the study of a wide range of time and space scales for a systematic analysis of a great number of samples obtained from a production unit. The thermal non-destructive evaluation methods are suitable in order to detect the evolution of cracks or delamination inside composite plates under mechanical stress. The temperature field processing also allows a great number of thermal parameters estimation even in the situations of convective transport and is helpful in order to analyze the fluid flows and thermal source terms at laboratory scale in microfluidic devices. Unfortunately, at industrial scale, the walls of molds or fabrication devices are often metallic and opaque to classical IR radiations. It is very difficult to use such contactless methods for the control inside the machines, but only with global wall surface measurements. However, in the near future, the study of terahertz radiations associated with the transparency properties of polymers or ceramics can be a way to set out new control methods. A great amount of data (about one 4- or 5D hypercube) will then be available with such methods and will open the way to new adaptative methods for the control of such processes.

10.5. Bibliography

[BAL 91] BALAGEAS D., DELPECH P., BOSCHER D. *et al.*, "New developments in stimulated infrared thermography applied to non destructive evaluation of laminates", in THOMPSON D.O., CHIMIENTI D.E. (eds), *Review on Progress in Quantitative Non-Destructive Testing*, Plenum Press, New York, 1991.

[BAM 09] BAMFORD M., BATSALE J.C., FUDYM O., "Nodal and modal strategies for longitudinal thermal diffusivity profile estimation. Application to the non-destructive evaluation of SiC/SiC composites under uniaxial tensile tests", *Infrared Physics and Technology*, vol. 52, no. 1, pp. 1–13, January 2009.

[BAM 08] BAMFORD M., BATSALE J.C., REUNGOAT D. *et al.*, "Simultaneous velocity and diffusivity mapping in the case of 3-D transient heat diffusion: heat pulse thermography and IR image sequence analysis", *QIRT Journal*, vol. 5, no. 1, pp. 97–126, 2008.

[BAR 01] BARDON J.P., "Mesure de température et de flux de chaleur par des méthodes par contact", *Ecole d'Hiver METTI*, University Press, Perpignan, vol. 1, pp. 25–30 January 1999, 2001.

[BAT 00] BATTAGLIA J.L., BATSALE J.C., "Estimation of heat flux and temperature in a tool during turning", *Inverse Problem in Science and Engineering*, vol. 8, pp. 435–456, 2000.

[BAT 07] BATSALE J.C., LASSERRE J.P., VARENNE M. *et al.*, "Indentification de la diffusivité thermique d'une cylindre tournant soumis à un flux de chaleur localisé", available at: www.sft.asso.fr/Local/sft/dir/user-3775/documents/ actes/.../374.pdf, 2007.

[BAT 10] BATTAGLIA J.L., KUSIAK A., SCHICK V. *et al.*, "Thermal characterization of the SiO_2-$Ge_2Sb_2Te_5$ interface from room tem-perature up to 400°C", *British Journal of Applied Physics*, vol. 107, no. 4, pp. 044314–6, 2010.

[BAT 15a] BATSALE J.C., FUDYM O., ORLANDE H., Inverse problems and regularized solutions, available at: www.metti.u-bordeaux.fr, February 2015.

[BAT 15b] BATSALE J.C., PRADERE C., Time/Space noise and temperature processing of a thermal signal, available at: www.metti.u-bordeaux.fr, February 2015.

[BEC 77] BECK J.V., ARNOLD K.J., *Parameter Estimation in Engineering and Science*, John Wiley & Sons, 1977.

[BOH 00] BOHAC V., GUSTAVSSON M.K., KUBICAR L. *et al.*, "Parameter estimations for measurements of thermal transport properties with the hot disk thermal constants analyzer", *Review of Scientific Instruments*, vol. 71, p. 6, 2000.

[BOU 00] BOUROUGA B., GOIZET V., BARDON J.-P., "Les aspects théoriques régissant l'instrumentation d'un capteur thermique pariétal à faible inertie", *International Journal of Thermal Sciences*, vol. 39, no. 1, pp. 96–109, January 2000.

[CAR 59] CARSLAW H.S., JAEGER J.C., *Conduction of Heat in Solids,* Clarendon Press, Oxford, 1959.

[CLE 10] CLERJAUD L., PRADERE C., BATSALE J.C. *et al.*, "Heterodyne method with an infrared camera for the thermal diffusivity estimation with periodic local heating in large range of frequencies (25 Hz to upper than 1 kHz)", *Quantitative InfraRed Thermography,* vol. 7, no. 1, pp 115–128, 2010.

[CRE 07] CRETIN B., GOMES S., TRANNOY N. *et al.*, *Scanning Thermal Microscopy. Microscale and Nanoscale Heat Transfer Topics*, Springer Verlag, Berlin Heidelberg, 2007.

[DAV 84] DAVIS W.R., *Hot-Wire Method for the Measurement of the Thermal Conductivity of Refractory Materials. Compendium of Thermophysical Property Measurement Methods*, in MAGLIC *et al.*, Plenum Press, New York, Londres, vol. 1, 1984.

[EUR 16] EURAMET-data-base, available at: http://archive.euramet.org/, 2016.

[FUD 05] FUDYM O., BATTAGLIA J.L., BATSALE J.C., "Measurement of thermophysical properties in semi-infinite media by random heating and fractional model identification", *Rev. of Sci. Instrum.*, vol. 76, no. 2, pp. 1–10, 2005.

[GAR 11] GARNIER B., LANZETTA F., "Measurements with contact in heat transfer: principles, implementation and pitfalls", available at: www.metti.u-bordeaux.fr, 2001.

[GAV 15] GAVERINA L., BATSALE J.C., PRADERE C. *et al.*, "Several considerations about a pulsed flying spot method implemented with IR thermography", *First QIRT Asia Conference*, available at http://qirt.gel.ulaval.ca/archives/qirtasia2015/papers/CP0030.pdf, 2015.

[GUS 91] GUSTAFSON D.E., "Transient plane source techniques for thermal conductivity and thermal diffusivity measurements of solid materials", *Review of Scientific Instruments*, vol. 62, p. 797, 1991.

[HAN 08] HANY C., PRADERE C., TOUTAIN J. *et al.*, "A millifluidic calorime-ter with infrared thermography for the measurement of chemical reactionenthalpy and kinetics", *QIRT Journal*, vol. 5, pp. 211–219, 2008.

[HAY 04] HAY B., FILZ J.R., BATSALE J.C., "Mesure de diffusivité thermique par méthode flash", *Techniques de l'ingénieur*, vol. R 2 955, pp. 1–12, 2004.

[JEN 01] JENSEN K.F., "Microreaction engineering-is small better?", *Chemical Engineering Science*, vol. 56, pp. 293–303, 2001.

[KRA 04] KRAPEZ J.C., SPAGNOLO L., FRIE B.M. *et al.*, "Measurement of in-plane diffusivity in non-homogeneous slabs by applying flash thermography", *International Journal of Thermal Sciences*, vol. 43, pp. 967–977, 2004.

[KUS 13] KUSIAK A., MARTAN J., BATTAGLIA J. *et al.*, "Using pulsed and modulated photothermal radiometry to measure the thermal conductivity of thin films", *Thermochimica Acta*, vol. 556, pp. 1–5, January 2013.

[LAD 00] LADEVIE B., FUDYM O., BATSALE J.C., "A simple device to estimatethermophysical properties of indulating materials", *International Communications in Heat and Mass Transfer*, vol. 27, no. 4, pp. 473–484, 2000.

[LAU 13] LAUNAY J., Analyse de l'homogénéisation thermique et cinétique d'un écoulement d'élastomère, PhD thesis, University of Nantes, France, March 2013.

[LAU 14] LAUNAY J., ALLANIC N., MOUSSEAU P. *et al.*, "Intrusive measurement of polymer flow temperature", *Polymer Science Engineering and Science*, vol. 54, no. 12, pp. 2806–2814, 2014.

[MAI 00] MAILLET D., ANDRE S., BATSALE J.C. *et al.*, *Thermal Quadrupoles – Solving the Heat Equation through Integral Transforms*, John Wiley, Chichester, 2000.

[MCG 88] MCGEE T., *Principles and Methods of Temperature Measurement,* John Wiley & Sons, 1988.

[NIS 90] NIST ITS-90, Thermocouple Database, available at: http://srdata.nist.gov, 1990.

[ORL 11] ORLANDE H., FUDYM O., MAILLET D. *et al.*, *Thermal Measurements and Inverse Techniques*, CRC Press, New York, 2011.

[OZI 00] OZISIK M.N., ORLANDE H., *Inverse Heat Transfer: Fundamentals and Applications*, Taylor and Francis, New York, 2000.

[MOL 04] MÖLLMANN K.P., LUTZ N., VOLLMER M., "Thermography of Microsystems", *Inframation Proceedings*, vol. 104 A, pp. 7–27, July 2004.

[MOU 01] MOURAND D., BATSALE J.C., "Sequential method for thermal diffusivities discrimination by infrared thermography", *High Temperatures - High Pressures*, vol. 33, pp. 127–134, 2001.

[PAR 61] PARKER W.J., JENKINS W., ABOTT J., "Flash method of determining thermal diffusivity, Heat capacity and thermal conductivity", *Journal of Applied Physics*, vol. 32, no. 9, pp. 1679–1684, 1961.

[PER 09] PEREZ L., LADEVIE B., TOCHON P. *et al.*, "A new transient thermal fouling probe for cross flow tubular heat exchangers", *International Journal of Heat and Mass Transfer*, vol. 52, nos. 1–2, pp. 407–414, 2009.

[PHI 95] PHILIPPI I., BATSALE J.C., MAILLET D. *et al.*, "Measurement of thermal diffusivity trough processing of infrared images", *Reivew of Scientific Instruments*, vol. 66, pp. 182–192, 1995.

[PRA 06a] PRADERE C., GOYHÉNÈCHE J.M., BATSALE J.C. *et al.*, "Thermal diffusivity measurements on a single fiber with microscale diameter at very high temperature", *International Journal of Thermal Sciences,* vol. 45, no. 5, pp. 443–451, 2006.

[PRA 06b] PRADERE C., JOANICOT M., BATSALE J.C. *et al.*, "Processing of temperature field in chemical microreactors with infrared thermography", *QIRT Journal,* vol. 3, no. 1, pp. 117–135, 2006.

[PRA 09] PRADERE C., BATSALE J.C. *et al.*, "Thermal properties of Carbon fibers at very high temperature", *Carbon*, vol. 47, no. 3, pp. 737–743, 2009.

[PRA 10] PRADERE C., CAUMES J.P., SALORT S. *et al.*, "Photothermal converters for quantitative 2D and 3D real-time terahertz imaging", *Quantitative Infrared Thermography,* vol. 7, no. 2, pp. 217–235, 2010.

[PRA 12] PRADERE C., CAUMES J.P., PALOMO E. *et al.*, "Use of SVD decomposition to increase signal and noise ratio on THz imaging measurements", *QIRT Conference*, available at http://qirt.gel.ulaval.ca/archives/qirt2014/ QIRT% 202014%20Papers/QIRT-2014-198.pdf, 2012.

[RAJ 02] RAJIC N., "Principal component thermography for flaw contrast enhancement and flaw depth characterisation in composite structures", *Composite Structures*, vol. 58, no. 4, pp. 521–528, December 2002.

[RAV 12] RAVEY C., PRADERE C., REGNIER N. *et al.*, "New temperature field processing from IR camera for velocity, thermal diffusivity and calorimetric non-intrusive measurements in microfluidics systems", *QIRT Journal*, vol. 9, pp. 79–98, 2012.

[RID 09] RIDES M., MORIKAWA J., HALLDAHL L. *et al.*, "Intercomparison of thermal conductivity and thermal diffusivity plastics", *Polymer Testing*, vol. 28, no. 5, pp. 480–489, 2009.

[ROM 15a] ROMANO M., NDIAYE C., DUPHIL A. *et al.*, "Fast infrared imaging spectroscopy technique (FIIST)", *Infrared Physics and Technology*, vol. 68, pp. 152–158, 2015.

[ROM 15b] ROMANO M., PRADERE C., SARRAZIN F. *et al.*, "Enthalpy, kinetics and mixing characterization in droplet-flow millifluidic device by infrared thermography", *Chemical Engineering Journal*, 2015.

[SAC 15] SACI A., BATTAGLIA J.L., DE I., "Accurate new methodology in scanning thermal microscopy", *IEEE Transactions on Nanotechnology*, vol. 14, no. 6, pp. 1035–1039, January 2015.

[SEN 99] SENKAN S., KRANTZ K., OZTURK S. *et al.*, "High-throughput testing of heterogeneous catalyst libraries using array microreactors and mass spectrometry", *Angewandte Chemie- International Edition*, vol. 38, pp. 2794–2799, 1999.

[SHE 06] SHEPARD S.M., HOU Y.L., AHMED T. *et al.*, "Reference-free interpretation of flash thermography data", *Insight*, vol. 48, no. 5, pp. 298–307, May 2006.

[SIE 71] SIEMENS W., "On the increase of electrical resistance in conductors with rise of temperatures", *The Bakerian Lecture (Royal Society)*, 1871.

[STA 99] STARK J., "New microthermoelectric devices based on bismuth telluride-type thin solid films", *18th International Conference on Thermoelectrics*, Baltimore, pp. 465–472, September 1999.

[THI 77] THIKONOV A., ARSENINE V., *Solutions for Ill Posed Problems*, Winston & Co, Washington, 1977.

[THO 08] THOMAS M., Propriétés thermiques de matériaux composites: caractérisation expérimentale et approche microstructurale, PhD thesis, University of Nantes, 2008.

[THO 10] THOMAS M., BOYARD N., LEFEVRE N. *et al.*, "An experimental device for the simultaneous estimation of the thermal conductivity 3-D tensor and the specific heat of orthotropic composite materials", *International Journal of Heat and Mass Transfer*, vol. 53, nos. 23–24, pp. 5487–5498, November 2010.

[THO 12] THOMAS W., KERLIN, JOHNSON M., *Practical Thermocouple Thermometry*, 2nd ed., International Society of Automation Publisher, 2012.

[THU 96] THUREAU P., "Fluxmètres thermiques", *Techniques de l'Ingénieur R2900-Mesures Physiques*, RC3 10, 1996.

[VIL 13] VILLIÈRE M., LECOINTE D., SOBOTKA V. *et al.*, "Experimental determination and modeling of thermal conductivity tensor of carbon/epoxy composite", *Composites: Part A*, vol. 46, pp. 60–68, 2013.

[YAG 15] EL YAGOUBI J., LAMON J., BATSALE J.C. *et al.*, "Multiscale thermal characterization of mechanically loaded ceramic matric composites, estimation of the diffusivity at mico/méso scale", *Experimental Mechanics*, vol. 55, pp. 783–794, 2015.

[ZHA 65] ZHANG X., DEGIOVANNI A., "Mesure de l'effusivité thermique de matériaux solides et homogènes par une méthode de sonde plane", *Journal de Physique III*, vol. 3, no. 6, pp. 1243–1265, 1965.

[ZHA 93] ZHANG X., DEGIOVANNI A., MAILLET D., "Hot-wire measurement of thermal conductiviy of solids: a new approach", *High Temperatures – High Pressures*, vol. 25, pp. 577–584, 1993.

11

Sensors for Heat Flux Measurement

Almost all forming processes involve heat transfer. Heat flux exchanged between the composite part and its environment provides us with information on physical phenomena occurring during the different steps of the process. This chapter details available heat flux sensor technologies and important characteristics to be taken into account (time constant, sensitivity, geometry, etc.) according to heat transfer mode and process environment. By way of illustrations, experimental heat flux data are presented and analyzed for several composite forming processes.

11.1. Motivations: heat flux sensor

Most of the time, temperature is the only tangible piece of information measured in forming processes. This variable is crucial but not sufficient to allow for the characterization of the thermal behavior of a system. Temperature indeed depicts the energy levels of the processed parts, but does not provide any information on how heat is transferred: inside the part, as well as between the part and the tools. Heat flux is therefore the relevant information that quantifies heat transfer in systems.

If we consider the electrical analogy, no electrical engineer would use only voltage (equivalent to temperature) without considering current (equivalent to heat flux) for circuit design. However, heat flux is seldom taken into consideration in industry in general, and in composite processing in particular. This is mainly due to lack of knowledge of existing technologies and often due to a misunderstanding of heat transfer. For these reasons, temperature measurements

Chapter written by Fabien CARA and Vincent SOBOTKA.

are generally preferred, interpretation of results seeming to be easier as well as the availability and variety of sensors. However, information given by heat flux sensors (HFSs) is very rich and useful for observing a wide range of phenomena. Since almost all physical and chemical phenomena imply heat generation or heat transfer, HFSs are good candidates to detect and quantify the following phenomena:

– real-time description of changes in material during filling phase: detection of flow front, quantification of heat exchanged and temperature of fluid in front of the sensor. Saturation degree might also be identified; however, this remains tricky because phenomena are coupled, notably through the dependence of thermal properties on the saturation;

– quantification of the consolidation of the material: exothermal peak characterization and possibly quantification of crosslinking degree for thermosets and crystallization degree for thermoplastics;

– quantification of heat exchanged between the mold and surroundings: ambient air, autoclave, oven, IR panels ... possible direct measurement of convective heat transfer coefficient and characterization of radiative component;

– quantification of the contact quality between the part and the molding cavity wall and detection of problems relative to lack of material, pressure or unsticking due to shrinkage;

– recording of surface temperature measurement of the molding cavity or of parts in movement.

In forming processes, the three heat transfer modes can be encountered as shown in Figure 11.1. We express below the heat exchanged at the surface of the composite part with its environment considering different cases or processes:

– for a part in a closed rigid mold, conduction occurs at the interface mold/part. In this case, the Fourier's law is used both in the part and in the mold as:

$$\vec{\phi}.\vec{n}_c = \left(-\lambda_{composite}.\vec{\nabla}T_{composite}\right)_{x=0}.\vec{n}_c = \left(-\lambda_{mould}.\vec{\nabla}T_{mould}\right)_{x=0}.\vec{n}_m \quad [11.1]$$

– convection: between the tool or the part and the surroundings (e.g. in autoclave or oven):

$$\vec{\phi}.\vec{n}_c = \left(-\lambda_{composite}.\vec{\nabla}T_{composite}\right)_{x=0}.\vec{n}_c = h\left[T_{composite}(x=0) - T_\infty\right] \quad [11.2]$$

– radiation: between the tool or the part and the walls of autoclave, for example or with radiant panels:

$$\vec{\phi}.\vec{n}_c = \left(-\lambda_{composite}.\vec{\nabla}T_{composite}\right)_{x=0}.\vec{n}_c = f\sigma\left[\varepsilon_{composite}T^4_{composite}\left(x=0\right) - \varepsilon_{source}T^4_{source}\right] \quad [11.3]$$

In these expressions, φ is the heat flux density (W/m²), n is the outer normal vector at the interface, λ is the thermal conductivity (W/m.K), h is the heat transfer coefficient (W/m².K), f is the form factor, σ is the Boltzmann constant (W/m².K^4) and ε is the surface emissivity.

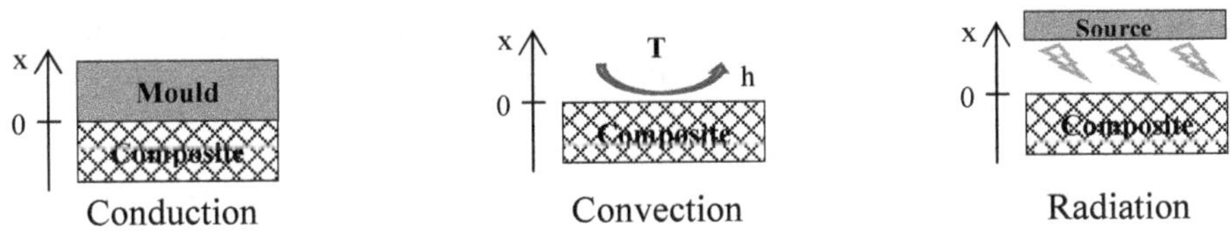

Figure 11.1. *Representation of the three heat transfer modes*

11.2. Principle of heat flux sensors

Whatever the involved transfer mode, technology of almost all heat flux sensors encountered in forming processes is based on conductive heat transfer. Thus, heat flux is most of the time deduced from temperature measurements inside the sensor.

There are different technologies for HFS, each with its advantages and disadvantages. The choice is made depending on the intended application. In this section, we present the main technologies encountered in commercial sensors. More details can be found in [THU 96]. Specific sensors are also developed for research applications. Some examples can be found in [ARA 96, LAN 99, SAI 04, MIT 12].

11.2.1. *Gradient heat flux sensor*

By considering the Fourier's law in 1D in a medium with known thermal conductivity, the measurement of heat flux is straightforward from the temperature gradient. Technology of gradient HFS is based on this principle.

11.2.1.1. *Normal gradient heat flux sensor*

For this sensor, the heat flux is determined by measuring the temperature difference between the two faces of a thin material whose thermal properties are known. In this configuration, the heat flux density is equal to:

$$\phi = \lambda \frac{T_{hot} - T_{cold}}{e} \qquad [11.4]$$

where λ is the thermal conductivity of the sensor, T_{hot} and T_{cold} are the temperatures of each face of the sensor and e is its thickness. From a theoretical point of view, only two temperatures are needed to determine the heat flux. In practice, these types of sensors use thermopiles, i.e. series connection of thermocouple junctions, and more precisely differential thermopiles. In this last case, the junctions are alternatively placed on both faces of the sensor (Figure 11.2).

This configuration increases the sensitivity of the sensor and allows measuring the heat flux with just a voltmeter.

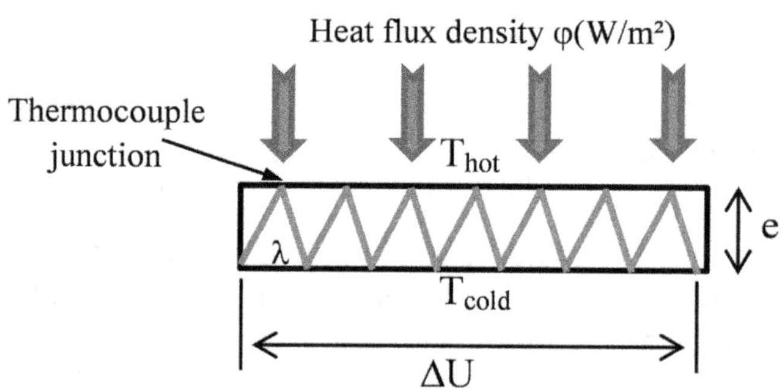

Figure 11.2. *Differential thermopile*

One limitation of this type of HFS is that only conductive term is considered in transfer inside the sensor. Its thermal inertia is ignored. This leads to consider sensors with short time constant in regard of the studied phenomena.

11.2.1.2. *Tangential gradient heat flux sensor*

A junction of two metals of different thermoelectric powers (Figure 11.3), submitted to a tangential temperature gradient, shows an electromotive force ΔU at the terminals of one of them. This property is used in the Théry-type [THE 80] heat flux sensors (Figure 11.4). In these sensors, tangential junctions are multiplied and placed in series on the surface of an insulating substrate. This type of HFS is used to measure normal heat flux on its surface. However, unlike a normal gradient sensor, the asymmetrical arrangement of the thermoelectric elements induces a tangential component of heat flux in each element. After calibration, measurement of voltage at the terminals of the sensors provides the value of normal heat flux. One advantage of this technology is that the measurement does not directly depend on the thickness of the sensors, contrary to normal HFS. These sensors being very thin, the uncertainty related to this parameters could indeed be important.

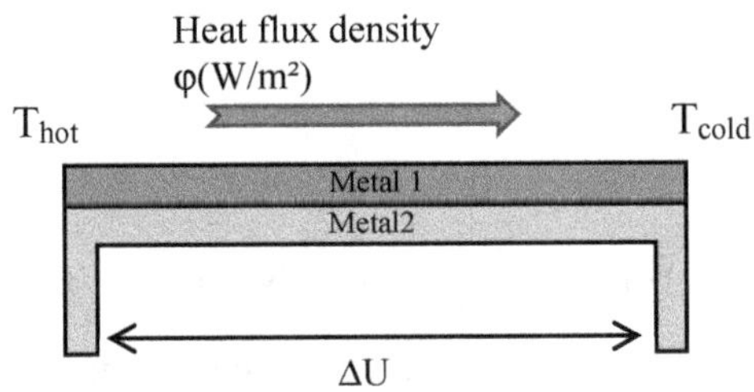

Figure 11.3. *Tangential heat flux principle*

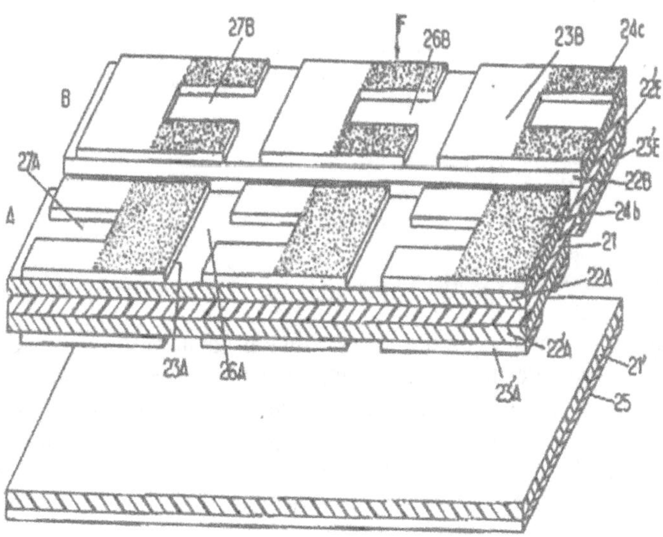

Figure 11.4. *Théry-type heat flux sensor*

11.2.2. *Inertial heat flux sensor*

Contrary to gradient heat flux sensors, inertial heat flux sensors are based on lumped thermal capacitance model. To meet the requirement of this model, the sensor must be composed of a thin element of conductive material so that temperature can be considered as uniform. Its temperature is measured due to a thermocouple.

The sensor initially at T_0 receives a heat flux density φ. It also exchanges heat with the surroundings at temperature T_∞ with a heat transfer coefficient h. In these conditions, the expression of its temperature evolution is equal to:

$$T = T_0 + \frac{\phi}{h}\left[1 - \exp\left(-\frac{t}{\tau}\right)\right] \qquad [11.5]$$

with $\tau = \frac{mCp}{hS}$ the time constant where m is the mass of the sensor, Cp is its heat capacity and S is its surface. The heat flux can be determined by considering long times:

$$\phi = h(T - T_0) \qquad [11.6]$$

However, the experimental determination of h is a difficult task. The heat flux can also be quantified by considering the derivative of equation [11.5] at the first times of the measurement. In this case, the heat flux is given by:

$$\left.\frac{dT}{dt}\right|_{t\to 0} = \frac{\phi S}{mCp} \qquad [11.7]$$

11.2.3. *Inverse heat flux sensor*

Another type of sensor uses the heat diffusion equation to estimate the normal heat flux at its surface. For this type of technology, both thermal inertia and thermal gradients are considered within the sensor. However, contrary to the previous types, the determination is not direct; this method requires an inverse estimation method. The principle is based on the measurement of several temperatures with microthermocouples in the thickness of the sensor.

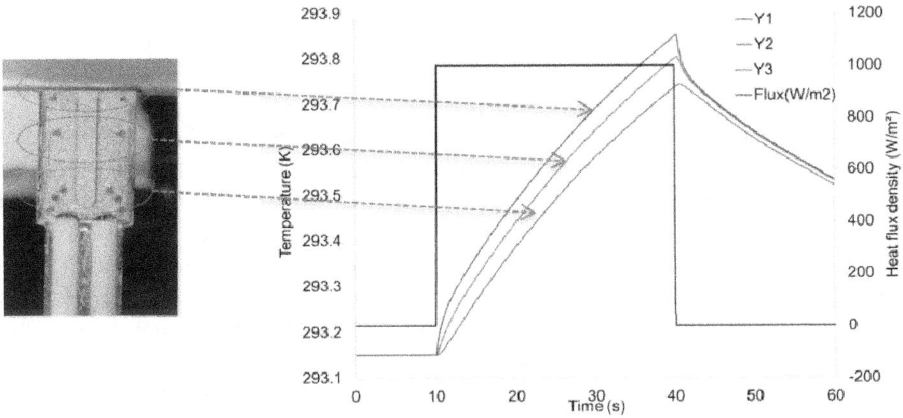

Figure 11.5. *Left: half HFS instrumented by three microthermocouples; right: temperatures measured by thermocouples when submitted to a rectangular heat flux. For a color version of the figure, see www.iste.co.uk/boyard/heat.zip*

Heat transfers within the depth of the sensor are described by the following set of equations:

$$\begin{cases} \rho Cp\dfrac{\partial T}{\partial t} = \lambda\dfrac{\partial^2 T}{\partial x^2}, 0 < x < L, 0 < t < t_f \\ -\lambda\dfrac{\partial T}{\partial x} = \phi(t), 0 < t < t_f \\ T = Y_3, x = L, 0 < t < t_f \\ T = T_0, 0 < x < L, t = 0 \end{cases} \qquad [11.8]$$

where ρ, Cp and λ are the thermophysical properties of the sensor; φ is the unknown heat flux density, Y_3 is the temperature given the thermocouple placed in the deepest position and T_0 is the initial temperature.

An inverse method based, for example, on sequential function specification method [BEC 85] is used to compute the heat flux as well as the surface temperature so as to make equal the computed temperatures with the ones measured by the thermocouples, the adjustable variable being the heat flux density φ.

To remain non-invasive during the measurement, this type of sensor is usually manufactured in the same material as the one of the medium in which it is inserted.

11.3. Main characteristics of HFS

11.3.1. *Invasiveness*

Invasiveness is the way a sensor disturbs heat transfer. If sensor's material thermal properties are different from its environment, the temperature and thus heat transfer will be modified. To point out this phenomenon, results from a simulation performed on a representative case are sketched in Figure 11.6. It consists of the curing of a composite part in a mold. The heat released by the transformation of the material is represented by a Gaussian function. An HFS is inserted inside the mold. Perfect thermal contacts between the sensor, mold and part are considered. Only thermal conductivity differs for the mold and the sensor. Temperatures and heat fluxes are plotted at two points at the surface of the composite part: one in front of the sensor (point 1) and one in front of the mold (point 2). Evolutions of temperatures and heat fluxes at these points are plotted in Figure 11.7. A difference of 12 K for an increase in 40 K and 4.5 kW/m² for a maximal peak value at 27 kW/m² is observed between the two points. The material of the sensor being different from the one of the mold, it modifies heat transfer and thus temperature in the composite part in non-negligible way. This behavior can also lead to defect in the part, thermal cycle and thus transformation cycle being non-uniform.

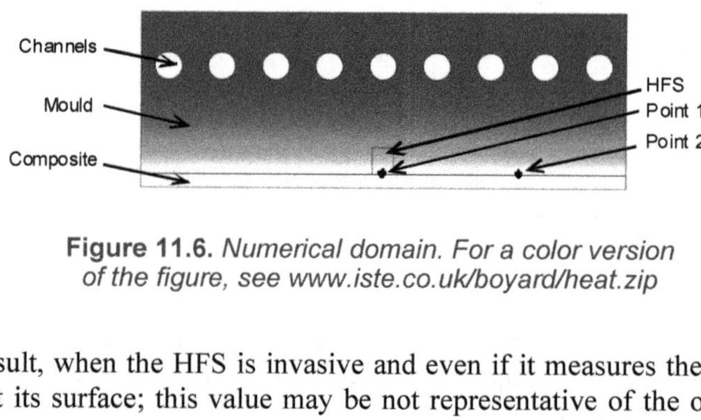

Figure 11.6. *Numerical domain. For a color version of the figure, see www.iste.co.uk/boyard/heat.zip*

As a result, when the HFS is invasive and even if it measures the right value of heat flux at its surface; this value may be not representative of the one that occurs without the presence of the sensor.

11.3.2. *Time constant*

Time constant characterizes the time a system requires to reach a new thermal balance when subjected to a sudden variation. Thus, by considering any point M in a system, the temperature evolution can be described by:

$$T(M,t) = T_F(M)+(T_0(M)-T_F(M))\exp\left(-\frac{t}{\tau}\right) \qquad [11.9]$$

With T_0 the initial temperature, T_F the final temperature and τ the time constant.

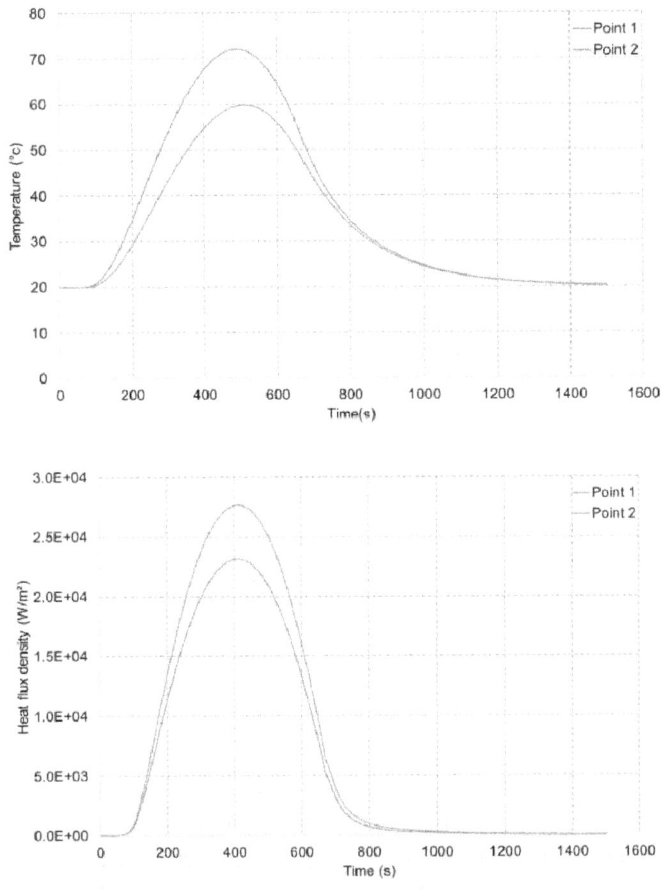

Figure 11.7. *Evolutions at points 1 and 2. Left: temperature; right: heat flux density. For a color version of the figure, see www.iste.co.uk/boyard/heat.zip*

This constant depends on the physical characteristics of the system, as well as on its boundary conditions. For example, in the case of a sensor of thickness e and subjected to imposed temperature on both sides, the time constant is equal to $\tau = \dfrac{e^2}{a\pi^2}$ where a is its thermal diffusivity. In the case of a capacitive HFS, as presented in section 11.2.2, the time constant is equal to $\tau = \dfrac{mCp}{hS}$.

As shown in Figure 11.8, 63.2% of the new regime is reached for t = τ and 99% for t = 5τ.

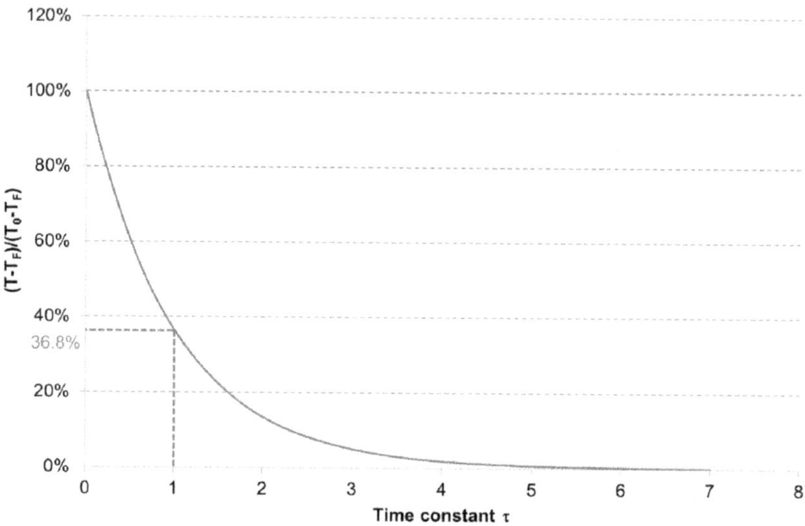

Figure 11.8. *Evolution of temperature according to time constant. For a color version of the figure, see www.iste.co.uk/boyard/heat.zip*

This constant is important because it allows choosing the appropriate sensor for the phenomenon to be observed and notably according to the time scale of this phenomenon. For example, when using a normal gradient sensor (section 11.2.1.1) based on Fourier's law, and thus neglecting the capacitive term, the measurement given by the sensor will be realistic only if variations of the observed phenomenon are slower than 5 times the time constant.

A comparison of different sensors on a log-time scale shows typical response time variation according to sensor technology (Figure 11.9).

From the Figure 11.9, time constant can be evaluated for three different types of sensors:

– radiative sensor: τ = 85 ms;

– convective/conductive flat sensor: τ = 130 ms;

– conductive non-invasive sensor: τ = 3.3 s.

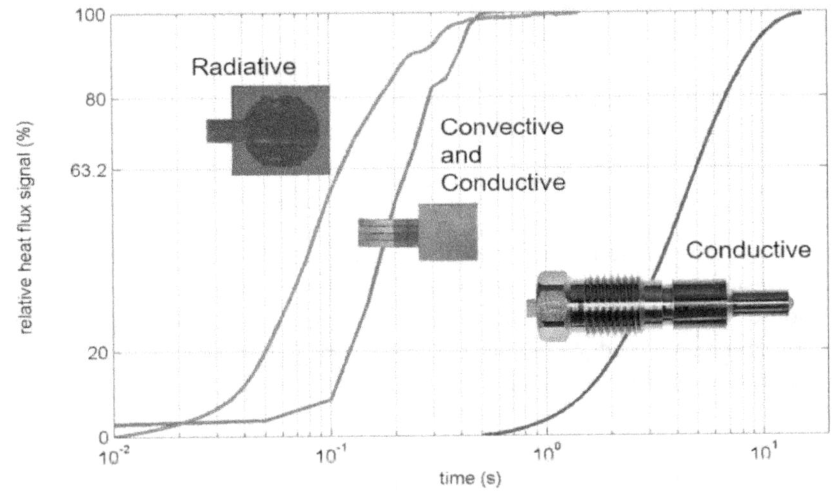

Figure 11.9. *Typical response to step input in heat flux for different sensors. For a color version of the figure, see www.iste.co.uk/boyard/heat.zip*

It is possible to find sensors with shorter and longer response time, the important fact is to have a sensor which is compatible with time scale of phenomenon to be observed. The conductive sensor with $\tau = 3.3$ s works perfectly on characterizing manufacturing processes with cycle time typically over 10 s.

Any thermal resistance at the interface between the sensor and the process will also affect response time. This is especially important when fast phenomena have to be observed. Response time for non-invasive sensors has to be considered relatively to sensor's location. A fast sensor also requires a fast data acquisition system which leads often to higher signal-to-noise ratio and more data to process.

Any mass inside the sensor or between the sensor and the material acts as a low-pass filter which slows down the signal as well as helps to reduce thermal noise. This is particularly clear on convection sensors, for instance, in autoclave where turbulent non-isothermal air or nitrogen flow can generate strong and fast variations of heat transfer, especially after setpoint changes.

11.3.3. *Calibration*

Signal processing is highly dependent on sensor's technology. Sensors based on thermoelectric effect generate directly a voltage, pyroelectric sensors generate an electric charge and multiple thermocouple-based sensors require dedicated measurement of each temperature and adequate processing.

For industrial control, a single calibration coefficient – for instance in $V/(W/m^2)$ for voltage-based sensors – is sufficient. The value is related to the sensitivity of the sensor. Typical values range from 1 $nV/(W/m^2)$ up to several $mV/(W/m^2)$ for Peltier-based sensors.

Heat flux sensors are normally supplied already calibrated in a steady state (under a static constant known heat flux). If thermal properties of constitutive materials of the sensor differ from surrounding ones in the application, it is necessary to perform an *in situ* characterization and calibration if exact heat flux values are requested. For instance, Kayme sensors are built specifically essentially with the same material as their environment, so that they therefore have good calibration values. Convective sensors are generally calibrated by conduction, and the value provided by the sensor will highly depend on how they are mounted, especially the thermal contact condition with the mold. Radiative sensors have their own reference and can be used directly. The signal measured will, however, be tightly linked to their temperature. In radiative measurements, the sensor is an active actor of the measurement, and it is difficult to approach the perfect "transparent" sensor.

Temperature dependence of a material's thermal properties and a non-constant Seebeck coefficient normally induces a drift according to temperature. For composite process monitoring, this effect can generally be neglected, but will be taken into account for the identification of characteristic properties of materials (enthalpy and thermal properties), especially when temperature variations are important.

It is also possible to control reactivity by detection of exothermal peaks without any calibration, where a heat flux sensor is used as an event sensor. Relative variations of heat flux are generally sufficient for most "end of cure" control applications. In this case, the lack of need for recalibration means subsequent savings.

It is highly advised to perform a temperature measurement together with a heat flux one. This complementary information can be used if necessary for calibration or simply in order to check the stability and repeatability of operating conditions of the sensor and, more interestingly, of the process to be controlled.

11.4. Type, positioning and use of heat flux sensors

A wide range of sensors has been developed in order to face multiple needs and use. There is unfortunately no universal sensor adapted to all uses. On the contrary to temperature sensors, the market is still a very small niche requiring a minimum of knowledge to choose the right model.

An ideal heat flux sensor would be thin, small and would be aligned with heat flux lines. Ideally, it would be tough, reusable, would not disturb heat transfer and provide a signal without cables. An inventory of main characteristics and existing models will help to select the right model.

11.4.1. *Commercial sensors*

Sensors can be found in various shapes. Depending on the application, environment and process constraints standard shapes may work fine but it is often necessary to design new sensors:

– thin foil shapes, some of which are flexible, can be used in many cases. They generally have good sensitivity and a fast response time. They are unfortunately fragile, sensitive to pressure and sometimes are difficult to place with their connections. Most of the time, their operating temperatures are also limited to temperatures lower than 250°C. Typical sizes range from 0.3 × 10 × 10 mm up to 0.5 × 300 × 300 mm (Captec); 0.18 × 28 × 35 mm (Newport);

Figure 11.10. *Flat foil-type heat flux sensors*

– thin sensors represent a new compromise with more robustness, and with a conveniently small shape: for example, 0.5 × 4.4 × 4.4 mm up to 2 × 30 × 30 mm (GreenTEG);

Figure 11.11. *Thin heat flux sensor*

– thicker plates or disks: less fragile, good handling of pressure on a wide temperature range. Typical sizes are 2.5 x ∅ 8 mm up to 600 x 600 mm (iTi); 1.5 × 7 × 10 mm, 2 × ∅ 19 mm up to 2.8 × 50 × 50 mm (Wuntronic); 5 × ∅ 80 mm (Hukseflux).

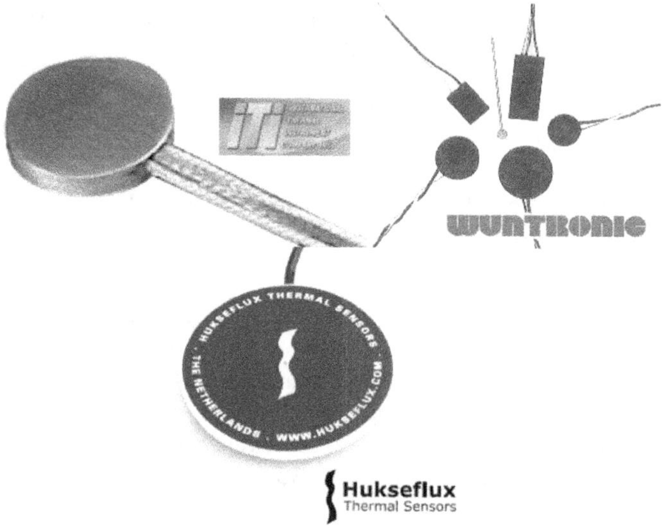

Figure 11.12. *Plates or disks heat flux sensors*

In this family, we can now add common Peltier thermoelectric cells which also perform as well as amplified heat flux sensors;

Figure 11.13. *Peltier module used in cooling devices*

– cylindrical shapes are often the assembly of a thin foil technology on a more robust mechanical part. Sizes range from ∅ 6.3 or 15.8 mm with or without integrated cooling (Vatell); ∅10 mm or ∅ 17.5 mm (TFX);

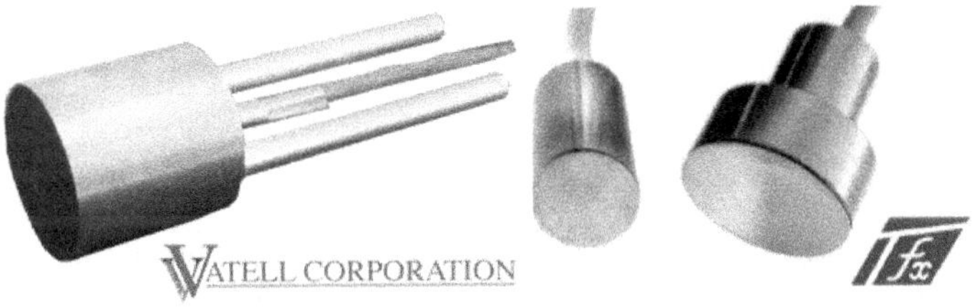

Figure 11.14. *Cylindrical shape heat flux sensor*

– needle or finger sensor shape with integrated screw is available for mounting inside walls or molds. The tip could be in contact (Kayme, Hukseflux) or not (TFX, Hukseflux).

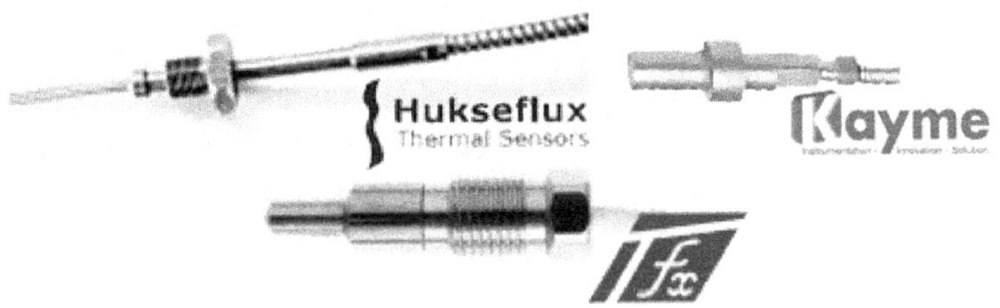

Figure 11.15. *Needle or finger-like heat flux sensor for in-mold mounting*

11.4.2. *Positioning of heat flux sensors*

The ideal position for heat flux sensors is at the interface between materials and exactly where the heat flux is to be measured. There are some specific models that can satisfy robustness and quality.

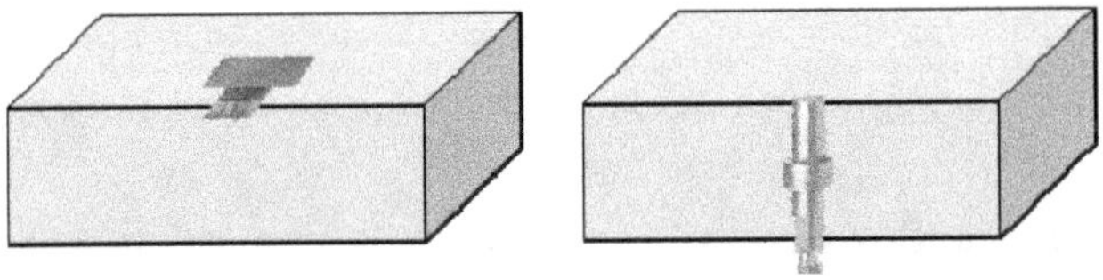

Figure 11.16. *Surface mounted sensor possibilities with flat sensors or access to the surface with screwed sensors*

With some tricks, it is also possible to access interesting interface areas in trial configurations with flat sensors, but at the expense of long preparations due to cable management, a work incompatible with industrial process cycle time and final quality of the part.

Unlike temperature sensors, there is no advantage to embedding a sensor inside the processed composite material. For example, a heat flux sensor placed at the centre of a composite in a symmetrical process will provide a constant zero value whatever the thermal cycle and reactivity of the material are. An exception would concern the study of sandwich structures; the sensor should be considered as consumable in this case.

For industrial process control, we can take advantage of heat transfer to avoid direct contact which is source of problems for the process (risk of vacuum/resin leakage, sensor marking on the part) and for the sensor itself (pressure, abrasive flow, mold cleaning, resin adhesion, infiltration, etc.). Sensors can be placed inside the tool typically at 1 mm beneath the surface and will use any remaining mold material as a transmitter. For systems associated with relative high level of flux (typically over 1 kW/m^2) found in most fast processes (active cycle time below 10 min), this is not a problem for time response and sensitivity.

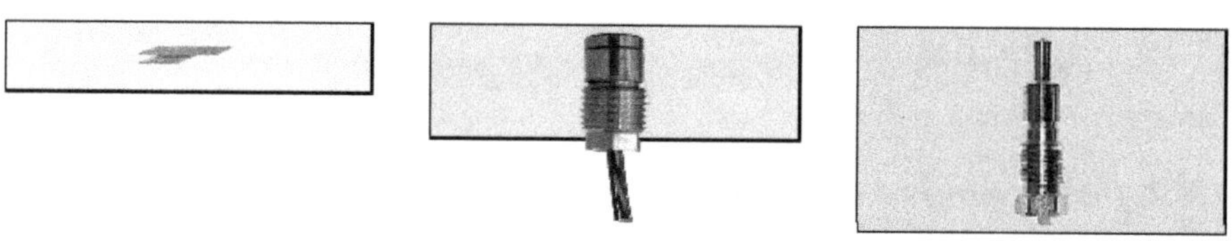

a) b) c)

Figure 11.17. *Non-invasive mounting possibilities with: a) flat sensor laminated in composite or sandwiched in metallic tool; b) sensor for thin 25 mm metallic tool; c) sensor for thick tools over 50 mm*

11.4.3. *Price*

The price of heat flux sensors is generally high compared to temperature sensors. The price increases with the size of the sensor and with its sensitivity. Prices range from hundreds to several thousands of euros according to models. One exception is the low-cost of Peltier devices, which can provide a heat flux signal for a few dollars.

11.5. Advantages and limitations of HFS compared to other *in situ* monitoring techniques

11.5.1. *Advantages*

– Easy access to a workable signal without the need of complex processing if precise calibration is not required;

– quantity of information contained in heat transfer signal. A lot of physical phenomena are associated with heat generation: chemical reactions, mechanical displacements and transfer phenomenon;

– possibility to avoid direct contact between the sensor and the part and use heat diffusivity of the tool to transfer the signal. This leads to really robust and reliable measurements;

– detection of heat transfer provided by the whole material. It allows a global cure control on a wide range of thicknesses (0.1 up to 100 mm).

11.5.2. *Limitation and care in using HFS*

– High sensitivity to thermal regulation. This is especially true if heating/cooling elements are close to the sensor. Regulation oscillations can also be a concern if their amplitude is in the same order of magnitude as the signal to be detected and if periodicity is closed to duration of studied phenomena;

– calibration of sensors is complex and should often be performed *in situ*. However, for a large number of applications involving process control and monitoring on a single sensor (event detection based on shape of the signal) a precise calibration is not necessary;

– isothermal processes are by definition characterized by no heat flux. It is, however, possible to use active sensors which generate their own calibrated heat flux in order to detect resin flow;

– detection of weak signal in the presence of important heat transfer in tools.

11.6. Examples

HFSs have been successfully used to control the following processes:

– compression molding (sheet molding compounds and wet molding);

– liquid composite molding (RTM and infusion);

– prepregs in autoclave;

– continuous composite molding;

– filament winding;

– tape placement;

– pultrusion.

The list is not exhaustive and all processes should be accessible according to available technologies. Some examples are described in the following sections.

11.6.1. *Compression molding*

This example concerns sheet molding compounds (SMCs), bulk molding compounds (BMCs), wet molding and more recently injection-compression molding. Reactive materials with thicknesses from 1 up to 100 mm, introduced in a high temperature (100–180°C) curing cycle between 1–30 min and relative high enthalpy (typically 100 J/g of composite) will have the same kind of thermal behavior. Non-invasive sensors are perfectly adapted to this family of processes and provide robust solutions with sensors lasting a mold's lifetime.

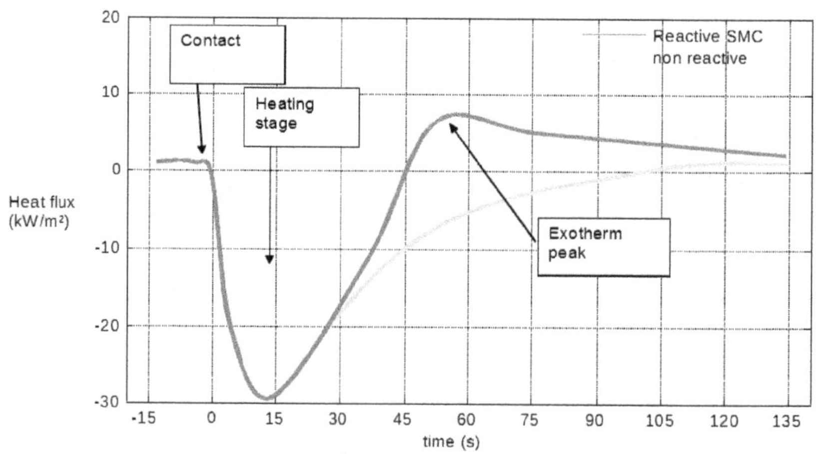

Figure 11.18. *Typical heat flux signal on standard SMC material. For a color version of the figure, see www.iste.co.uk/boyard/heat.zip*

The entire cycle is clearly described by a sensor's signal: for people used to the thermal signature of these cycles, the shape of heat flux measured inside the mold is closer to the temperature value inside the composite than the actual surface heat flux:

– *initial value* of heat flux depends on how the sensor is mounted and where it is placed according to the mold's heating circuits. Positive values are generally a sign of back cooling of the sensor, typical of long screws. This offset has no effect on process monitoring except if it is not stable;

– *contact* between the cold material and the hot mold is seen as a drop of the signal. If the sensor is on the bottom mold, in the material deposition area, it is possible to detect variabilities in timings by this method;

– during the compression phase, it is possible to detect heat transfer rate variations linked to internal *pressure changes*;

– when the material heats up, the heat flux is negative since the mold provides heat to the material. The negative amplitude is a characteristic of temperature difference between the material's initial temperature and the mold. It is a convenient way to access the *prepreg's temperature* changes during production;

– when the temperature is high enough, the reaction starts and reaches a maximum corresponding to exothermal peak. The heat flux becomes positive, meaning that the heat is transferred from the material toward the mold. The amplitude of exothermal peak is directly related to internal temperature reached inside the composite, and it is possible to estimate for each part produced the *maximal temperature reached by the composite* in a very repeatable manner;

– at the end of cure, the heat flux signal returns to a stable value. This value may differ from initial values due to the fact that the mold is closed and heat transfer through the cured composite will occur. This is especially true if both sides of mold are not at the same temperature.

The exact *end of cure* of the composite is difficult to detect precisely for two reasons:

– most thermoset materials see their curing rate decreasing as they approach complete conversion, mainly due to diffusion limitation at the molecular scale. This means that the end of cure is an asymptotic curve which has no end;

– the heat of reaction is not transferred directly to the mold, a part being stored by the thermal capacity of the composite part. Exothermal heat flux signal in the mold corresponds therefore to apparent cooling of the composite which is heated by the reaction. Due to material's heat capacity and poor thermal conductivity (compared to the mold), it takes time to cool down the composite, slowing down apparent curing signal seen by the sensor.

Taking into account these observations, it is possible to use heat flux signals as a robust indicator of reactivity and use a relative heat flux level as a reference for actual reactivity of the part inside the mold. A single non-invasive sensor (compatible with A-class parts) coupled with a smart electronic control device can generate over 10% of productivity increase by safely releasing each part at its optimal cure time. Such a system also helps to prevent undercure due to thermal regulation issues or problems in resin formulation.

11.6.2. *Resin transfer molding flow front detection*

Besides curing aspects, heat transfer is often associated with mass transfer in convection mechanism. This can be used to detect and characterize resin flow front during injection process. When injection is isothermal, there is no heat transfer associated with filling. A heat flux sensor is therefore useless except if the sensor can generate its own heat flux and characterize the efficiency of heat transfer linked to convection or thermal properties changes in front of the sensor.

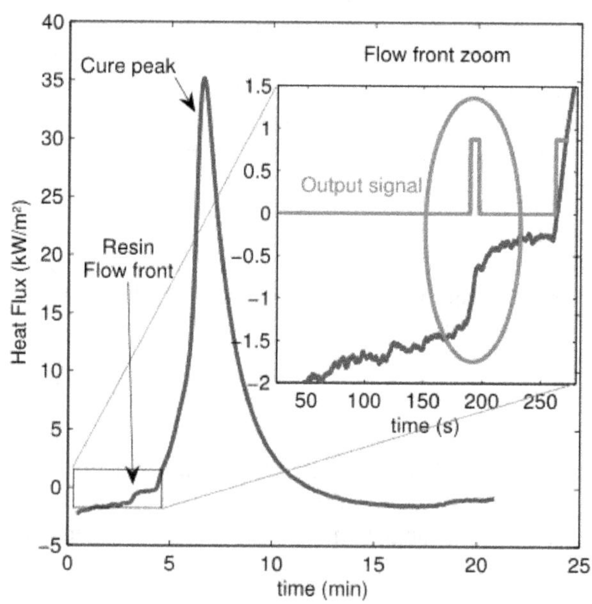

Figure 11.19. *Heat flux observed in a reaction injection molding (RIM) on a metallic tool, illustration of flow front detection for automation purpose. For a color version of the figure, see www.iste.co.uk/boyard/heat.zip*

Heat flux levels associated with resin flow front require good sensitivities and an adapted signal process in order to remove background perturbations, such as baseline drift and other artifacts. It is, however, possible to use the heat flux signal for flow front detection (signal level change), flow rate changes or flow stop. In Figure 11.19, the first step circled in red corresponds to fresh resin arrival. After 1 min of injection, the flow stops and the signal increases strongly due to exothermic reaction.

11.6.3. *Resin transfer molding: influence of mold temperature*

In this section, the influence of the mold temperature in the RTM process is shown as well as notably the phenomena that can be deduced from heat flux measurement from an HFS positioned into a mold in contact with the composite part. To point out this influence, four parts were molded by keeping the temperature of the polyester resin constant at 35°C and by setting the mold temperature at four temperatures, 25, 31, 46 and 50°C. The heat fluxes measured during the four cycles are gathered in Figure 11.20. In this figure, injection is numbered 1 and curing is numbered 2:

– *injection phase*: injection being non-isothermal, the HFS can be used to detect the filling front. The temperature of the resin being constant for the four experiments, the heat fluxes observed during the injection phase only differ by the temperature of the mold which can be higher or lower than the temperature of the resin. It is thus negative for the experiments at 46 and 50°C and positive for the ones at 25 and 31°C;

– *curing phase*: during this step, HFS can be used to measure the conversion degree by integrating the measured heat flux. For the proposed experiments, the crosslinking time is 4 times shorter at 50°C rather than at 25°C. This observation may suggest that using higher temperatures can substantially reduce the cycle time.

However, the shape of heat fluxes changes during crosslinking according to the mold temperature. The kinetics present two spread peaks for the experiment at lower temperatures, and only one peak at higher temperature. This evolution is due to the presence of two distinct chemical reactions with different temperature dependencies. By integrating the heat flux during crosslinking, we can deduce the enthalpy is released by the resin. By comparing the moldings at 25 and 50°C, we can observe that the degree of cure is smaller at higher temperatures, which is not obvious and may increase the risk of post-shrinkage. This result is due to the end of one of the two chemical reactions (copolymerization) that prevents the second one (homopolymerization) from finalizing. More details can be found in [DUP 05].

11.6.4. *Internal temperature prediction during infusion*

The process of infusion is relatively simple to instrument. A simple flat heat flux sensor can be taped onto the top of the vacuum bag. In this experiment, we have measured the heat flux and embedded temperature of the sensor and the temperature provided by a thermocouple placed inside the composite.

The temperature of the material inside the vacuum bag can be easily predicted by the following formula:

$$T_{mat} \approx T_{sensor} + k.\phi \qquad [11.10]$$

In this equation, T_{sensor} is the temperature measured inside the heat flux sensor, φ is the heat flux density and the coefficient *k* is the thermal resistance of the composite part.

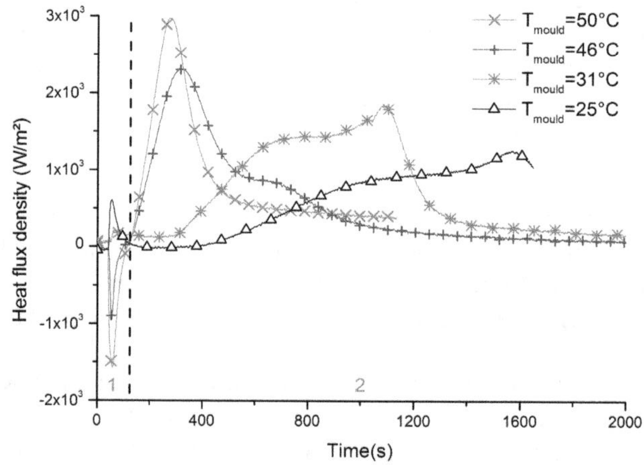

Figure 11.20. *Influence of mold temperature on heat transfer on RTM process. For a color version of the figure, see www.iste.co.uk/boyard/heat.zip*

In this example, *k* is estimated from one measurement from a thermocouple inside the material used as reference for calibration. The obtained *k* value can be used in production for a given part in order to monitor the exothermal peak inside the part without touching the composite.

We can see in Figure 11.21 that temperature as given by equation [11.10] (red curve) fits well with the temperature measured by the thermocouple inside the composite part (dotted red curve). This method works when heat transfers occur in thermal sliding regime, i.e. when inertial term $\rho Cp\frac{\partial T}{\partial t}$ remains constant at each instant.

This method can also be applied in prepregs cured in autoclave and has been proven to work through caul plates and even through 5 mm thick rubber membranes. The same approach can also be performed with in-mold sensors.

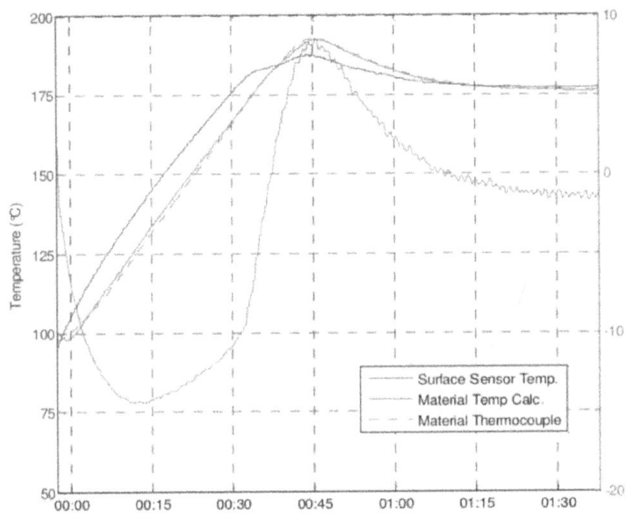

Figure 11.21. *Infusion cycle with a flat heat flux placed on top of vacuum bag. For a color version of the figure, see www.iste.co.uk/boyard/heat.zip*

11.6.5. *Glass mat transfer: internal temperature monitoring*

Heat flux can also be useful for monitoring thermoplastic composites. The following example performed successively on three different materials (PP, PPS and PA6) shows how a heat flux sensor can be used to monitor forming and cooling stages when it is not possible to measure the internal temperature of the part.

In this experiment, a thermocouple is placed in the composite and the temperature is measured during heating stage in an infrared oven. Then, the hot material is transferred inside the mold. The mold is closed which cuts the thermocouple wires. They become unusable.

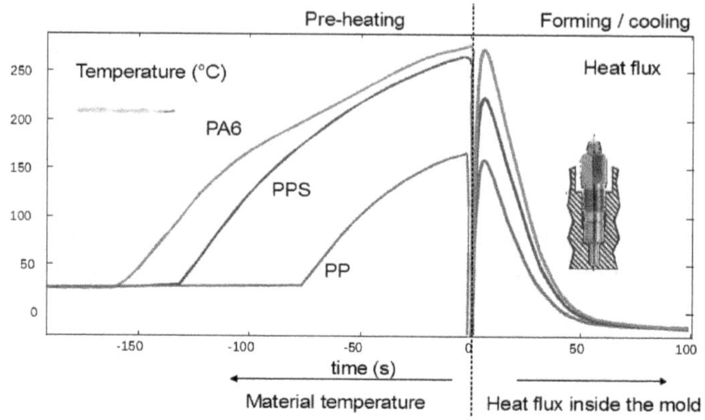

Figure 11.22. *Two-step GMT processing of three different materials monitored by in-material thermocouple during preheating and by non-invasive in-mold heat flux sensor during forming and cooling. For a color version of the figure, see www.iste.co.uk/boyard/heat.zip*

For t > 0, the signal displayed in Figure 11.22 is the heat flux provided by a non-invasive sensor. Without calibration nor post-processing, it is possible to see that heat flux signal represents the cooling of the part accurately and provides an interesting point of view on the process.

Further work on this subject and similar processes (thermoforming and extrusion blowing) shows that this technique allows to directly monitor the material's temperature after calibration and is also capable of showing mass or thickness variations.

11.7. Conclusions

Applications of HFSs in the composite industry are recent but today remain little known to engineers and process engineering companies. Placed at the crossroads of material science, processing technology and thermal engineering, HFSs can enlighten the understanding of composite processing, optimize composites manufacturing and provide reliable processes.

11.8. HFS suppliers

– Captec: www.captec.fr

– Heimann Sensor: www.heimannsensor.com

– Hukseflux: www.hukseflux.com

– GreenTeg: www.greenteg.com

– ITI: www.thermalinstrumentcompany.com

– Kayme: www.kayme.eu

– Newport: www.newportus.com

– Rdf: www.rdfcorp.com

– Sequoia: www.sequoia.co.uk

– TFX: www.thermoflux.ch

– Vatell: www.vatell.com

– Wuntronic: www.wuntronic.de

11.9. Bibliography

[ARA 96] ARAI N., MATSUNAMI A., CHURCHILL S.W., "A review of measurements of heat flux density applicable to the field of combustion", *Experimental Thermal and Fluid Science*, vol. 12, pp. 452–460, 1996.

[BEC 85] BECK J.V., BLACKWELL B., CLAIR J.C., *Inverse Heat Conduction: Ill-Posed Problems*, Wiley-International, 1985.

[DUP 05] DUPUY J., ADAMI J., MAAZOUZ A., *et al.*, "Kinetic modeling of an unsaturated polyester resin using two complementary techniques: near infrared spectroscopy and heat flux sensors", *Polymer Engineering and Science*, vol. 45, pp. 846–856, 2005.

[LAN 99] LANGLEYA L.W., BARNESA A., MATIJASEVICB G. *et al.*, "High-sensitivity, surface-attached heat flux sensors", *Microelectronics Journal*, vol. 30, pp. 1163–1168, 1999.

[MIT 12] MITYAKOVA A.V., SAPOZHNIKOV S.Z., MITYAKOV V.Y. *et al.*, "Gradient heat flux sensors for high temperature environments", *Sensors and Actuators A*, vol. 176, pp. 1–9, 2012.

[SAI 04] SAIDI A., KIM J., "Heat flux sensor with minimal impact on boundary conditions", *Experimental Thermal and Fluid Science*, vol. 28, pp. 903–908, 2004.

[THE 80] THERY P., MARECHAL J.E., "Etude et caractérisation d'un nouveau fluxmètre calorifique", *Journal of Physics E: Scientific Instruments*, vol. 13, pp. 860–865, 1980.

[THU 96] THUREAU P., "Fluxmètres thermiques", *Techniques de l'ingénieur*, r2900, 1996.

12

Thermal Radiative Properties of Polymers and Associated Composites

The knowledge of thermal radiative of materials is crucial to accurately compute the temperature fields and/or the radiative fluxes within manufacturing processes or engineered systems in which heat transfers are mainly governed by thermal radiation. For cases of polymers matrices, reinforced or not with fibers structures (composed of glass or carbon), it appears that these properties are less known than for other advanced textured materials such as refractory ceramics. However, more and more shaping processes use infrared sources or laser sources for heating these polymer composites. According to the volumetric fraction, size and chemical composition of the embedded fibers and also to the composites thicknesses, these materials can be considered either optically thick or optically thin. In turn, this means that solving radiative transfer problems can be dealt with by only considering heat exchanges on their surfaces or by also being extended to their whole volumes. This chapter recalls what is meant by thermal radiative properties as the main physical concepts that define them. Some measurements performed on typical samples by infrared spectroscopy will be provided and finally discussed.

12.1. Introduction

Until now, research aiming to measure and determine consecutively the pertinent parameters that govern the spectral dependence of the thermal radiative properties of polymer composites [ADV 12, EVE 07] is rather uncommon. According to the heating device (infrared or laser sources) used in a manufacturing process, spectral information going from the far infrared range up to the visible range [1,000–0.4 μm] is required to finely compute fractional or total thermal radiative properties. These properties can be used afterwards to model heat and mass transfers occurring during the shaping processes of polymer composites [HAK 14]. On the one hand, few experimental set-ups today allow the fine characterization of the thermal radiative

Chapter written by Benoit ROUSSEAU.

properties from the room-temperature up to the mid-temperature ($T_{maximal}$~ 700 K). $T_{maximal}$ corresponds here to the highest temperatures where some improved polymer matrices can reach their melting point [BAS 88]. On the other hand, connecting the thermal radiative properties with the textural organization of the matter within these composites for several length scales (from nanometer to millimeter), knowing that this organization is itself imposed by the heating and cooling steps involved during the elaboration processes, appears today as a real open question [ROU 11, HAK 14]. But, why is an answer to this latter question so important? Simply, because it is well known that the textural features of heterogeneous materials have a noticeable impact on their radiative properties. Indeed, the radiative properties of a 1D-slab made with a homogeneous semi-transparent material are not the same as those of the same slab, but in which heterogeneities have been introduced [ROU 07, ROU 10]. The term "texture" stands for the spatial organization and the size, shape and orientation distribution of heterogeneities dispersed within a host matrix. Since polymer composites can be described by the association of a polymeric matrix in which reinforcements (fibers, particles, etc.) are spatially arranged, conferring them with superior physical properties than when they are individually taken, the question of the exact role played by the texture is therefore clearly stated. Once this role is known, we can propose to later establish robust behavioral laws where the radiative properties will be expressed from chemical (optical indices of the host polymer and of the embedded reinforcements) and zero-order textural parameters (volumetric fraction of fibers, for example) [ROU 11]. Such laws can be useful for anticipating the radiative behavior of the composites under severe conditions of employment when experimental data are scarce. Before achieving this tricky task, let us first deal with the measurement of the thermal radiative properties.

In the following, the main thermal radiative properties are therefore shown. Let us state that a clear distinction must be made between the properties that can be directly measured (reflectance, transmittance, absorptance and/or emittance) and the homogenized ones which required, for textured semi-transparent composite, an inversion process to be properly estimated (volumetric absorption and scattering coefficient, scattering phase function) [HOW 03]. Fundamentals of thermal radiation will be firstly developed. Then, secondly, a method aiming at the computation of the radiative properties of any homogeneous slabs with optically polished faces will be described. Practical results with polypropylene and silica will be shown. Once this framework is settled, we discuss with appropriate examples the effect of fibers structure on the radiative properties of advanced thermoplastic composites: unidirectional E-glass fibers reinforcing polypropylene (PP) rods commercially known as Roving TWINTEX® and carbon fiber reinforcing a polyetheretherketone (PEEK) tape.

12.2. Fundamental requisites concerning thermal radiation

This section presents the main physical concepts allowing us to hand lethermal radiative properties. For more detailed explanations, readers can consult unavoidable and recognized books written, for example, by Siegel, Howell and Menguc [HOW 03] or Modest [MOD 03]. Let us specify that readers have often to take time to make the correspondences between all the notations (space coordinates, directions, solid angles, etc.), knowing moreover that French and English researchers and engineers can describe the same energetic quantity with different symbol letters (luminance in French which is denoted by L is the same quantity as the radiative intensity in English but is denoted by I). Moreover, at given time t, since radiative quantities depend on the wavelengths, on the position of the matter in space and on the direction of propagation, we can simultaneously consider spectral and directional expressions which are also function of the temperature. This means that the denomination of quantities can quickly become cumbersome, knowing that it appears more practical, for heat and mass transfer problems, to work with total (integration of spectral properties over the whole spectral range) and hemispherical (integration of directional properties over a 2π solid angle) expressions.

12.2.1. *Spectral range of thermal radiation*

The term "thermal radiation" is commonly used to describe the heat transfers caused by electromagnetic waves. More exactly, the propagation of radiative energy can be explained from two viewpoints: quantum mechanics and electromagnetic waves theory. In quantum mechanics, massless particles called photons explain energy exchanges at the atomic scale. This concept is appropriate for describing the radiative properties of gases. For solid or liquid matter, the electromagnetic wave theory is more useful. Having said that, in 1924, De Broglie proposed a practical relationship, based on the wave–particle duality that conciliates the concepts of waves and quantum particles. The wavelength, λ, of a given wave is associated with momentum of particle, p, through the Planck constant, h ($h = 6.626\ 10^{-34}$ J.s) by:

$$\lambda = \mathrm{h}/p \quad [12.1]$$

Moreover, a wave transports with it an amount of energy E defined by:

$$E = h\nu \quad [12.2]$$

where ν is the frequency. Electromagnetic waves propagate within any medium with a high velocity c, which is generally lower than the speed of the light in a vacuum, c_0 (with c_0 = 2.998 10^8 m/s). The effect enacted by the matter is depicted by n ($n = c/c_0$) known as the refractive index that is the real part of its complex refractive index, $\tilde{n}$. The refractive index is higher than 1 for almost all natural or synthetic materials in the infrared and the visible spectra. Thus, the average refractive index computed in the visible spectral range is about 1.49 for poly(methyl methacrylate) (PMMA), 1.64 for polyethylene terephthalate (PET), 1.50 for polyethylene (PE) [EVE 07]. Note that for E-glass fibers mean average of n is about 1.54 [WAL 01] and for carbon fibers n is about 2.32 [STA 93]. The imaginary part of $\tilde{n}$ corresponds to the index of absorption, k, and indicates how an electromagnetic wave can be attenuated during its propagation through the material. Electromagnetic waves can be classified according to their frequency, ν, wavelength, λ, wavenumber, σ and angular frequency, ω. These four quantities are linked to each other through the relationship:

$$\nu = \omega/2\pi = c/\lambda = c\sigma \qquad [12.3]$$

Practically, wavelengths can be converted into wavenumbers by using $\sigma = 10000/\lambda$ where λ in micron gives σ in cm^{-1}. Since the energy of a wave is conserved during a propagation between two adjacent materials, the frequency of radiation does not evolve. Figure 12.1 shows the spectral repartition of electromagnetic waves according to great well-known categories (gamma rays, X rays, ultraviolet, visible, infrared and microwave) that circumscribe the spectral range corresponding to the thermal radiation.

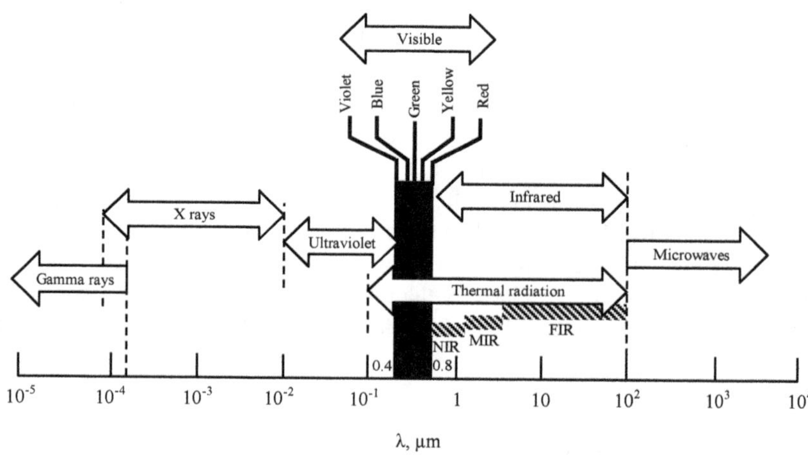

Figure 12.1. *Electromagnetic wave spectrum adapted from [INC 90]*

Note that the spectral domain of the thermal radiation can be subdivided into several wavelength subdomains: ultraviolet and visible (λ = 0.1–0.8 µm), near infrared (λ = 0.8–2 µm), mid infrared (λ = 2–4 µm) and far infrared (λ = 4–100 µm) spectra.

12.2.2. *Radiant energy, radiant flux, radiative flux density and radiative intensity*

Electromagnetic waves with wavelengths spanning in the [0,1–100 μm] spectral range are directly involved in radiative heat transfers. When such a wave propagates through a first medium (air or polymer composite) and encounters a second medium, it may be reflected at the interface and the non-reflected part will therefore penetrate into the new medium. The wave may be attenuated in the new medium and if the attenuation is complete, the wave is totally absorbed. According to the microscopic mechanisms involved in the attenuation of the energy carried out by the incoming waves, we can distinguish opaque, semi-transparent and transparent medium. The first two terms correspond to the cases where the crossed medium can totally or partially attenuate the waves. The latter term refers to the case where any attenuation occurs. To quantify these energetic exchanges, we can first introduce the radiant energy, $\mathcal{E}_e$ (J), which depicts the energy of a given electromagnetic wave. The radiant flux or radiant power, Q(W), corresponds then to to the radiant per unit time that is reflected, transmitted or absorbed by a medium.

It is also interesting to consider the radiative flux density, denoted by q (W/m^2), that is the amount of radiant flux, Q, transported through a given area, A. q acts as a generalization of the heat flux density well known in the field of thermal science. When q impinges toward a surface, it is often called irradiance written in textbooks as G. Contrarily, when q is emitted from a surface it may call radiant exitance or radiant emittance. Now that q and Q have been introduced, let us concentrate on a crucial quantity that is the radiative intensity, I, to treat radiative transfer exchange. Figure 12.2(a) recalls first the equivalence between the Cartesian coordinate and the spherical coordinate system that is useful to describe the elementary solid angle, Ω. The radiative intensity, I, is depicted in Figure 12.2(b). I is defined as the differential radiant flux (derivative order 2), d^2Q, emitted from an elementary surface localized at $\vec{r}$, dA, per the elementary solid angle, $d\Omega$ around $\vec{u}$.

$$I(\vec{r},\vec{u}) = \frac{d^2Q(\vec{r},\vec{u})}{\cos\theta dA d\Omega} \quad [12.4]$$

with $\cos\theta = \vec{\Omega}.\vec{n}$. In the spectral case, the previous relationship becomes:

$$I_\lambda(\vec{r},\vec{u}) = \frac{d^2Q_\lambda(\vec{r},\vec{u})}{\cos\theta dA d\Omega\, d\lambda} \quad [12.5]$$

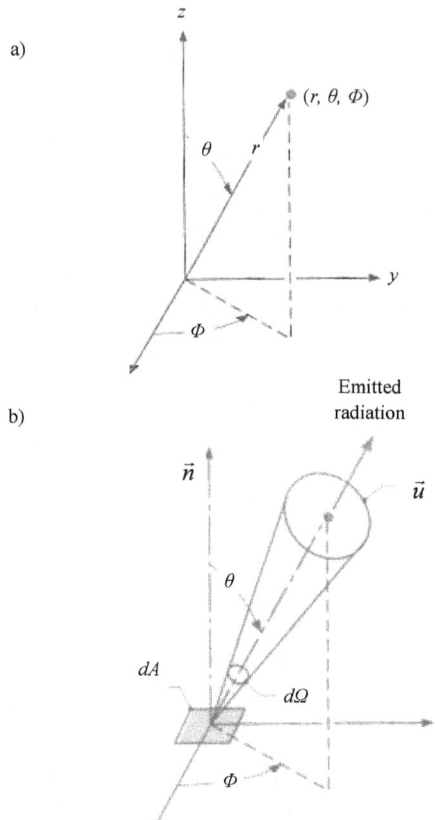

Figure 12.2. *a) Spherical coordinate system. θ is the zenith angle and ϕ is the azimuthal angle b) Radiation intensity emitted from an infinitesimal surface dA into an infinitesimal solid angle $d\Omega$ where $d\Omega = \sin\theta \, d\theta d\phi$ [INC 90]*

It appears that the spectral radiative intensity is a function of seven independent parameters since we have three space coordinates (x, y, z) along $\vec{r}$, two angular coordinates (θ, ϕ) for direction $\vec{u}$, one wavelength, λ, and one time, t. The unit of I_λ is Wm^{-2}. $\mu m.sr^{-1}$. The total radiative intensity, I_T, that is the integration of the spectral radiative intensity over the whole spectral range is defined by:

$$I_T(\vec{r}, \vec{u}) = \int_0^\infty I_\lambda(\vec{r}, \vec{u}) \, d\lambda \qquad [12.6]$$

Practically, this integration is performed on a finite wavelength domain $[\lambda_{min}, \lambda_{max}]$ imposed by the temperature of the medium. This aspect will be detailed in the next section.

To go one step further, we can also play with the quantity called the spectral radiative flux density, $q_\lambda(\vec{r})$. If we consider, the thermal radiation subdued within an incoming infinitesimal solid angle around the direction $\vec{u}_{in}$that hits a surface and the thermal radiation that outgoes from the same surface with a direction $\vec{u}_{out}$, the spectral net radiative flux density, q_{net} is defined by:

$$q_{\lambda,net}(\vec{r}) = q_{\lambda,out}(\vec{r}) - q_{\lambda,in}(\vec{r}) \qquad [12.7]$$

with

$$q_{\lambda,in}(\vec{r}) = \int_{cos\theta_{in}<0} I_\lambda(\vec{r},\vec{u}_{in}) cos\theta_{in}\, d\Omega_{in} \qquad [12.8]$$

and

$$q_{\lambda,out}(\vec{r}) = \int_{cos\theta_{out}>0} I_\lambda(\vec{r},\vec{u}_{out}) cos\theta_{out}\, d\Omega_{out} \qquad [12.9]$$

Finally, the spectral net radiative heat flux density is evaluated as the flux going in the positive $\vec{n}$ direction:

$$q_{\lambda,net}(\vec{r}) = \vec{q}_\lambda(\vec{r}).\vec{n} = \int_{4\pi} I_\lambda(\vec{r},\vec{u})\vec{n}\,.\vec{u} d\Omega \qquad [12.10]$$

The total radiative heat flux density at the surface is given by integration of equation [12.8] over the whole spectrum:

$$q_{T,net}(\vec{r}) = \int_0^\infty \int_{4\pi} I_\lambda(\vec{r},\vec{u})\vec{n}\,.\vec{\Omega} d\,\Omega\, d\lambda \qquad [12.11]$$

Let us specify that unity of $q_{T,net}$ is in W.m^{-2}. For a surface, [12.10] allows us to compute the spectral emissive power E_λ, at location point, $\vec{r}$, that is defined as the rate as which radiant energy of wavelength λ is emitted in all directions pointing away from the surface per unit wavelength $d\lambda$ about λ per unit time and per unit surface area.

$$E_\lambda(\vec{r}) = \int_{2\pi} I_\lambda(\vec{r},\vec{u})\vec{n}\,.\vec{\Omega} d\Omega \qquad [12.12]$$

$E_\lambda(\vec{r})$ obviously depends on the local temperature and on the optical properties of the surface. The notion of spectral emissive power is interesting to introduce the concept of blackbody as we can see in the next section.

12.2.3. *Blackbody spectral emissive power*

In the early 19th Century, many scientists have attempted to theoretically predict the Sun's emission spectrum (Lord Rayleigh, Sir James Jeans and Wilhelm Wien). The most famous of this group was Max Planck who published works on quantum mechanics in 1901. In particular, he found the spectral emissive power of a blackbody commonly known as Planck's law. Thus, for a black surface surrounded by a transparent medium with a refractive index, n, we have:

$$E_{\lambda b}(T) = \frac{2\pi h c_0^2}{n^2\lambda^5\left[exp\left(\frac{hc_0}{nk_B\lambda T}\right)-1\right]} = \frac{\pi C_1}{n^2\lambda^5\left[exp\left(\frac{C_2}{n\lambda T}\right)-1\right]} \quad [12.13]$$

Two universal constants are used in equation [12.12]: Planck's constant, h, and Boltzmann's constant, k_B where k_B = 1.3806504 10^{23} J/K. Two auxiliary constants are introduced $C_1 = 2hc_0^2$ and $C_2 = hc_0/k_B$. Recall that for a blackbody, its radiative intensity is uniform in all directions. According to Lambert's cosine law, it follows that $E_{\lambda b}$ is independent of θ and ϕ. This means that the blackbody is a perfect diffuse emitter. By using equation [12.12], we can easily demonstrate that:

$$E_{\lambda b}(T) = \pi I_{\lambda b}(T) \quad [12.14]$$

equations [12.12] and [12.13] are afterwards combined to retrieve the spectral distribution of Planck's law that is depicted in Figure 12.3. It can be seen that the spectral radiative intensity increases when the temperature increases. Moreover, the maximum of the spectral radiative intensity shifts toward lower wavelengths as temperature increases. For a temperature of T of 700 K, which corresponds to the highest temperature where advanced polymer matrices can be ever solid, emission is predominantly in the infrared region of the electromagnetic spectrum. It allows us to define λ_{max} given by the Wien's displacement law:

$$\lambda_{max} = C_3 T \quad [12.15]$$

with C_3 the third radiative constant equal to 2897.8 μm.K. When the operating temperatures of the infrared sources involved in a manufacturing processes are known, equation [12.15] indicates the wavelength λ_{max} that mainly defines the spectral range where the radiative flux densities must be computed.

Let us now specify what happens when total blackbody intensity and emissive power are calculated when n is assumed to be constant:

$$I_{Tb}(T) = \int_0^\infty \frac{C_1}{n^2\lambda^5\left[exp\left(\frac{C_2}{n\lambda T}\right)-1\right]} d\lambda = n^2 \frac{\sigma_B}{\pi} T^4 \quad [12.16]$$

where σ_B is the Stefan–Boltzmann constant which depends on $C_1$and C_2 since:

$$\sigma_B = \frac{C_1 \pi^5}{15\, C_2^4} \qquad [12.17]$$

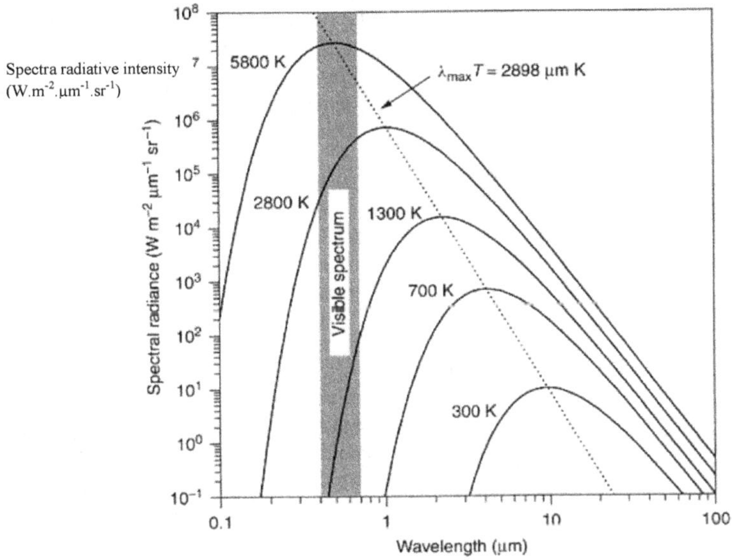

Figure 12.3. *Spectral radiative intensity, $I_{\lambda b}$, of a blackbody at several temperatures [INC 90]*

The numerical value of σ_B is 5.670 10^{-8} $W.m^{-2}.K^{-4}$. This means that the total emissive power can be given by:

$$E_{Tb}(T) = \pi I_{Tb}(T) = n^2 \sigma T^4 \qquad [12.18]$$

Equation [12.17] is known as the Stefan–Boltzmann law where the total spectral emissive power is given in $W.m^{-2}$. It enables computation of the radiative heat flux density in all directions and over all wavelengths simply by knowing the temperature of the blackbody.

Most of the time, it is practical to compute the emissive power of the blackbody in a finite wavelength interval or band. For a given temperature and the interval from 0 to λ, the corresponding fraction is determined as follows:

$$F_{(0\to\lambda)}(T) = \frac{\int_0^{\lambda} E_{\lambda b}(T) d\lambda}{\int_0^{\infty} E_{\lambda b}(T) d\lambda} = \int_0^{n\lambda T} \frac{E_{\lambda b}(T)}{n^3 \sigma T^5} d(n\lambda T) \qquad [12.19]$$

When the case of a band between λ_1 and λ_2 is considered, we obtain:

$$F_{(\lambda_1 \to \lambda_2)}(T) = \frac{\int_0^{\lambda_2} E_{\lambda b}(T)d\lambda - \int_0^{\lambda_2} E_{\lambda b}(T)d\lambda}{\int_0^{\infty} E_{\lambda b}(T)d\lambda} = F_{(0 \to \lambda_2)}(T) - F_{(0 \to \lambda_1)}(T) \qquad [12.20]$$

$F_{(0 \to \lambda)}(T)$ is a function of $n\lambda T$ and is tabulated. From equation [12.17] and by using suitable tabulated data, we can show that 96% of the spectral emissive power at a fixed temperature, T, is emitted for a spectral domain going from $0.5\lambda_{max}$ to $5\lambda_{max}$. In other words, if the temperature, T, of a polymer composite exposed to surrounding thermal radiations is known, we can define the largest spectral window where the spectral absorbed radiative heat flux must be considered.

12.2.4. *Radiative properties at interfaces*

In the previous section, the concept of a blackbody has been briefly introduced. A blackbody is considered as the ideal emitter, radiating equivalently in all directions. But in reality, materials and therefore polymer composites cannot radiate energy as the blackbody can do perfectly. In the following, a blackbody will be a reference in describing the emission of a real surface which envelops a given material.

Now, just before describing the radiative properties, since materials can behave as transparent, semi-transparent or opaque media, it appears judicious as Siegel, Howell and Menguc explained in the fifth edition of the famous book "Thermal Radiation Heat Transfer" [HOW 03] to make the distinction between the quantity ending by –ivity and by –ance. The –ivity ending refers to intensive properties such as thermal conductivity and diffusivity, whereas the –ance ending is reserved for extensive properties such as conductance. More exactly, extensive properties are proportional to the amount of material. For an opaque material using the expression emittance for describing the surface emission properties appears inappropriate since they are not governed by the material thickness. In contrary for a semi-transparent medium, the expression emittance is more pertinent since the spatial organization of the matter contained within its volume acts directly on the property. Once this framework exposed, let us describe the main spectral properties.

12.2.4.1. *Directional spectral emissivity*

Consider the radiative energy leaving an opaque elementary surface dA at a temperature T, itself function of the vector position $\vec{r}(x, y, z)$, per unit time in the interval spectral domain $d\lambda$ and within the solid angle $d\Omega$ where $d\Omega = \sin\theta d\theta d\phi$. The differential spectral radiant flux, d^2Q_λ is given by:

$$d^2Q_\lambda(\theta,\phi,T)d\lambda = I_\lambda(\theta,\phi,T)dA\cos\theta d\Omega\, d\lambda \qquad [12.21]$$

However, for the blackbody, reference $I_{\lambda b}(T)$ is independent of (θ,ϕ) and:

$$d^2Q_{\lambda b}(\theta,\phi,T)d\lambda = I_{\lambda b}(T)dA\cos\theta d\Omega\, d\lambda \qquad [12.22]$$

The directional spectral emissivity, $\epsilon(\theta,\phi,T)$ is then defined as follows:

$$\epsilon_\lambda(\theta,\phi,T) = \frac{d^2Q_\lambda(\theta,\phi,T)d\lambda}{d^2Q_{\lambda b}(\theta,\phi,T)d\lambda} = \frac{I_\lambda(\theta,\phi,T)}{I_{\lambda b}(T)} \qquad [12.23]$$

Once this capital spectral quantity is known, we can compute its hemispherical total expression as follows:

$$\epsilon_{\cap,T}(T) = \frac{\int_\cap \int_0^\infty [\epsilon(\theta,\phi,T)I_{\lambda b}(T)d\lambda]\cos\theta d\Omega}{\sigma_B T^4} \qquad [12.24]$$

We specify that:

$$\int_0^{2\pi}\int_0^{\pi/2}\epsilon(\theta,\phi,T)sin\theta d\theta d\phi = \int_\cap \epsilon(\theta,\phi,T)d\Omega \qquad [12.25]$$

12.2.4.2. *Directional spectral absorptivity*

Imagine a thermal source radiating thermal radiation toward an opaque elementary surface dA in the direction (θ_i,ϕ_i). The incident differential spectral radiant flux, d^2Q_λ is given by:

$$d^2Q_{\lambda,i}(\theta_i,\phi_i,T)d\lambda = I_{\lambda,i}(\theta_i,\phi_i,T)d\Omega_i dA\cos\theta_i d\lambda \qquad [12.26]$$

The directional spectral absorptivity is defined as the fraction of incident directional spectral radiant flux absorbed on the surface at T.

$$\alpha_\lambda(\theta_i,\phi_i,T) = \frac{d^2Q_{\lambda,a}(\theta_i,\phi_i,T)d\lambda}{I_{\lambda,i}(\theta_i,\phi_i,T)d\Omega_i dA\cos\theta_i d\lambda} \qquad [12.27]$$

If the surrounding radiative energy comes from a blackbody source at temperature T, we obtain:

$$\alpha_\lambda(\theta_i,\phi_i,T) = \frac{d^2Q_{\lambda,a}(\theta_i,\phi_i,T)d\lambda}{I_{\lambda b}(T)d\Omega_i dA\cos\theta_i d\lambda} \qquad [12.28]$$

Starting from equations [12.22] and [12.23], we can depict the directional spectral emitted radiant flux in the direction (θ_i, ϕ_i):

$$d^2Q_{\lambda,e}(\theta_i, \phi_i, T)d\lambda = \epsilon_\lambda(\theta_i, \phi_i, T)I_{\lambda b}(T)d\Omega_i dA \cos\theta_i d\lambda \qquad [12.29]$$

Kirchhoff's first law states that if an elementary surface dA at temperature T inserted in an enclosure with blackbody features at the same temperature T, we have for a general incidence (θ, ϕ) :

$$\epsilon_\lambda(\theta, \phi, T) = \alpha_\lambda(\theta, \phi, T) \qquad [12.30]$$

Equation [12.30] means that if the conditions for local thermodynamic equilibrium are fulfilled, microscopic optical mechanisms within the matter that act on the light absorption are the same as those that act on the light emission. Moreover, as for emissivity, hemispherical and total absorptivity can be computed without modifying Kirchhoff's law.

12.2.4.3. *Spectral bidirectional reflectivity distribution function*

Consider an incident pencil of light impinging on an elementary surface dA characterized by its spectral radiant flux $d^2Q_\lambda(\theta_i, \phi_i, T)$ from a direction (θ_i, ϕ_i). A part of spectral radiant flux is reflected in the direction (θ_r, ϕ_r). The spectral directional irradiance $dG_\lambda(\theta_i, \phi_i, T)$ is given by:

$$dG_\lambda(\theta_i, \phi_i, T) = \frac{d^2Q_\lambda(\theta_i, \phi_i, T)d\lambda}{dAd\lambda} = I_\lambda(\theta_i, \phi_i, T)\cos\theta_i d\Omega_i \qquad [12.31]$$

$dG_\lambda(\theta_i, \phi_i, T)$ corresponds to differential radiant flux impinging dA from (θ_i, ϕ_i) per unit solid angle $d\Omega_i$. $I_{\lambda,r}(\theta_r, \phi_r, \theta_i, \phi_i, T)$ is the reflected spectral radiative intensity in (θ_r, ϕ_r) due to $dG_\lambda(\theta_i, \phi_i, T)$. The spectral bidirectional reflectivity distribution function [NIC 77], $f_\lambda(\theta_r, \phi_r, \theta_i, \phi_i, T)$, is then defined as:

$$f_\lambda(\theta_r, \phi_r, \theta_i, \phi_i, T) = \frac{I_{\lambda,r}(\theta_r, \phi_r, \theta_i, \phi_i, T)}{I_\lambda(\theta_i, \phi_i, T)\cos\theta_i d\Omega_i} \qquad [12.32]$$

The term $f_\lambda(\theta_r, \phi_r, \theta_i, \phi_i, T)$ is also abbreviated as the spectral BRDF and it may take any value from zero to infinity. The unit of BRDF is sr^{-1}. The BRDF can be used to express the reflected quantities by an opaque surface. Note that BRDF is measured with hardness due to the very small value of $d\Omega_r$. The BRDF obeys to Helmholtz reciprocity principle that stipulates how a ray of light and its reverse ray encounter matched optical events, in particular for the case of reflections at an interface. This means that:

$$f_\lambda(\theta_r,\phi_r,\theta_i,\phi_i,T) = f_\lambda(\theta_i,\phi_i,\theta_r,\phi_r,T) \quad [12.33]$$

With simple mathematical handling, we can determinate the spectral directional hemispherical reflectivity, $\rho_\lambda^\cap(\theta_i,\phi_i,T)$

$$\rho_\lambda^\cap(\theta_i,\phi_i,T) = \int_\cap f_\lambda(\theta_r,\phi_r,\theta_i,\phi_i,T)\cos\theta_r\, d\Omega_r \quad [12.34]$$

For an opaque surface, we can also show a practical relationship for defining the spectral directional absorptivity

$$\alpha_\lambda(\theta_i,\phi_i,T) = 1 - \rho_\lambda^\cap(\theta_i,\phi_i,T) \quad [12.35]$$

From equation [12.34] and for the case of a surface optically behaving as a perfect diffuse reflector:

$$\rho_\lambda^\cap(T) = \pi f_\lambda(T) \quad [12.36]$$

In the opposite case of a mirror-like surface, the Fresnel laws apply leading to θ_i=θ_r and ϕ_r=$\phi_i + \pi$. For angles different from $\theta_i$and ϕ_i, the spectral BRDF is equal to zero so that :

$$\rho_{\lambda,s}(\theta_i,\phi_i,T) = f_\lambda(\theta_r,\phi_r,\theta_i,\phi_i,T)\cos\theta_i d\Omega_i \equiv f_\lambda(\theta_i,\phi_i,T)\cos\theta_i d\Omega_i \quad [12.37]$$

12.2.4.4. *Spectral bidirectional transmissivity distribution function*

When the surface is not opaque, the spectral radiant flux $d^2Q_\lambda(\theta_i,\phi_i,T)$ from a direction (θ_i,ϕ_i) and contained in the elementary solid angle $d\Omega_i$ can cross it. Therefore, a fraction of $d^2Q_\lambda(\theta_i,\phi_i,T)$ can be transmitted in a direction (θ_t,ϕ_t) through the single interface. As for the spectral BRDF, the spectral BDTF is defined by the relation:

$$f_\lambda(\theta_t,\phi_t,\theta_i,\phi_i,T) = \frac{I_{\lambda,t}(\theta_t,\phi_t,\theta_i,\phi_i,T)}{I_\lambda(\theta_i,\phi_i,T)\cos\theta_i d\Omega_i} \quad [12.38]$$

As in the case of the reflectivity, the spectral directional hemispherical transmittivity is connected to the spectral BDTF as follows:

$$\tau_\lambda^\cap(\theta_i,\phi_i,T) = \int_\cap f_\lambda(\theta_t,\phi_t,\theta_i,\phi_i,T)\cos\theta_r\, d\Omega_r \quad [12.39]$$

12.2.5. *Radiative properties of semi-transparent slabs*

At the beginning of section 12.2.2, we have briefly recalled the concept of irradiance. A parallel can be drawn with equation [12.12] that defines what the spectral emissive power, $E_\lambda(\vec{r})$, is. Indeed, the spectral irradiance, $G_\lambda(\vec{r})$, at the location point, $\vec{r}$, corresponds to the spectral radiative flux arriving on an elementary surface, dA, from the whole hemisphere above it.

$$G_\lambda(\vec{r}) = \int_{2\pi} I_\lambda(\vec{r},\vec{u})\vec{n}.\vec{u}d\Omega = \int_{2\pi} G'_\lambda(\vec{r},\vec{u})\, d\Omega \qquad [12.40]$$

In equation [12.40], $G'_\lambda(\vec{r},\vec{u})$ is the spectral directional irradiance. $G_\lambda(\vec{r})$ is often associated with another spectral quantity that is the spectral radiosity, $J_\lambda(\vec{r})$. The latter quantity is used to describe all the spectral radiative flux leaving a surface. More details concerning this quantity can be found in [HOW 03] and [MOD 03]. The total expressions of $G_\lambda(\vec{r})$ and $J_\lambda(\vec{r})$ are often involved in the net-radiation method that had been developed to quantify the radiative exchange in any enclosed domain where exchanges between two opaque surfaces are treated due to geometric configuration factors [HOW 03]. For the case of a 1D semi-transparent window of thickness, d, and with two opposite optically polished slabs surrounded by air, the net-radiation method was modified for taking account of its semi-transparency. Moreover, for this case, radiative quantities are obviously independent of azimuthal angle, ϕ. For a single direction of incidence, θ_i, the following relations [SIE 73] are successively found:

$$J_{\lambda,1} = \rho_{\lambda,\theta_i} G_{\lambda,1} + \left(1 - \rho_{\lambda,\theta_i}\right) G_{\lambda,2} \qquad [12.41a]$$

$$J_{\lambda,2} = \left(1 - \rho_{\lambda,\theta_i}\right) G_{\lambda,1} + \rho_{\lambda,\theta_i} G_{\lambda,2} \qquad [12.41b]$$

$$J_{\lambda,3} = \rho_{\lambda,\theta_i} G_{\lambda,3} + \left(1 - \rho_{\lambda,\theta_i}\right) G_{\lambda,4} \qquad [12.41c]$$

$$J_{\lambda,4} = \left(1 - \rho_{\lambda,\theta_i}\right) G_{\lambda,3} + \rho_{\lambda,\theta_i} G_{\lambda,4} \qquad [12.41d]$$

with, $G_{\lambda,1} = 1$, $G_{\lambda,4} = 0$, $G_{\lambda,2} = J_{\lambda,3}\tau_\lambda$ and $G_{\lambda,3} = J_{\lambda,2}\tau_\lambda$. Combining the whole equations yields:

$$\alpha_\lambda(\theta_i) = \frac{\left(1-\rho_{\lambda,\theta_i}\right)(1-\tau_\lambda)}{1-\rho_{\lambda,\theta_i}\tau_\lambda} \qquad [12.42]$$

where $\alpha_\lambda(\theta_i)$ is the absorptance at the temperature, T. In equation [12.42], ρ_{λ,θ_i} is given by the Fresnel law for an oblique incidence [BOR 99] and $\tau_\lambda$is the attenuation of the radiation along the path $d/\cos\theta_2$. The angle θ_2 is given by the generalized Snell–Descartes law. For the particular case of the normal incidence ($\theta_i = 0$), we obtain:

$$\alpha_\lambda(\theta_i = 0) = \frac{\left(1-\rho_{\lambda,\theta_i=0}\right)\left(1-e^{-4\pi k_\lambda d/\lambda}\right)}{1-\rho_{\lambda,\theta_i=0}e^{-4\pi k_\lambda d/\lambda}} \qquad [12.43]$$

where

$$\rho_{\lambda,\theta_i=0} = \frac{(n_\lambda-1)^2+k_\lambda{}^2}{(n_\lambda+1)^2+k_\lambda{}^2} \qquad [12.44]$$

Let us specify that the lab is composed of a homogeneous, isotropic and non-magnetic material. Equations [12.43] and [12.44] are particularly important because they directly connect the absorptance to the complex refractive of the matter, $\tilde{n}_\lambda = n_\lambda + i\,k_\lambda$ contained within the slab. It means that when the spectral normal reflectance $R_\lambda(\theta_i = 0)$ and/or the spectral normal transmittance $T(\theta_i = 0)$ are computed due to equation [12.41] and if they are measured in parallel, the determination of $\tilde{n}_\lambda$ is straightforward. However, since $\tilde{n}_\lambda = \sqrt{\tilde{\varepsilon}_\lambda}$, where $\tilde{\varepsilon}_\lambda$ is the complex dielectric function that probes the vibrational activity of the matter at the atomic scale.

Finally, recall that according to Kirchhoff's second law which comes down a radiation balance within the semi-transparent material:

$$\alpha_\lambda(\theta_i) = \varepsilon_\lambda(\theta_i) = 1 - R_\lambda(\theta_i) - T_\lambda(\theta_i) \qquad [12.45]$$

$R_\lambda(\theta_i)$ and $T_\lambda(\theta_i)$ are the spectral directional reflectance and transmittance. These two quantities depend on the volumetric absorption of the thermal radiation within the slab and on the multiple specular reflections occurring between the two opposite parallel boundaries separated by the distance, d.

12.3. Prediction of the radiative properties of homogeneous semi-transparent slabs: case of the isotactic polypropylene

Until now, the main spectral radiative properties have been described. All the quantities are linked to the spectral radiative intensity, $I_\lambda(\vec{r},\vec{u})$. For the case of a 1D homogeneous semi-transparent slab (with azimuthal symmetry), we are able to define the spectral directional absorptance, $\alpha_\lambda(\theta_i)$, the spectral directional reflectance, $R_\lambda(\theta_i)$ and the spectral directional transmittance, $T_\lambda(\theta_i)$, all these properties being connected by the Kirchhoff's laws.

12.3.1. *Intrinsic optical properties for a homogeneous, isotropic and non-magnetic medium*

In the framework of the electromagnetic waves theory, the resolution of Maxwell's equations at the interfaces of an isotropic non-magnetic homogeneous semi-transparent slab shows that, finally, $\alpha_\lambda(\theta_i)$, $R_\lambda(\theta_i)$ and $T_\lambda(\theta_i)$ are straightforwardly connected to its spectral complex refractive index $\tilde{n}_\lambda = n_\lambda + i\, k_\lambda$, and of course, to its thickness, d. The knowledge of $\tilde{n}_\lambda$ is therefore crucial. We do not spend time, here, to detail all the methodologies developed in the field of solid state physics to retrieve $\tilde{n}_\lambda$. Briefly, for polymer matrices, $\tilde{n}_\lambda$ may be first obtained by modeling the complex dielectric function, $\tilde{\varepsilon}_\lambda$, with harmonic oscillator models adapted either for crystallized (Lorentz model [GER 02, KIT 05]) or for amorphous materials [MEN 05, MEN 14]. For inorganic semi-transparent samples (such as single crystals, thin layers, multi-layers based on heteropolar oxides), these approaches [GER 02, GER 73, GER 83, BRU 13] are now currently used on a large domain of temperature [4-2,500 K]. Secondly, we may use the law of causality, which states that an effect may not precede its cause, to obtain general relationships between n_λ, k_λ: this defines the well-known Kramers–Kronig relations [KRO 26, KRA 27]. In the case of normal incidence, the Kramers–Kronig relations may also be used to retrieve n_λ and k_λ from $R_\lambda(\theta_i = 0)$ [KIT 05, BRE 92]. Finally, other estimation processes are based on the maximum entropy method [BRU 01] or on piecewise polynomial functions [DES 07a, DES 07b].

Figure 12.4 depicts the complex refractive indices obtained at T = 20°C of pure silica [ROU 07], polystyrene [REM 09] and amorphous polypropylene [HAK 15]. n_λ and k_λ for pure silica are typical of the optical response of heteropolar semi-transparent materials dominated by lattice vibration absorption mechanisms in the Far Infrared Range (FIR) spectral range (ν_3 phonon band ~9.8 μm), multi-phonon absorption mechanisms at the beginning of the Mid Infrared Range (MIR) and no after. For polystyrene and polypropylene, the spectral dependence of the indices of absorption is rather the same knowing that weak absorption mechanisms due to their respective organic chains can be observed. Spectroscopists often use the viewed bands to identify the chemical composition of the studied polymer [EVE 07]. For the Near Infrared Range (NIR) range, the values of n_λ for the two polymers are comparable with the one of silica. The major difference appears for k_λ where two decades distinguish the attenuation of light for the two polymers and silica. It means that for a same physical thickness (typically 5 mm), a slab of polypropylene will be rather opaque whereas slab of silica will be contrarily transparent for the MIR spectral range.

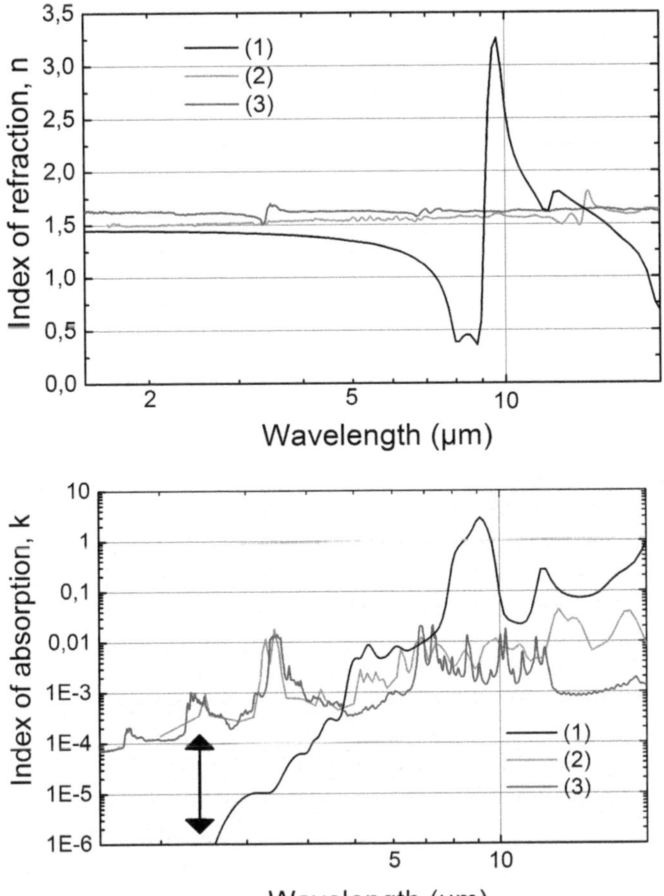

Figure 12.4. *Spectral dependence [1.5–20 µm] at T = 20°C of the complex refractive indices,* $\tilde{n}_\lambda = n_\lambda + i\, k_\lambda$*, of silica 1), of polystyrene 2) and of isotactic polypropylene 3). For a color version of the figure, see www.iste.co.uk/boyard/heat.zip*

12.3.2. *Prediction of the radiative properties of polypropylene slabs at 20°C*

Now, $\tilde{n}_\lambda$ for polypropylene will be injected in equation [12.42] to compute $\alpha_\lambda(\theta_i = 0)$, $R_\lambda(\theta_i = 0)$, $T_\lambda(\theta_i = 0)$ for a slab of 1 mm thick with moreover:

$$R_\lambda(\theta_i) = \rho_{\lambda,\theta_i}\left[1 + \frac{\left(1-\rho_{\lambda,\theta_i}\right)^2 \tau_\lambda{}^2}{1-\left(\tau_\lambda\, \rho_{\lambda,\theta_i}\right)^2}\right] = \rho_{\lambda,\theta_i}[1 + \tau_\lambda T_\lambda(\theta_i)] \qquad [12.46]$$

and

$$T_\lambda(\theta_i) = \frac{\left(1-\rho_{\lambda,\theta_i}\right)^2 \tau_\lambda}{1-\left(\tau_\lambda\, \rho_{\lambda,\theta_i}\right)^2} \qquad [12.47]$$

Let us describe briefly the absorption bands viewed on the polypropylene slab (from $\alpha_\lambda(\theta_i = 0)$ and/or from $T_\lambda(\theta_i = 0)$). The constitutional repeat unit of polypropylene is $-(CH_2\text{-}CH(CH_3)\text{-})_n$ so that the mechanisms of absorption are governed, here, by skeleton vibration of the organic chains in the spectral range from 6.67 up to 13.4 µm. In particular, C-C stretching vibrations can act in this spectral domain. Two peaks at ~3.40 and 3.45 are assigned to the symmetrical CH_3 stretching vibrations and a band at 3.47 µm to the overtone CH_2 bending vibrations. Two other bands at 1.72 and 2.38 µm come from superior order combinations of the primary modes [SUN 96]. These two bands are important since they may explain why the index of absorption of polypropylene in the MIR spectral range is higher of two decades than the one of pure silica (see Figure 12.5). Here, pure silica is representative of what happens in this spectral range for most of pure heteropolar semi-transparent materials (alumina, magnesia, mullite, etc.). If pure silica is doped with hydroxyl groups (-OH) during its elaboration, a broad band can appear in the MIR domain of the absorptance spectrum [ROU 07]. In other words, a high value of the term $4\pi k_\lambda d/\lambda$ in equation [12.43] which is known as the optical thickness $K_\lambda d/\lambda$, explicates directly a high level for $\alpha_\lambda(\theta_i = 0)$. Recall that $K_\lambda = 4\pi k_\lambda/\lambda$ is the absorption coefficient (mm^{-1}). Thus, Figure 12.6 reports the evolution of $\alpha_\lambda(\theta_i = 0)$ for a polypropylene slab with growing thicknesses going from 0.1 to 10 mm. The higher the thickness is, the lower the sharpness of the bands is. Note that for thicknesses higher than 5 mm, the slab behaves as a gray optically thick medium with $\alpha_T(\theta_i = 0) \sim 0.95$ at T = 20°C.

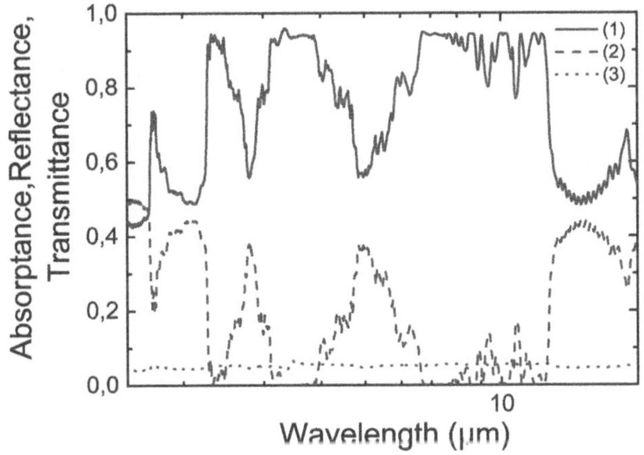

Figure 12.5. *Spectral dependence of the absorptance (1), transmittance (2) and reflectance (3) of slab of 1 mm thick composed of isotactic polypropylene at T = 20°C*

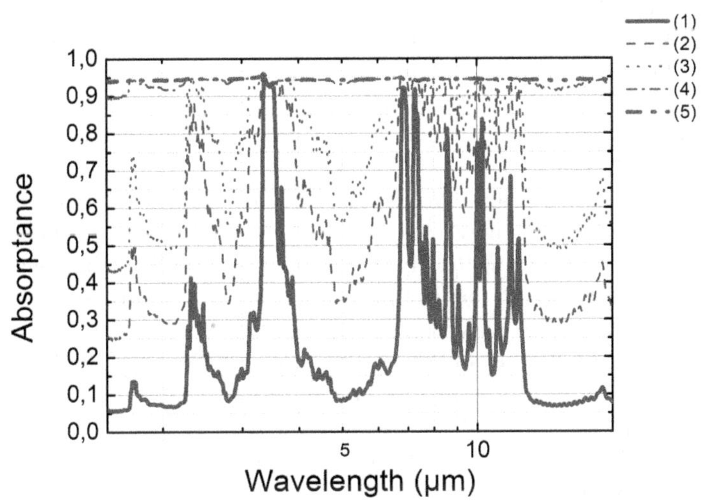

Figure 12.6. *Spectral dependence of the absorptances of several slabs of isotactic polypropylene at T = 20°C: (1) 0.1 mm, (2) 0.5 mm, (3) 1 mm, (4), 5 mm and (5) 10 mm*

If the surface roughness of thick (> 5 mm) polypropylene slabs can be increased, therefore $\alpha_T(\theta_i = 0)$ may tend to 1 [TA 10]. This would mean that a polypropylene slab classically viewed as translucent (amorphous state) or whitish (semi-crystalline state) for the visible spectral range may be considered, surprisingly, as a nearly blackbody like material in the infrared range. This information is crucial for the determination of the temperature of surface during a heating process with contactless methods (infrared pyrometer or optical sensor). Generally, a small amount of black carbon powder (1–4% in mass fraction) is added to polypropylene pellets during the shaping process to color the original whitish sample in black [TAR 12], the blackness often being associated with a high spectral directional emissivity.

12.4. Radiative properties of polymer composites with fiber structures

Since the main parameters (θ_i,n_λ, k_λ, d) required to tune the normal spectral absorptances of polymer matrices at a given temperature, T, have been set, we now turn to the case of polymer composites. It is well known that advanced polymer composites, developed for either aeronautic or automotive application, are made up of high-strength fibers embedded in a host polymeric matrix. The volume fraction of fibers, $f_{v,fibres}$, is often in the range 0.5–0.7, such that high fraction volumes conferring to the composite high mechanical properties in terms of stiffness and toughness. In general, fibers are linearly aligned or interlaced. When fibers are woven, it leads to two- or three-dimensional interlocked structures [GUT 97]. In the case of one-dimensional arrangement, the spacing between fibers, δ, is instead law. For $f_{v,fibres} = \pi/4$ and $\emptyset_{fibres}$ = 7 µm, δ~0.9 µm. According to the incident wavelength and to the nature of fibers (glass or carbon), the understanding of the radiative properties may be rather tricky in light of plausible light-dependent scattering [LEE 94, BAI 00, COQ 06]. In other words, should the radiative properties of the composites be interpreted by simply mixing the radiative properties of the constituents taken independently or not? And what is the real role played by the absorbing polymeric matrix [YIN 06]? These points constitute open questions that must be encompassed in the future. Moreover, in the following, $\alpha_\lambda(\theta_i = 0)$ for two composites with one-directional aligned fibers will be discussed. Specifying that the two considered composites are based on thermoplastic matrices and the discussion would also be extended to the case of thermoset matrices [GUT 97, CHE 95a, CHE 95b].

12.4.1. *Normal spectral absorptance of Roving TWINTEX®*

In Figure 12.7, the normal spectral absorptances (T = 20°C) of three slabs composed, respectively, of pure silica, amorphous polypropylene and polypropylene composite reinforced with E-glass fibers are depicted. The commercial name of the composite is Roving TWINTEX® (Owens Corning, Toledo, Ohio, USA) and is made up of commingled E-glass fibers and polypropylene filaments. Since silica is one of the major constituents of E-glass, its absorptance spectrum can be used as a useful guideline for the final discussion of this section. For both the silica and the polypropylene slabs, $\alpha_\lambda(\theta_i = 0)$, were computed (through equations [12.43] and [12.44]) with the complex refractive indices exposed in Figure 12.4 and for a thickness of 1.6 µm. The shape of the absorbance curve for the silica slab is typical of those obtained for heteropolar dielectric compounds [MEN 14]. For the spectral range going from 4 up to 20 µm, the absorptance is rather high (~0.8) whereas it falls to zero the range going from 1.5 to 2 µm. For the polypropylene slab, $\alpha_\lambda(\theta_i = 0)$ is rather high on the whole spectral range and particularly in the spectral window where the silica slab is obviously transparent. However, the more noticeable growth concerns the absorptance spectrum of the Roving TWINTEX® with $f_{m,fibres} = 0.6$ shows the highest values whatever the wavelength is. Qualitatively, this means that, for this composite the sample is optically thick and its spectrum results from the combination of the strongest absorption mechanisms of the E-glass and the polypropylene. Therefore, it clearly appears that in the MIR spectral range the absorbance spectrum of the roving may be rather similar than the one imposed by the polypropylene slab.

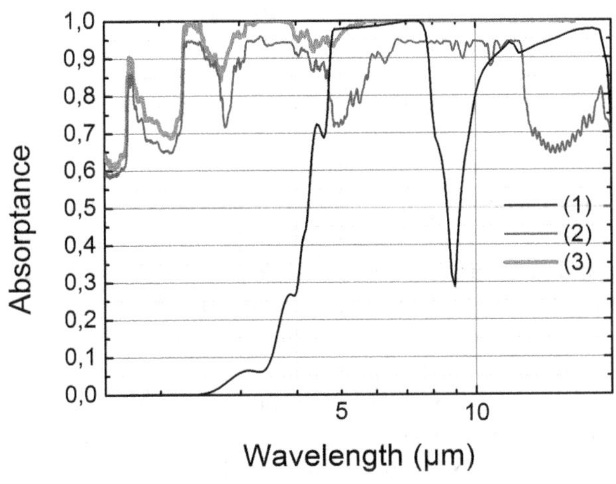

Figure 12.7. *Normal spectral absorptance at T = 20°C of slabs of 1.6 mm thick of (1) pure silica, (2) amorphous polypropylene (3) Roving TWINTEX®. For a color version of the figure, see www.iste.co.uk/boyard/heat.zip*

Slight differences may be explained by a local organization of the polymeric chains at the local scale ranging from an amorphous state for the polypropylene slab to a semi-crystalline state for the Roving TWINTEX® slab.

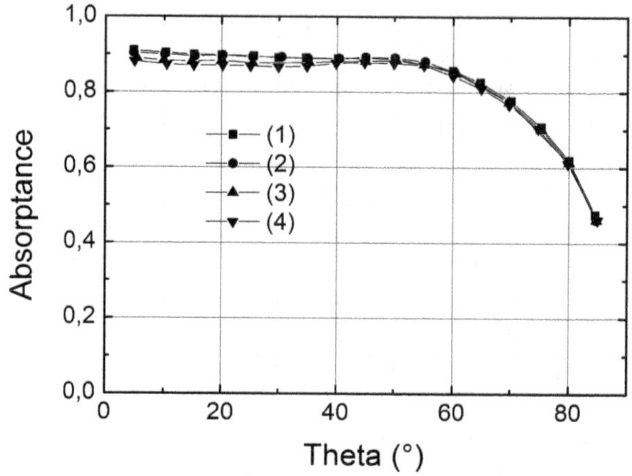

Figure 12.8. *Directional spectral absorptance at T = 20°C of a slab of PEEK reinforced with AS4 carbon fibers [STO 15] (1) 0.5 μm, (2) 1 μm, (3) 1.5 μm (4) 2 μm*

12.4.2. *Normal spectral absorptance of sheet of PEEK with carbon fibers*

Since the previous composite is based on glass fibers, its optical response is dominated in the MIR spectral range by the optical contribution of the polymeric matrix, and we can ask the question of what happens for the case of carbon fibers. Figure 12.8 shows the spectral directional absorptance of a PEEK matrix reinforced with carbon fibers. Straightforwardly, the optical response in the MIR/NIR range is here dominated by the presence of the carbon fibers [STO 15]. Recall that, in this example, the volumetric fraction of unidirectional carbon fibers occupies 55% of the composite [STO 15]. The main interesting effect lies in the diminution of the absorbance for growing angle of incidence when the illuminating beam ranges from 5 to 85° (here $\alpha_\lambda(\theta_i)$ is given by equation [12.35] see section 12.2.4.2). $\alpha_\lambda(\theta_i = 85°)$ drops on 44% that is spectacular. This means that for manufacturing processes such as automated tape placement (ATP) where an oblique laser heating is used, data obtained at the normal incidence are insufficient to perform exact predictions of the heat and mass transfers. The decreasing of $\alpha_\lambda(\theta_i)$ at high incidence angle is typical of what happens for dielectric materials [HOW 03].

12.5. Conclusion

This chapter proposes some robust basis for handling concepts and quantities useful in the field of radiative heat transfer. Special attention has been paid to the discussion of the presentation of these quantities such as spectral directional intensity of any materials and the blackbody with specific comments focused on polymer composites or their manufacturing processes. The radiative properties (polypropylene and silica glass) of scholar-type materials have been discussed and allow us to give a quick picture on the microscopic absorption mechanisms involving in the radiative properties of the resulting composites. To go one step further in the interpretation of the properties either spectral or angular, today we need to apply physical modeling well known in the solid state physics to the specific case of polymer composites with fibers structures. Due to the micrometric size of the fibers (~ 7 μm), the electromagnetic wave theory must be employed in the future to deal with this new field of investigation.

12.6. Bibliography

[ADV 12] ADVANI S.G., HSIAO K.-T., *Manufacturing Techniques For Polymer Matrix Composites (PMCs),* Elsevier, 2012.

[BAI 00] BAILLIS D., SACADURA J.-F., "Thermal radiation properties of dispersed media: theoretical prediction and experimental characterization", *Journal of Quantitative Spectroscopy and Radiative Transfer*, vol. 67, pp. 327–363, 2000.

[BAS 88] BASSETT D., OLLEY R., AL RAHEIL I., "On crystallization phenomena in PEEK", *Polymer*, vol. 29, pp. 1745–1754, 1988.

[BOR 99] BORN M., WOLF E., *Principles of Optics: Electromagnetic Theory of Propagation, Interference and Diffraction of Light*, 7th ed., Cambridge University Press, 1999.

[BRE 92] BREWSTER M.Q., *Thermal Radiative Transfer and Properties*, John Wiley & Sons, 1992.

[BRU 01] BRUN J.-F., DE SOUSA MENESES D., ROUSSEAU B. *et al.*, "Dispersion relations and phase retrieval in infrared reflection spectra analysis", *Applied Spectroscopy*, vol. 55, pp. 774–780, 2001.

[BRU 13] BRUN J., DEL CAMPO L., MENESES D. *et al.*, "Infrared optical properties of α-alumina with the approach to melting: γ-like tetrahedral structure and small polaron conduction", *Journal of Applied Physics*, vol. 114, p. 223501, 2013.

[CHE 95a] CHERN B.-C., MOON T., HOWELL J., "Thermal analysis of in-situ curing for thermoset, hoop-wound structures using infrared heating: part I—predictions assuming independent scattering", *Journal of Heat Transfer*, vol. 117, pp. 674–680, 1995.

[CHE 95b] CHERN B.-C., MOON T., HOWELL J., "Thermal analysis of in-situ curing for thermoset, hoop-wound structures using infrared heating: part II—dependent scattering effect", *Journal of Heat Transfer*, vol. 117, pp. 681–686, 1995.

[COQ 06] COQUARD R., BAILLIS D., "Radiative properties of dense fibrous medium containing fibers in the geometric limit", *Journal of Heat Transfer*, vol. 128, pp. 1022–1030, 2006.

[DES 05] DE SOUSA MENESES D., GRUENER G., MALKI M. *et al.*, "Causal Voigt profile for modeling reflectivity spectra of glasses", *Journal of Non-Crystalline Solids*, vol. 351, pp. 124–129, 2005.

[DES 07a] DE SOUSA MENESES D., ROUSSEAU B., ECHEGUT P. *et al.*, "Piecewise polynomial dielectric function model and its application for the retrieval of optical functions", *Applied Spectroscopy*, vol. 61, pp. 644–648, 2007.

[DES 07b] DE SOUSA MENESES D., ECHEGUT P. *et al.*, "Retrieval of linear optical functions from finite range spectra", *Applied Spectroscopy*, vol. 61, pp. 1390–1397, 2007.

[DES 14] DE SOUSA MENESES D., ECKES M., DEL CAMPO L. *et al.*, "Phase transformations of crystalline SiO2 versus dynamic disorder between room temperature and liquid state", *Journal of Physics: Condensed Matter*, vol. 26, p. 255402, 2014

[EVE 07] EVERALL N.J., CHALMERS J.M., GRIFFITHS P.R., *Vibrational Spectroscopy of Polymers: Principles and Practice*, John Wiley & Sons Ltd., 2007.

[GER 73] GERVAIS F., Anharmonicité des modes de vibration infrarouge dans le quartz-alpha et quelques silicates: interaction phonon-phonon entre 300 et 1500 K, Thesis, University of Orléans, 1973.

[GER 83] GERVAIS F., "High-temperature infrared reflectivity spectroscopy by scanning interferometry", *Infrared and Millimeter Waves*, vol. 8, pp. 279–339, 1983.

[GER 02] GERVAIS F., "Optical conductivity of oxides", *Materials Science and Engineering: R: Reports*, vol. 39, pp. 29–92, 2002.

[GUT 97] GUTOWSKI T.G.P., *Advanced Composites Manufacturing*, John Wiley & Sons, 1997.

[HAK 14a] HAKOUME D., DOMBROVSKY L., DELAUNAY D. *et al.*, "Effect of processing temperature on radiative properties of polypropylene and heat transfer in the pure and glass fibre reinforced polymer", *Proceedings of the 15th International Heat Transfer Conference*, IHTC-15, Kyoto, Japan, August 10-15, 2014.

[HAK 14b] HAKOUME D., DOMBROVSKY L., DELAUNAY D. *et al.*, "Spectroscopic diagnostics of morphological changes arising in thermal processing of polypropylene", *Applied Optics*, vol. 53, pp. 2702–2710, 2014.

[HAK 15] HAKOUME D., Influence des conditions de mise en forme sur les propriétés radiatives de polymères semi-cristallins, Thesis, Ecole Centrale de Nantes, 2015.

[HOW 03] HOWELL J.R., SIEGE R.L., MENGUC M.P., *Thermal Radiation Heat Transfer*, CRC Press, 2010.

[INC 90] INCROPERA F.P., DEWITT D.P., *Fundamentals of Heat and Mass Transfer*, 3rd ed., John Wiley & Sons, Inc., Singapore, pp. 1–27, 1990.

[KIT 05] KITTEL C., *Introduction to Solid State Physics,* Wiley, 2005.

[KRA 27] KRAMERS H.A., *La diffusion de la lumière par les atomes*, Como, 1927.

[KRO 26] KRONIG R.d., "On the theory of dispersion of x-rays", *JOSA*, vol. 12, pp. 547–556, 1926.

[LEE 94] LEE S.-C., "Dependent vs independent scattering in fibrous composites containing parallel fibers", *Journal of Thermophysics and Heat Transfer*, vol. 8, pp. 641–646, 1994.

[MOD 03] MODEST M., *Radiative Heat Transfer,* Academic Press, 2003.

[NIC 77] NICODEMUS F.E., RICHMOND J.C., HSIA J.J. *et al.*, Geometrical considerations and nomenclature for reflectance, US Department of Commerce, National Bureau of Standards Washington, DC, 1977.

[REM 09] RÉMI C., DOMINIQUE B., DANIEL Q., "Radiative properties of expanded polystyrene foams", *Journal of Heat Transfer*, vol. 131, p. 012702, 2009.

[ROU 07a] ROUSSEAU B., CANIZARES A., DE SOUSA MENESES D. *et al.*, "Direct simulation of the high temperature optical behaviour of a porous medium based on a CT image", *Colloids and Surfaces A: Physicochemical and Engineering Aspects*, vol. 300, pp. 162–168, 2007.

[ROU 07b] ROUSSEAU B., DE SOUSA MENESES D., ECHEGUT P. *et al.*, "Characterization of the thermal radiative properties of heterogeneous porous materials up to T= 2 500 K by a spectroscopic device", *EUROTHERM*, vol. 81, p. 00, 2007.

[ROU 07c] ROUSSEAU B., DI MICHIEL M., CANIZARES A. *et al.*, "Temperature effect (300-1500 K) on the infrared photon transport inside an x-ray microtomographic reconstructed porous silica glass", *Journal of Quantitative Spectroscopy and Radiative Transfer,* vol. 104, pp. 257–265, 2007.

[ROU 10] ROUSSEAU B., GOMART H., ZANGHI D. *et al.*, "Synchrotron x-ray µ-tomography to model the thermal radiative properties of an opaque ceramic coating at T = 1000 K", *Journal of Materials Research*, vol. 25, pp. 1890–1897, 2010.

[ROU 11] ROUSSEAU B., DE SOUSA MENESES D., ECHEGUT P. *et al.*, "Textural parameters influencing the radiative properties of a semitransparent porous media", *International Journal of Thermal Sciences*, vol. 50, pp. 178–186, 2011.

[SIE 73] SIEGEL R., "Net radiation method for transmission through partially transparent plates", *Solar Energy*, vol. 15, pp. 273–276, 1973.

[STA 93] STAGG B., CHARALAMPOPOULOS T., "Refractive indices of pyrolytic graphite, amorphous carbon, and flame soot in the temperature range 25 to 600 C", *Combustion and Flame*, vol. 94, pp. 381–396, 1993.

[STO 15] STOKES-GRIFFIN C., COMPSTON P., "Optical characterisation and modelling for oblique near-infrared laser heating of carbon fibre reinforced thermoplastic composites", *Optics and Lasers in Engineering*, vol. 72, pp. 1–11, 2015

[SUN 96] SUNDELL T., FAGERHOLM H., CROZIER H., "Isotacticity determination of polypropylene using FT-Raman spectroscopy", *Polymer*, vol. 37, pp. 3227–3231, 1996.

[TA 10] TA M.T., ROLLAND J.Y., ECHEGUT P. *et al.*, "Prediction of thermal radiative properties (300--1000 K) of La[sub 2]NiO[sub 4 + delta] ceramics", *Applied Physics Letters*, vol. 97, pp. 181917–181913, 2010.

[TAR 12] TARDIF X., Cristallisation et transferts thermiques dans un polymère thermoplastique semi-cristallin en refroidissement rapide sous pression, Thesis, University of Nantes, 2012.

[WAL 01] WALLENBERGER F.T., WATSON J.C., LI H., *Glass Fibers*, ASM International, Materials Park, Ohio, pp. 27–34, 2001.

[YIN 06] YIN J., PILON L., "Efficiency factors and radiation characteristics of spherical scatterers in an absor95bing medium", *JOSA A*, vol. 23, pp. 2784–2796, 2006.

13

Infrared Radiation Applied to Polymer Processes

The infrared heating of polymers is often a crucial step during polymer processing, because radiative heat transfer can be very efficient in comparison to conductive or convective heat transfer. Indeed, the thermal conductivity of polymers is really low ($0.1 \leq k \leq 0.6$ W/m/K) and consequently optimizing the heating and/or cooling steps remains a challenge. The results presented in this chapter are based on intensive research work, conducted at the ICA Institute over the past 10 years.

After a brief introduction, the authors present basic concepts dealing with radiative properties and more particularly infrared. The main radiative properties (emissivity, reflectivity, absorptivity, etc.) both for emitters and polymers are measured and analyzed. In addition, different methods to determine relevant parameters, such as filament lamp temperatures, are detailed. The development of accurate surface temperature measurements using an infrared camera is subsequently described. This powerful non-contact measurement device can also be applied for estimating heat transfer coefficients as well as anisotropic thermal conductivities of composites.

The infrared heating modeling of polymer preforms, typically used in the stretch blow molding process, is then presented. The assumptions of a cold and non-scattering medium suitable for thermoplastic polymers allow us to use a simplified radiative transfer equation. For radiation modeling, the Beer–Lambert law together with the ray tracing method is well adapted to simulate heating of semi-transparent polymers. Different examples are then presented, dealing with injection blow molding of P.E.T. bottles and curing of composites made of epoxide thermoset resin.

Future work and perspectives conclude this short but complete description of infrared radiation of polymers.

Chapter written by Yannick LE MAOULT and Fabrice SCHMIDT.

13.1. Introduction

Radiant energy has numerous familiar applications. A very common one is that of microwave heating, nowadays adopted for cooking at home. Another application is the laser, most known by the public for surgical applications. Halogen lamps are appreciated for home lighting, as well as automotive lighting systems, and the drying of paints in the automotive industry is also a well-known application. The focus of this chapter, however, is on infrared heating of polymers.

13.1.1. *Why use infrared heating for polymers?*

In the industrial context of polymer processing, radiant energy has gained increasing markets due to low thermal conductivity of most polymers (0.1–0.6 W/m/K) [FRA 02]. Different energy transfer means are used in industrial manufacturing processes:

– ultraviolet (UV);

– infrared (IR);

– high-frequency dielectric heating (HF);

– microwave dielectric heating (MO or UHF);

– laser beam.

The main advantages of IR heating include [ADE 02, WIL 14]:

– direct energy transfer between the source and the product to be treated (can penetrate into polymers giving a volumetric heating effect);

– high-power density;

– radiation focusing;

– control flexibility (low thermal inertia);

– low investment cost;

– adaptable and expandable.

13.1.2. *Application of radiative transfers in polymer processing*

In Table 13.1, some applications of infrared heating are presented. Polymer processes such as injection stretch-blow molding (ISBM) [ROS 04] and thermoforming [THR 99] are commonly used for manufacturing packaging products (bottles, cups, etc.). Both need a preliminary and efficient IR heating step before forming. In addition, development of IR heating technologies has gained recent interest for curing thermoset composites [WIL 14, NAK 11] since it is necessary to optimize manufacturing time and costs, and to provide the performances of these composites. Repair operation on damaged composite structures, such as an airplane fuselage, is another area in which infrared technologies are displacing conventional heating mats and conductive methods [WIL 14].

In the case of the ISBM process for manufacturing polyethylene terephthalate (PET) bottles, a tube-shaped preform previously injected is heated above the glass transition temperature $\left(T_g \cong 80°\text{C}\right)$ before stretching and inflating inside a bottle mold. It is well known that initial temperature distribution of the preform will have a major impact on mechanical and structural properties of the final bottle [BOR 09].

Industry	*Activity*	*Elementary operation can be processed by IR*
Automotive	Body	Gluing, varnishing, painting, printing
	Equipment	Surface treatment, gluing, forming, molding, embossing Flocking, varnishing, painting, printing
Thermoplastics, composites	Technical parts	Thermoforming, injection blow molding, gluing, welding, curing
	Flooring	Varnishing, coating, polymerization, foaming
Textile	Finishing	Drying, dehydration, polymerization, thermo-fixing, coating
	Technical textiles	Drying, heat treatment, gluing, polymerization, thermo-fixing, forming, molding, graining, coating

Table 13.1. *Some industrial applications of infrared heating [ADE 02]*

In addition, the rate of bottle manufacture can exceed 60,000 bottles per hour for a current ISBM machine. It is this high production rate and the production method that exposes the ISBM bottle manufacturing process as a high-energy consumption process with around half the total energy of a consumable bottle of water lending itself to manufacturing (Figure 13.1).

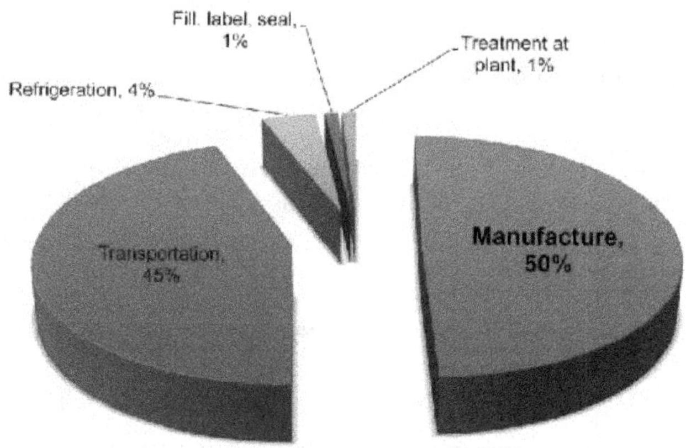

Figure 13.1. *Energy required producing bottled water [NIX 15]. For a color version of the figure, see www.iste.co.uk/boyard/heat.zip*

Optimizing temperature distribution while reducing energy consumption are major motivations to understand and model the interaction between IR radiation and the polymeric semi-products (preforms, sheets, plates, etc.).

13.2. Infrared radiation characteristics

Electromagnetic waves are in charge of heat transfer by radiation traveling at the speed of light ($c = 2.998 \times 10^8$ m/s in vacuum), which is called electromagnetic radiation. It is a kind of energy emitted as photons and exhibits as electromagnetic wave (Figure 13.2), which is characterized by wavelength λ and frequency ν:

$$\lambda = \frac{c}{\nu} \quad [13.1]$$

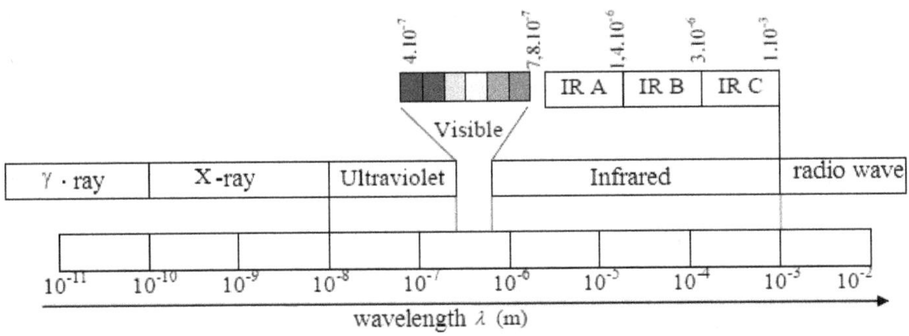

Figure 13.2. *Electromagnetic wave spectrum [IEC 87]. For a color version of the figure, see www.iste.co.uk/boyard/heat.zip*

According to most of the authors [MOD 03, SIE 02, MON 04], the characteristic infrared bandwidth is from 0.78–1,000 µm. Three main wavelength bands are used in sensing techniques or in the processing of materials: IR A [0.78–1.4 µm] corresponding to near-infrared, IR B [1.4–3 µm] corresponding to short-wavelength infrared (SW) and long-wavelength (LW) [8–15 µm].

13.2.1. *Radiative properties (basis and main definitions)*

The radiative heat flux Φ^i emitted by infrared lamps can be either reflected (ρ_λ), absorbed (α_λ) or transmitted (τ_λ) by the polymer (Figure 13.3).

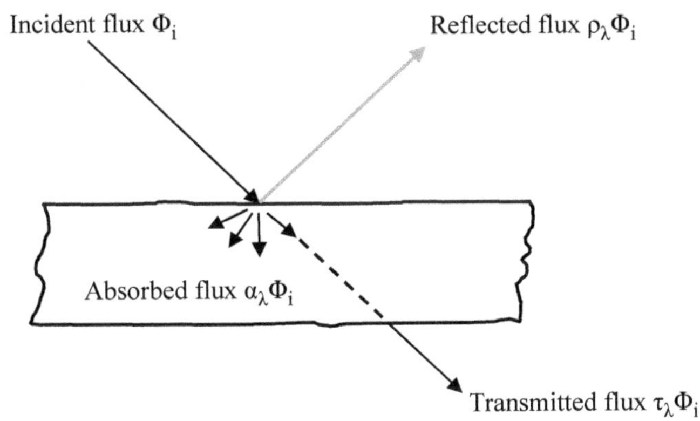

Figure 13.3. *Radiative flux distribution*

Thus, Kirchhoff's first law simply expresses the energy balance:

$$\alpha_\lambda + \tau_\lambda + \rho_\lambda = 1 \qquad [13.2]$$

In addition, using Kirchhoff's second law, we can assume the energy absorbed is equal to energy emitted:

$$\varepsilon_\lambda = \alpha_\lambda = 1 - \tau_\lambda - \rho_\lambda \qquad [13.3]$$

where ε_λ is the spectral emissivity coefficient. In addition, the reflection is related to the value of the surface roughness. Specular or diffuse reflection is characterized by a specific ratio given by $\frac{\sigma}{\lambda}$ where σ is the root mean square (RMS) roughness of the surface and λ is the wavelength of the incident radiation. If $\frac{\sigma}{\lambda} \ll 1$, then the surface is assumed to be specular; on the other hand, if $\frac{\sigma}{\lambda} \gg 1$, the surface is assumed to be a diffuse one. For most polymers parts considered here, the reflection is mainly diffuse. In addition, if the surface is assumed to be opaque, there is no radiation transmitted ($\tau_\lambda = 0$). The measurement of all these optical parameters is of prime importance in order to optically and spectrally characterize infrared heating of polymers and also to measure surface temperature distribution (mean emissivity coefficient is needed for infrared camera measurements).

13.2.1.1. *IR-spectrometer measurement (transmission and reflectivity)*

For the determination of the intrinsic transmittivity (τ_λ) and reflectivity (ρ_λ) coefficients of semi-transparent polymers, transmission factor (T_λ) and reflection factor (R_λ) are measured using FT-IR spectrometer $(0.4 - 25\ \mu\text{m})$ together with an integrating sphere in the case of diffuse surface as sketched in Figure 13.4.

From the literature [SIE 02], expressions of transmission and reflection factors are well known:

$$\begin{cases} T_\lambda = \dfrac{(1-\rho_\lambda)^2 \tau_\lambda}{1-(\rho_\lambda \tau_\lambda)^2} \\ R_\lambda = \rho_\lambda \left[1 + \tau_\lambda^2 \dfrac{(1-\rho_\lambda)^2 \tau_\lambda^2}{1-(\rho_\lambda \tau_\lambda)^2} \right] \end{cases} \qquad [13.4]$$

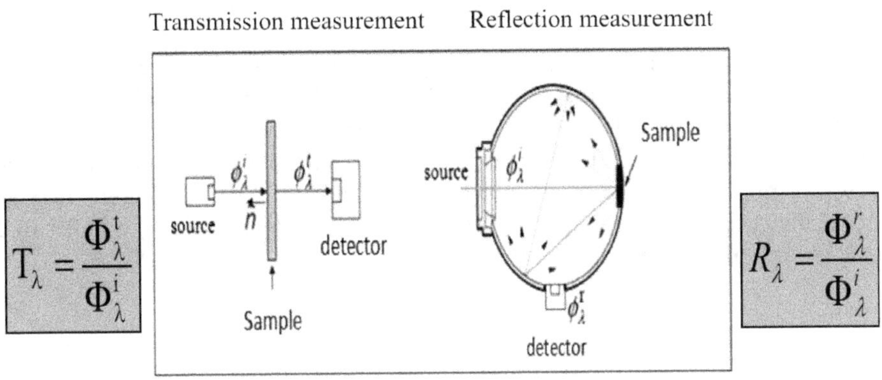

Figure 13.4. *Radiative flux distribution [AKU 15]*

Although a full resolution is possible using a numerical method [AKU 15], a simplified experimental approach, which has already been proven in several studies [BOR 09, MON 04], considers the following assumptions $\rho_\lambda \langle\langle 1$ and $\rho_\lambda \tau_\lambda \langle\langle 1$ for polymers material [SIE 02]:

$$T_\lambda \cong \tau_\lambda \frac{1-\rho_\lambda}{1+\rho_\lambda} \quad \Rightarrow \quad -\ln T_\lambda \cong -\ln \tau_\lambda + \ln \frac{1+\rho_\lambda}{1-\rho_\lambda} \qquad [13.5]$$

If we consider *non-scattering and cold medium assumptions* (see section 13.3), transmittivity τ_λ is given according to the Beer–Lambert law:

$$\tau_\lambda = e^{-\kappa_\lambda t} \qquad [13.6]$$

Equation [13.5] may then be simplified as follows:

$$-\ln T_\lambda \cong \kappa_\lambda t + C_\lambda \qquad [13.7]$$

where k_λ is the intrinsic absorption coefficient *independent* of the sample thickness t and C_λ is a value corresponding to calculation of the $-\text{In}T_\lambda$ intercept of the experimental line. The C_λ value permits us to calculate the reflectivity coefficient, in addition to reflection mode measurements that can be performed.

13.2.1.2. *Example of transmission and reflection spectra for some polymers*

As mentioned before, PET is extensively used for the manufacture of thin-walled containers that are used for a multitude of applications, especially beverage containers. In order to achieve accurate transmission measurements, it is necessary to obtain thin samples at different thicknesses from amorphous PET injected-molded plaques or preforms. More details are described in [MON 01]. In Figure 13.5, the spectral transmission is plotted for two PET Tergal T74F9 (IV = 0.74) samples (0.24 and 0.34 mm thicknesses).

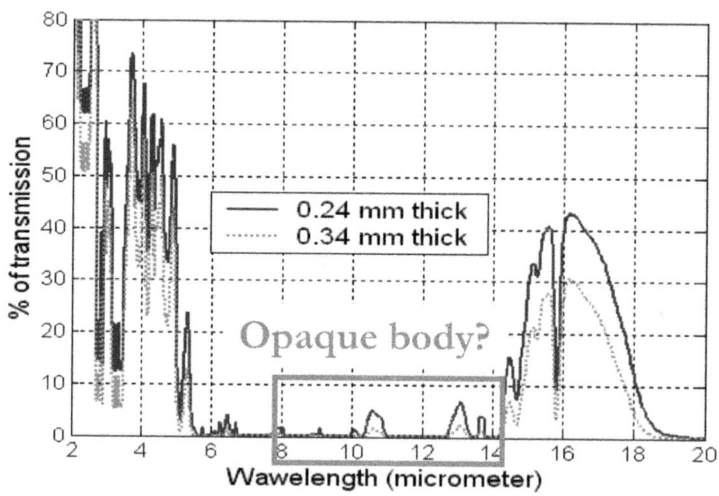

Figure 13.5. *Spectral transmission coefficient of PET [BOR 07]. For a color version of the figure, see www.iste.co.uk/boyard/heat.zip*

We note that for $\lambda \leq 7$ µm, the transmission is very important. At the opposite, transmission is low in the spectral band (8 – 14 µm). This means that PET can be considered as opaque in this specific wavelength band. This is not the case for all polymers. For example, HDPE is not opaque in this spectral band. In addition, reflection measurements for PET have been performed with 2 mm thick samples (Figure 13.6), confirming a low average value (approximately 7%).

Further values for other polymers will be presented in the next section in correlation with lamp temperatures and the surface colors.

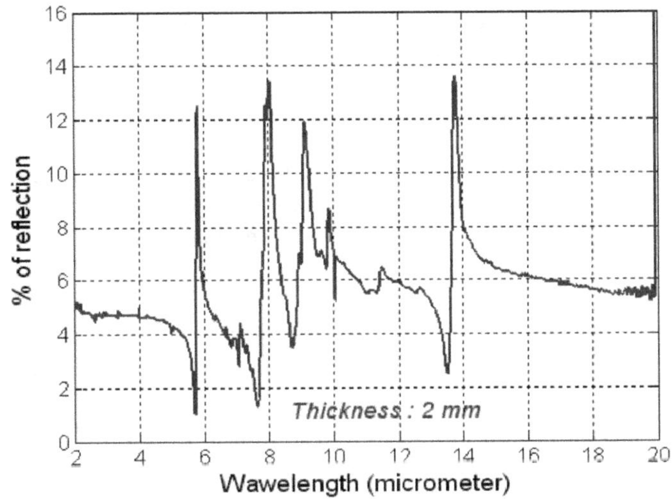

Figure 13.6. *Spectral reflection coefficient of PET [BOR 07]*

13.2.2. *Infrared emitters: characterization*

Infrared emitters vary mainly by the technology used, their geometry and temperature T_L. The role of temperature is predominant because it affects the maximum power and wavelength range associated with the element. The main problem is the choice of the wavelength range for radiative interactions between infrared lamps and polymers.

In addition, reflectors may change the spatial distribution of radiative flux. Other parameters, such as inertia during heating and cooling, can be critical depending on the type of application. As different technologies [KNI 96] exist on the market (metallic radiant tubes, quartz tubes, ceramics, vitrified ceramic radiant panels, catalytic and non-catalytic gas, etc.), we will focus here only on specific electrical emitters mainly used for ISBM and thermoforming processes: halogen, quartz tube and ceramic lamps.

The choice of emitters will be conditioned according to Wien's law giving the value of wavelength λ_m for which spectral intensity I_λ is maximal [SIE 02]:

$$\lambda_m = \frac{2897.8}{T_L} \quad (T_L \text{ in Kelvin}) \qquad [13.8]$$

Thus, the wavelength range associated with an emitter may be calculated by knowing that 95.6% of energy is between $0.5\times\lambda_m$ and $5\times\lambda_m$.

13.2.2.1. *Emitter temperature*

In Table 13.2, temperatures of typical emitters (halogen, quartz tube and ceramic lamps) are referenced as well as the related wavelength range. Note that for halogen lamps, most of the energy is emitted in the near-infrared (NIR) and short-wavelength infrared (SW).

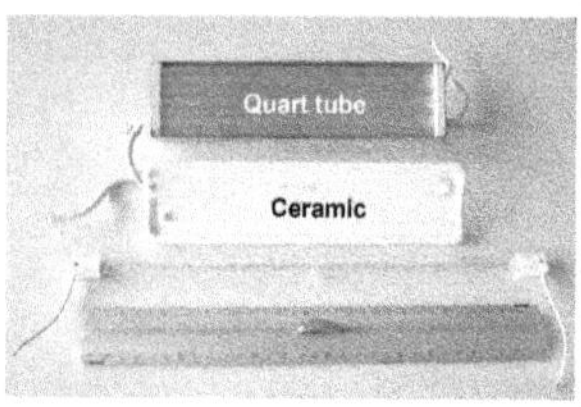

	Heater temperature (K)	Wavelength range $\Delta\lambda$ (μm)
Quartz tube	1,140	1.3. – 12.5
Ceramic	970	1.5 – 15
Halogen lamp	2,470	0.6 – 6

Table 13.2. *Wavelength ranges for the different emitters [AND 05]*

Using the previous measures of intrinsic spectral absorption coefficient (Figure 13.5) allows a mean penetration depth of radiation energy $\overline{e}(T)$ to be computed, which is the inverse of mean absorption coefficient $\overline{\kappa}(T)$:

$$\overline{e}(T) = \frac{1}{\overline{\kappa}(T)} = \frac{\int_{\Delta\lambda} I_\lambda^0(T)\,d\lambda}{\int_{\Delta\lambda} \kappa_\lambda I_\lambda^0(T)\,d\lambda} \qquad [13.9]$$

where $I_\lambda^0(T)$ is the blackbody spectral intensity given by Planck's law [MOD 03]. The spectral band $\Delta\lambda$ = [0.4, 25] μm corresponds to the typical range of the spectrometer Bruker Vertex 70 available in our laboratory. In Table 13.3, penetration depths are calculated for different thermoplastic polymers (ABS, PS and PP), natural, black (N) or white (B), considering temperatures of lamps (see Table 13.2). Indeed, most of the values are very low (a few tens of micrometer) and of the same order of magnitude for ceramic and quartz tube emitters. The highest value is for natural PP (622 μm) together with halogen lamp. In comparison (Figure 13.5), transmission radiation is much more important for PET polymer and particularly for NIR.

$\bar{e}$ (μm)	ABSB	ABSN	PS	PSB	PSN	PP	PPB	PPN
Ceramic	69	29	195	64	39	429	157	65
Quartz tube	64	26	215	64	35	451	157	59
Halogen	41	18	233	79	29	622	128	44

Table 13.3. *Mean penetration depths [AND 05]*

The effect of color can be further highlighted by calculating the mean reflectivity coefficient $\bar{\rho}(T)$ on the same spectral band:

$$\bar{\rho}(T) = \frac{\int_{\Delta\lambda} \rho_\lambda I_\lambda^0(T)\, d\lambda}{\int_{\Delta\lambda} I_\lambda^0(T)\, d\lambda} \qquad [13.10]$$

Most of the values are less than 10%, as expected (section 13.2.1.2). But for white samples with halogen lamps (NIR and SW), the mean reflectivity coefficient may be more than 40%. These results confirm that infrared heating using halogen lamps, will be affected by the color of the surface, particularly by white.

$\bar{\rho}(T)$ (%)	PS	PSB	PSN	PP	PPB	PPN	ABSB	ABSN
Ceramic	2.9	3.9	2.7	3.9	4.1	3.5	4.6	4.1
Quartz tube	6.3	10.0	3.6	7.9	10.2	3.9	11.4	4.9
Halogen	19.8	41.5	6.3	19.0	41.3	5.2	45.6	6.5

Table 13.4. *Mean reflectivity coefficient [AND 05]*

13.2.2.2. *Emission spectrum of halogen lamps*

Considering now the halogen lamp, a coiled tungsten filament is enclosed into a quartz tube filled with neutral gas. In order to calculate the power emitted by the lamp P_L, we need to know the spectral emissivity of the tungsten filament versus temperature $\varepsilon_\lambda^F(T_F)$. Using data issued from the literature [MON 01] and completing for $\lambda \geq 2.6$ µm using Hagen–Rubens law [SIE 02], allows an estimation of the spectral emissivity for pure tungsten (Figure 13.7). Indeed, we observe strong variations versus wavelength and temperature that prevents assuming this parameter constant. Thus, we can estimate the power emitted by the lamp as follows:

$$P_L = \sigma_{SB}\overline{\varepsilon}^F(T_F)T_F^4 S_F \quad [13.11]$$

where σ_{SB} is the Stefan–Boltzmann constant [SIE 02], S_F is the area of filament, T_F is the temperature of the filament and $\overline{\varepsilon}^F$ is the mean emissivity, defined following the same way as the mean reflectivity coefficient (equation [13.10]).

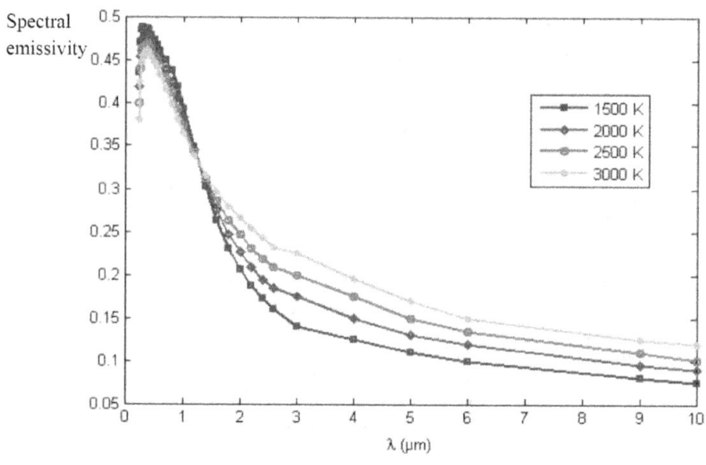

Figure 13.7. *Spectral emissivity of pure tungsten [MON 01]. For a color version of the figure, see www.iste.co.uk/boyard/heat.zip*

Of course, the filament is not made only of pure tungsten, but of an alloy. For $T_F = 2470$ K (i.e. the temperature of a halogen lamp), the mean emissivity is equal to 0.27. In addition, the quartz tube also emits and contributes to 9% of total emitted power [MON 01].

13.2.2.3. *Directional emission of typical lamps used in forming processes*

An in-lab experimental set-up using a thermopile (for full details, see [MON 01] and [AND 05]) has been developed in order to evaluate directivity of each emitter and to provide data for computational methods described later in this chapter. In Figure 13.8, polar emissions of different heaters are plotted and compared to Lambertian source (isotropic emission).

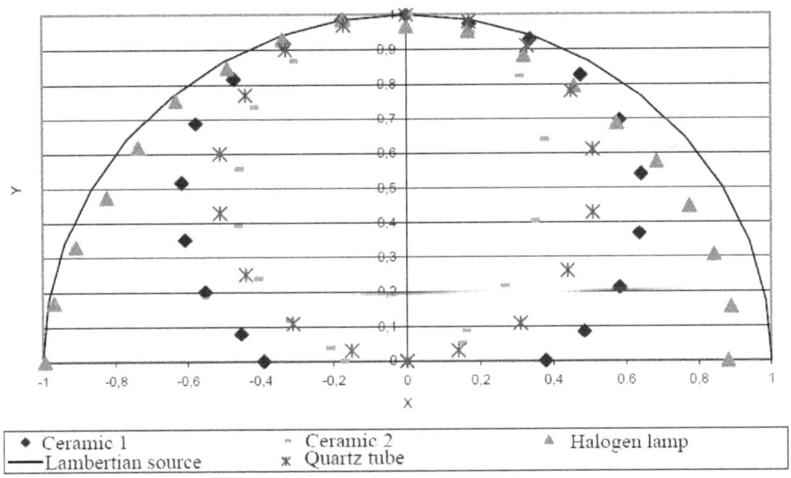

Figure 13.8. *Polar emission of different heaters [AND 05]. For a color version of the figure, see www.iste.co.uk/boyard/heat.zip*

We note that a “clear” halogen lamp (i.e. with no ceramic coating on the back surface of the quartz tube) is similar to a Lambertian source. This is not the case for all the other emitters.

13.2.3. *Infrared camera measurements*

One of the key points in polymer processing is the control of surface temperature of parts or preforms during the heating step and, as a consequence, the optimization of the volumic heat source inside the material to ensure the best efficiency of heating. Due to the high production rate needed for plastic bottles, as seen before, this step is mainly provided by IR lamps.

Some tests have been conducted by researchers to evaluate the ability of thermocouples as a temperature sensor on the surface or inside the polymer heated [BAR 94]. These authors have shown that the response of the embedded thermocouples in a semi-transparent material such as PET can be drastically affected by the IR radiation traveling through the material toward the thermoelectric junction. This leads to errors in the temperature measurements which can be very different from the true local temperature of the polymer. This drawback implies rigorous corrections (that means a specific model) of the measurement process. Moreover, the size of the thermocouple has to be chosen very carefully to avoid conduction into the wires. Such a method is not suited to quick experiments on industrial machines where processing time is a critical parameter.

For these reasons, non-contact temperature measurement has been tested in our approach. Some authors proposed in the past to use infrared pyrometers [HAI 94] and gave comparisons depending on spectral properties of the material observed. To be more complete, we have chosen temperature field measurement with an infrared camera instead of single spot analysis. In all these situations, we have to take into account a radiometric equation related to different parameters which are necessary to convert raw digital values obtained at the output of the camera into true temperatures. The radiometric equation is given by:

$$DL^0\left(T_{eq}\right)=\tau_{atm}\left(T_{atm}\right)\left(\varepsilon_{polym}DL^0\left(T_{polym}\right)+\rho_{polym}DL^0\left(T_{env}\right)\right)+DL^0\left(T_{atm}\right) \qquad [13.12]$$

– $DL^0\left(T_{eq}\right)$ corresponds to the digital level (DL) delivered by the camera, T_{eq} is an apparent blackbody temperature;

– τ_{atm} represents the atmospheric transmission along the optical path between the camera and the polymer semi-product (preform, sheet, etc.);

– $DL^0\left(T_{polym}\right)$ corresponds to the signal delivered by the camera when the target seen is a blackbody at the temperature of the polymer semi-product;

– $DL^0\left(T_{env}\right)$ is the signal related to an environment term which will be discussed later;

– $DL^0\left(T_{atm}\right)$ is a contribution given by an equivalent blackbody level and corresponding to the atmospheric contribution for long distance measurement.

To obtain these different terms, we need to investigate a typical experimental situation encountered in our problem. This situation is depicted in Figure 13.9 and shows our in-lab blow molding machine including an infrared oven on the right side with six lamps (1,000 W each), a reflector behind (can be an aluminum plate or a specific white ceramic) and on the left side the preform and the mold dedicated to the blowing step of a 0.5 l bottle after the heating step.

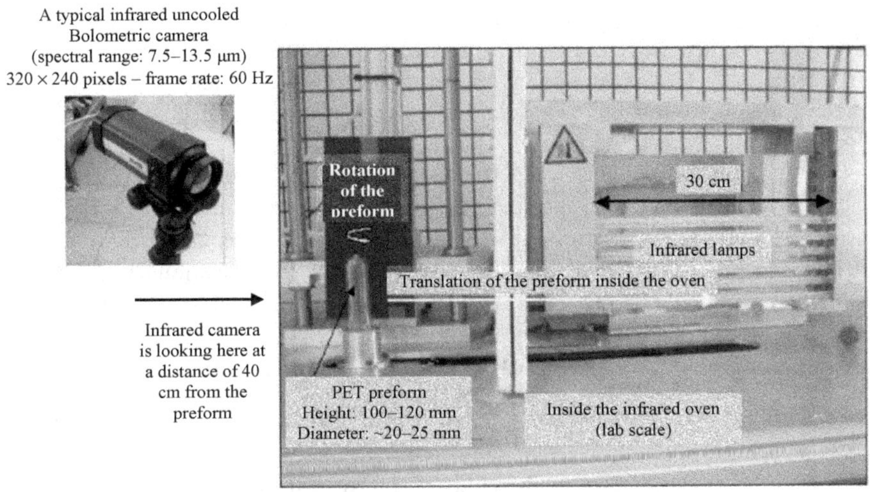

Figure 13.9. *In-lab infrared oven*

At first, it is important to choose the best spectral range for the camera to ensure that the measurement done corresponds to a surface temperature. With the help of equation [13.9] and choosing specific spectral ranges (short wave, medium wave and long wave), it is easy to calculate the penetration depth (see section 13.2.2.1). For example, if we consider PET, the results for e(T), discussed in [MON 04], are summarized in Table 13.5.

spectral range / e(T)	1–2.5 µm	2.5–5µm	5–20 µm
T = 2,400/2,500 K	3.2 mm	0.4 mm	0
T = 400 K	0.5 mm	0.25 mm	30. 10^{-3} mm

Table 13.5. *Penetration depths for T = 2,400/2,500 K and T = 400 K*

It can be seen that the 5–20 µm wave range corresponds to the opaque region: the incident radiative flux coming from the lamps is considered as a boundary condition at the surface and regarding the emission of the preform, the flux received by an infrared sensor can be assumed as a surface emission (not a volume emission).

For this reason, the best compromise here is to choose the 7–14 µm window available among the different ranges for infrared cameras. Most used windows are 7.5–13.5 µm or 8–12 µm depending on the supplier of the camera, which also fits to the best atmospheric transmission window for the infrared [WOL 89].

Note also that for the lamp temperature and the short wave range, e(T) is equal to 3.2 mm, a value greater than the real thickness of our preform (3 mm). This shows that for the maximum of emission of the lamp (≅ 1.2 µm), the radiative energy contributes to a volumic heating of the preform (see section 13.2.2). As a conclusion, by choosing the 7–14 µm spectral range, a real surface temperature measurement can be performed. To ensure that this temperature is a true temperature, we now have to consider the other terms of [13.12]:

– τ_{atm} (atmospheric transmission): for the spectral range 7–14 µm, a simple test done with a blackbody shows that there is no significant attenuation of the signal received by the camera when the blackbody is moved from a very short distance up to several meters. As the preform is viewed at less than 0.5 m, the τ_{atm} parameter is assumed to be equal to 1;

– as a result of Table 13.5, PET is an opaque material in the 7–14 µm spectral region). Then, a reflectance measurement done on an FTIR spectrometer gives $\bar{\rho} = 0.07$ as a mean value of ρ_λ between 7 and 14 µm, this leads to $\bar{\varepsilon} = 0.93$ for our PET Tergal T74F9 (IV= 0.74) (this value can vary slightly according to the grade of PET used);

– contribution of the atmosphere in the emission ($DL^0(T_{atm})$ term): due to the very short distance between the camera and the preform and the spectral range used in our measurements, this term is assumed to be negligible;

– $DL^0(T_{env})$: this term requires special attention depending on the geometric configuration of the oven used. For an "open" oven: the environment term is related to a radiative contribution coming from the wall of the room where the oven is installed (a "room" temperature). Then, a measurement made with the infrared camera on a homogeneous screen beside the heating area, measurement made at room temperature (every part of the oven is in radiative equilibrium with the surroundings) gives the $DL^0(T_{env})$ and in a second step T_{env} by inverting the calibration curve DL^0. However, such a situation can correspond only to a simple lab situation. In a more realistic situation, the oven is composed of different walls and reflectors in front of the panel of lamps or ventilation system as depicted in Figure 13.10.

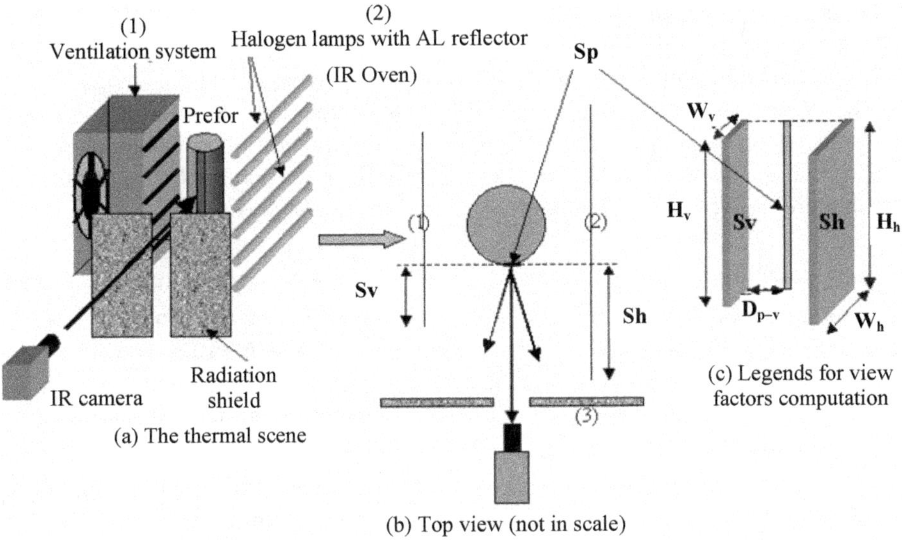

Sp: selected area (vertical strip) on the preform for the temperature measurements
Sv: restricted ventilation support area visible by Sp
Sh: restricted halogen lamps area visible by Sp

Figure 13.10. *A more realistic configuration (our system)*

Each element, depending on its own temperature and optical properties, can contribute to the energy measured by the infrared camera on a vertical strip S_p on the front of the preform as shown in Figure 13.10 (the size of this vertical strip is taken greater than a given number of pixels depending on the spatial resolution of the camera used). The reflectors and grid of the cooling fan are considered flat surfaces (i.e. here S_v and S_k) for which optical properties such as reflectivity have been analyzed with an infrared spectrometer, and temperatures have been recorded with a specific thermocouple fixed on the inner part of each plate. With these data, the energy exchanged, in a first approach, between the ventilation system and the preform and between the halogen lamps with a back surface plane has been obtained via a view factors computation (e.g. plate to plate or strip to plate).

As in a radiosity method, we have then computed the sum of these last two terms incoming on the camera lens (i.e. irradiance E_s in W/m^2). Then, a numerical calibration curve has been provided involving $T_{bb} = f(E_b)$ where E_b was the irradiance in W / m^2 when the preform is assumed to be a perfect blackbody, for different levels of irradiance. The irradiance E_s has also been plotted, with measured optical properties of the preform and the different components of our oven, for different levels of temperature of the preform reproducing the heating stage. Lastly, the two curves were compared to quantify the discrepancy induced by the effect of the walls of the oven. By considering that the computation of E_s is another way to establish [13.12], we tested in the computations an equivalent environment temperature T_{env} (global contribution of reflections on the preform).

This parametric study leads to $T_{env} \cong 25°C$, giving good agreement with the "blackbody curve": $\Delta T \cong 1°C$ around 120°C. An experimental approach has also been tested: an image difference technique between the preform viewed by the infrared camera out of the oven and the preform just as the entrance of the oven: the preform is still in equilibrium with the room temperature but the reflections of the new environment can be deduced; this other approach shows a T_{env} more close to 27/28°C but the difference between computation and experiment had no significant effect on the non-contact measurement done on the preform in the oven (1 or 2°C around 120°C). As a conclusion, a temperature map on the PET preform was successfully achieved inside the infrared oven, generally considered as difficult conditions. An example of such a map is given in Figure 13.11.

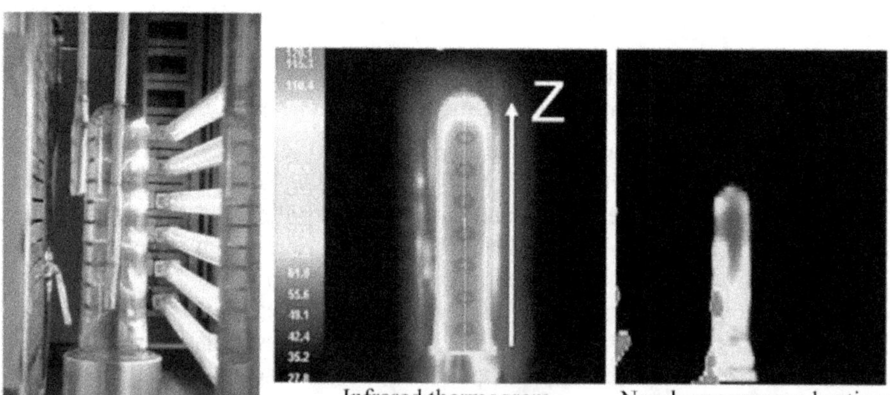

PET preform inside the lab infrared oven

Infrared thermogram showing markers on the height of the preform (temperature spots)

Non-homogeneous heating: IR thermography as an optimization tool for the process

Figure 13.11. *Quantitative infrared thermography inside the lab infrared oven. For a color version of the figure, see www.iste.co.uk/boyard/heat.zip*

This section has shown the different steps to use an infrared camera, not only for thermal imaging but also to achieve quantitative infrared thermography. As mentioned before, these measurements had been used to obtain temperature spots of profiles along the height of the preform helping to adjust the oven set-up regarding the variation of the temperature on the PET.

Other applications, such as validation of lamp temperatures via infrared thermography measurements on quartz tubes (indirect validation of filament temperature) [MAO 05] or heat transfer convection coefficient along the height (Z) of the preform [MON 04] helping to complete the dataset needed for a complete simulation of the heating step, are described in [MON 01].

13.3. Modeling of infrared radiation

As previously shown, characterizing and measuring temperature of infrared polymers is essential before developing radiation simulations. In section 13.2.2, we highlighted that thermoplastics can be opaque or semi-transparent, depending on the type of emitters. In addition, in some kind of composite materials, due to the presence of fibers reinforcement and particularly carbon fibers, it is the same for thermoplastic or thermoset composites [NAK 11]. In the following section, we will consider both opaque and semi-transparent cases.

13.3.1. *Opaque medium: surface to surface methods*

Entire surfaces (tungsten filament, reflectors, composites or polymers, etc.) are assumed opaque in the infrared bandwidth, and behave like a Lambertian source (isotropic emission). In addition, only integrated optical properties should be considered. An ambient medium (for example, air) is assumed to be a non-participative radiative medium.

Thus, surface to surface methods could be applied in order to account for radiation interactions. Then, radiation heat transfer between opaque surfaces is based upon radiosity which is defined as the total radiative flux leaving a surface. The radiosity method subdivides a scene into different areas acting as Lambertian emitters and subsequently determines a condition in which the radiative transfer between the areas leads to an energetic equilibrium [SIE 02]. The fraction of radiative energy absorbed by the composite (or opaque polymer) may be calculated via view factor computations.

Different analytical approaches exist for simple view factor calculations [SIE 02]. However, most of the time, numerical methods are necessary to account for complex geometries. The main advantage of this approach is the reduction of computer time in comparison to more general methods such as ray tracing and Monte Carlo. Figure 13.12(a) shows the geometry of a 3D composite part heated by one halogen heater. The global error between heat fluxes computed using hemicube numerical method [COH 85] and ray tracing method (Figure 13.12(b)) is less than 10%. In addition, CPU time is reduced approximately by 7 (Intel Core 2Duo 2.1 Ghz 2.99 Go).

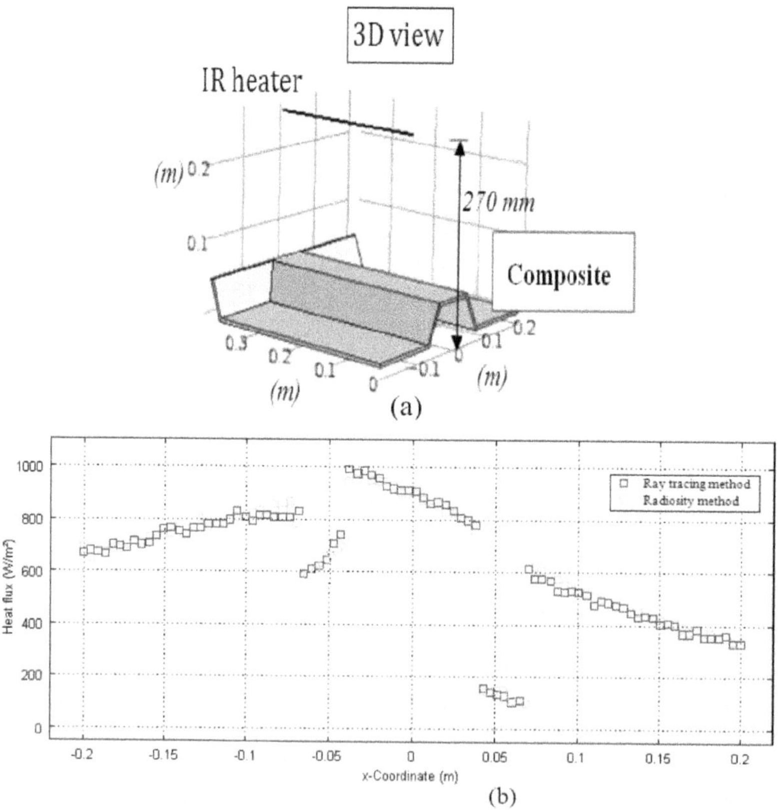

Figure 13.12. *Heat flux along width [NAK 13]. For a color version of the figure, see www.iste.co.uk/boyard/heat.zip*

In order to develop optimization procedures for improving surface temperature distributions, computation times for solving a radiative transfer problem should be greatly minimized. It should be noted that this simplified approach is not useable for infrared heating of semi-transparent polymers or thermoplastic composites, where absorption or scattering has to be accounted for.

13.3.2. *Semi-transparent medium*

The propagation of radiation through a semi-transparent medium is affected by absorption, emission and scattering processes. The radiative transfer equation describes the interaction of photons inside the matter, that is to say: losses of energy related to absorption, gains energy by emission and scattering.

13.3.2.1. *Radiative transfer equation*

The fundamental quantity which describes a field of radiation is the so-called spectral intensity $I_\lambda\left(\vec{s},\vec{\Omega}\right)$ depending on the position vector $\vec{s}$ and the direction vector $\vec{\Omega}$. Thus, the differential form of the equation for radiative transfer is as follows:

$$\frac{dI_\lambda\left(\vec{s},\vec{\Omega}\right)}{ds}+\underbrace{\left(\kappa_\lambda+D_\lambda\right)I_\lambda\left(\vec{s},\vec{\Omega}\right)}_{absorption}=\underbrace{\kappa_\lambda I_\lambda^0\left(T(s)\right)}_{emission}+\underbrace{\frac{D_\lambda}{4\pi}\int_0^{4\pi}\psi\left(\vec{\Omega},\vec{\Omega}'\right)I_\lambda\left(\vec{s},\vec{\Omega}'\right)d\Omega'}_{scattering} \quad [13.13]$$

where D_λ is the scattering coefficient, ψ is the phase function, Ω' is the solid angle and $\vec{\Omega}'$ is the scattering direction vector.

Radiation scattering may occur during infrared heating of semi-transparent thermoplastics for which optical properties are influenced by the crystalline microstructure (morphology and/or degree of crystallinity). Another interesting case is the radiation heating of a semi-transparent thermoplastic composite (Figure 13.13) using a laser beam. The difference between refractive indices of the polycarbonate matrix and the glass fibers will induce light scattering and attenuation of the intensity. In this case, accurate measurement of the scattering coefficient D_λ will be crucial.

13.3.2.2. *Cold and non-scattering material assumption: Beer–Lambert law*

For many thermoplastic polymers, it is possible to neglect scattering during infrared heating. This is typically the case for PET during the heat conditioning step of the ISBM process. Degree of crystallization after injection molding of PET preforms is less than 5% [MON 01]. This assumption is no more possible for ISBM of mineral filled PET where particles may aggregate with a size up to 0.5 µm [BIL 12].

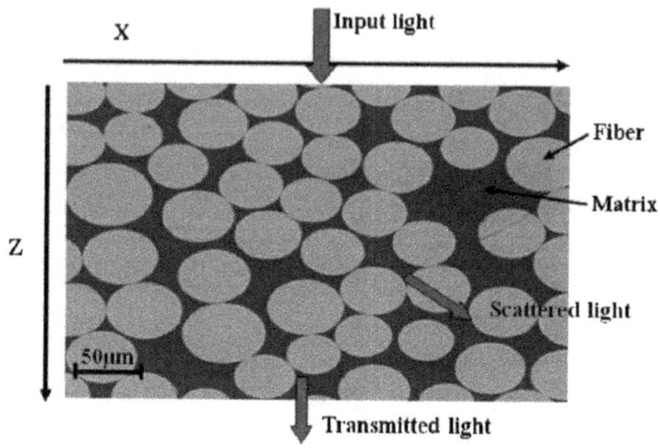

Figure 13.13. *UD composite microstructure (polycarbonate matrix and 40% glass fibers) [AKU 15]*

Another important assumption is the so-called "cold medium". In fact, self-emission of most of the heated polymers may be neglected during the heating step, considering that the polymer bulk temperature ($T_{PET} \leq 400$ K) is very low in comparison to the source temperature ($T_F \leq 2000$ K) for tungsten filament). Therefore, the assumption of a cold material is convenient [SCH 03].

Using these assumptions greatly simplifies the radiative transfer equation and allows integrating easily:

$$\frac{dI_\lambda\left(\vec{s},\vec{\Omega}\right)}{ds} = -\kappa_\lambda I_\lambda\left(\vec{s},\vec{\Omega}\right) \Rightarrow I_\lambda\left(\vec{s},\vec{\Omega}\right) = I_\lambda\left(\vec{s}_o,\vec{\Omega}\right)e^{-\kappa_\lambda(s-s_o)} \qquad [13.14]$$

where s_o can be, for example, the position vector at the polymer surface. We obtain a similar relation to Beer–Lambert law previously used in section 13.2.1.1.

For the case of a semi-transparent material, the heat balance equation may be written as follows:

$$\rho c_p \frac{dT}{dt} = \vec{\nabla}\cdot\left(k\vec{\nabla}T\right) - \vec{\nabla}\cdot\vec{q}_r \qquad [13.15]$$

where ρ is the polymer density, c_p is the specific heat and $\vec{q}_r$ is the radiative flux. In this case, the radiative flux is simply approximated using the Beer–Lambert law [COS 11]:

$$\vec{\nabla}\cdot\vec{q} \cong -\int_0^{\infty} \kappa_\lambda e^{-\kappa_\lambda(s-s_o)} \int_0^{4\pi} I_\lambda\left(\vec{s}_o,\vec{\Omega}\right) d\Omega\, d\lambda = -\int_0^{\infty} \kappa_\lambda M_\lambda\left(\vec{s}_o\right) e^{-\kappa_\lambda(s-s_o)} d\lambda \qquad [13.16]$$

where $M_\lambda(\vec{s}_o)\left(W.m^{-2}.\mu m^{-1}\right)$ is the incident spectral emissive power (for example, from the lamps to the polymer part). $M_\lambda(\vec{s}_o)$ is given by the Planck's law [MOD 03] and represents the power received by the skin/surface of the part from the lamp.

13.3.3. *Ray tracing method*

Different numerical methods may be used to compute the source term $\vec{\nabla}\cdot\vec{q}$. The most efficient are the ray tracing [COS 11] and Monte Carlo [HOW 98] methods. The ray tracing method is commonly used in 3D image rendering. This method is very close to the physics of light propagation, as a ray can represent the path of a photon (or a packet of photons).

Ray tracing allows for the simulation of a wide variety of optical effects, such as reflection, refraction and absorption. In addition, this method enables us to take into account most of constitutive elements of an IR oven such as multiple lamps (various geometries) and reflectors (ceramic or metallic). The general principle of the method is to discretize, into a set of rays, the radiative heat flux emitted by each lamp. Then, each ray is followed inside the oven, from its emission point and throughout the thickness of the polymer. The ray tracer computes the optical path of each ray, accounting for specular or diffuse reflections, refractions, etc.

In addition, assumptions should be made for the different optical properties of lamps, reflectors and part. For example, the tungsten filament of a halogen lamp can be assumed to be a Lambertian emitter. This assumption provides the definition of ray direction vectors (Figure 13.14) for rays coming from the filament.

The direction vector is defined by two parameters: θ and φ, respectively, defined in the ranges $\left[0,\frac{\pi}{2}\right]$ and $[0,2\pi]$. Notations are illustrated in Figure 13.14. The computation of θ and φ has a strong effect on the ray tracing accuracy. Determinist discretization of the emission space could lead to errors due to the ray effect. Computing the two angles by stochastic variables avoids this [LI 97]. Another advantage of the stochastic method is that it is also possible to control the convergence of the computation. In addition, the direction change of a ray that crosses a polymer-air or matrix-fiber interface (for composite) is given by the Snell–Descartes law (Figure 13.15):

$$n_1 \sin\theta_1 = n_2 \sin\theta_2 \quad [13.17]$$

where n_1, n_2 are, respectively, the refractive indices of mediums 1 and 2.

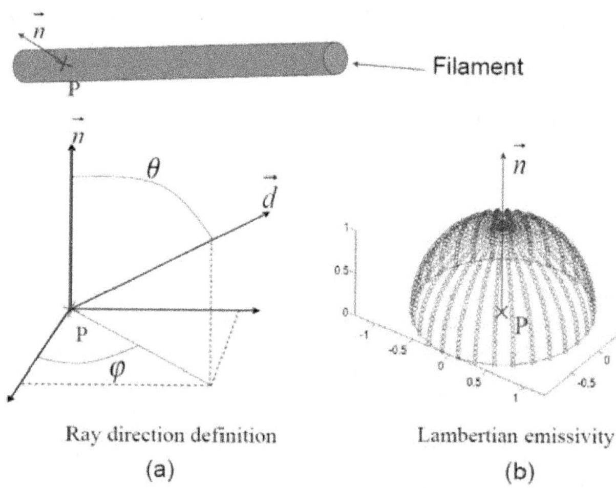

Figure 13.14. *Ray definition for ray tracing [COS 11]*

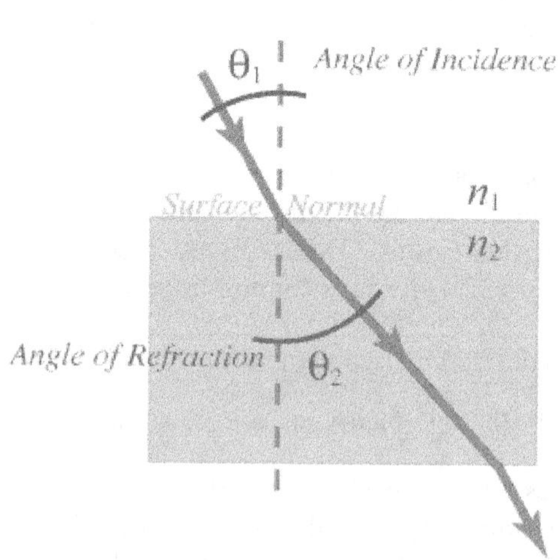

Figure 13.15. *Ray refraction at interface crossing. For a color version of the figure, see www.iste.co.uk/boyard/heat.zip*

As a result, the ray tracing method can be time-consuming. It is therefore very important to optimize the number of rays as well the stochastic computation of polar emissions [COS 11].

13.4. Polymer processing applications

There are many applications of infrared heating in polymer processing (see section 13.1). In this section, we will focus on two promising approaches: one dealing with IR heating of thermoplastic material, "Optimization of preform temperature for ISBM process" (this study was conducted within the framework of 6th EEC framework. STREP project APT_PACK; NMP-PRIORITY) and the other one with thermoset curing "Optimal infrared composite curing" (this research work was supported by TOSHIBA LIGHTING company).

13.4.1. *Optimization of preform temperature for the ISBM process*

As described previously, the IR heating step is crucial for bottle forming. The important question is: what is the optimal preform temperature distribution before inflating? Answering this question is not easy because the main interest of the manufacturer tends to the quality of the final product (thickness distribution, mechanical properties, bottle transparency, etc.). Thus, if we define the problem in terms of an optimization algorithm, we understand that the relevant objective function is related to the forming step (for example, bottle thickness, top-load, etc.) while the optimization variable (temperature) deals with the IR heating step. This means that coupling between both steps (heating and forming) is necessary. In order to reduce CPU time, we have decided to split the optimization problem into two steps:

– First, using an interpolation function of the temperature versus length of the preform, we have computed the optimal values of temperatures located at different heights of the preform due to blow-molding numerical simulations. Full details of the procedure are given in [BOR 09]. The objective function is defined as the standard deviation of the computed bottle thicknesses. The preform geometry was a 18.5 g preform-PET T74F9 (IV = 0.74). The in-lab blowing machine including an IR oven and a 50 cl bottle mold is represented in Figure 13.16.

Figure 13.17 illustrates the temperature distribution along the preform length before and after optimization. Initial conditions were chosen in order to apply a uniform temperature on the preform. In addition, the optimal temperature distribution determined by the industrial partner (Logoplast Company) is plotted. Note that there is a good agreement in the trends between the temperature profile experimentally determined (trial-and-error method) within industrial conditions, and the temperature distribution computed using the optimization method.

The uniform temperature distribution leads to a strongly non-uniform thickness distribution for the bottle, as illustrated by Figure 13.18. After optimization, there is a temperature gradient along the preform length, which provides a more uniform thickness.

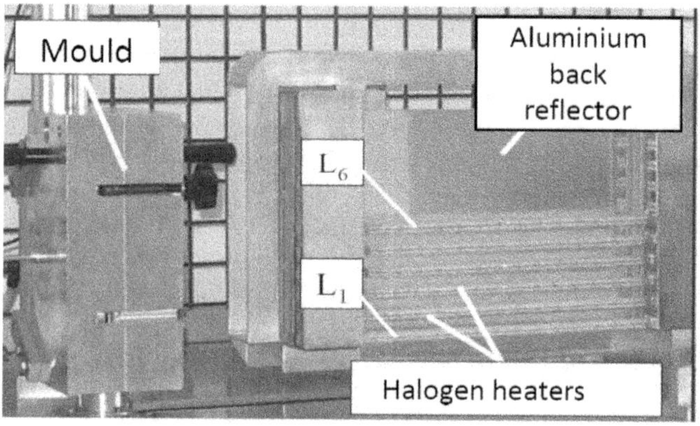

Figure 13.16. *In-lab blow molding machine*

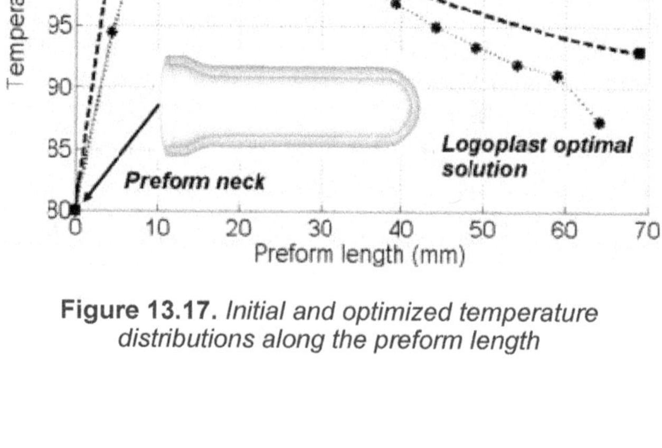

Figure 13.17. *Initial and optimized temperature distributions along the preform length*

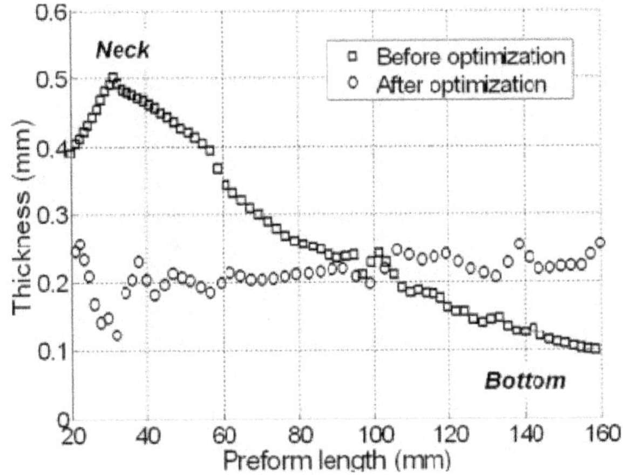

Figure 13.18. *Thickness distribution of the bottle before and after optimization*

– Then, when the optimal temperature distribution has been computed, a second optimization procedure is applied for computing the IR oven optimal parameters. The semi-transparent infrared model has been described in section 13.3. It is important to note that, in order to account for preform rotation, the source term has to be computed as a periodic function versus time in the numerical model. Figure 13.19 shows an example of periodic source term, plotted versus time and distributed inside the preform mesh (Figure 13.20). The computations have been performed using ray tracing in-lab software (so called RAYHEAT) and then coupled to commercial FEM software such as ABAQUS and COMSOL Multiphysics®.

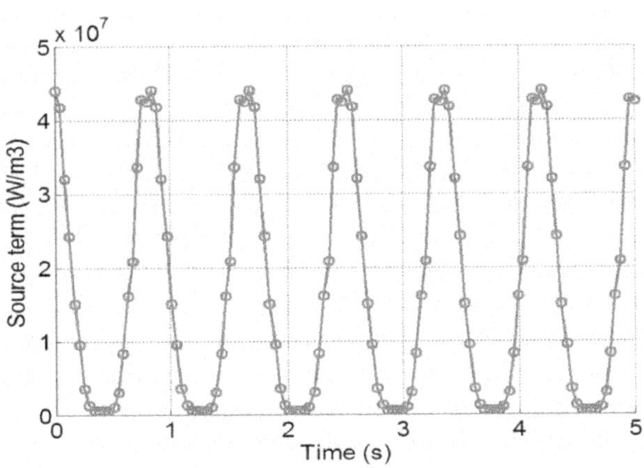

Figure 13.19. *Periodic source term versus time*

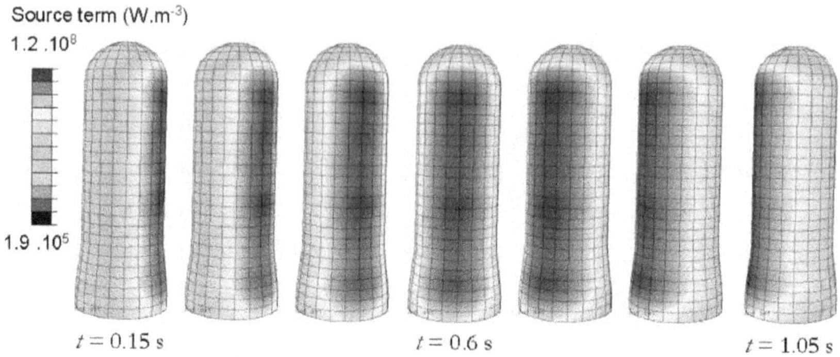

Figure 13.20. *Source term distribution inside preform during time. For a color version of the figure, see www.iste.co.uk/boyard/heat.zip*

In order to illustrate the accuracy of the IR heating model, a comparison is proposed between PET preform simulations and IR camera measurement (Agema 880 [8–12] μm). Six halogen lamps of Philips 2 kW are set to power percentages referenced in Table 13.6.

P_1 (%)	P_2 (%)	P_3 (%)	P_4 (%)	P_5 (%)
100	100	20	5	60

Table 13.6. *Percentage power of each halogen emitter*

In addition, the distance between the lamps and the preform, as well as the mesh for the heaters and preform, is represented in Figure 13.21. For this case, the rotation speed of preform is 1.1 s^{-1}. The heating time is 25 s while cooling time is 10 s. For the polar emission of each lamp, 1.5 million rays have been used. CPU time for computation of source term is 90 min (T9500 2.6 GHz, 3 Go RAM). In addition, the preform is meshed with 6,000 tetrahedral nodes (20 hoop nodes, 10 thickness nodes and 30 height nodes). Figure 13.22 shows the excellent agreement between computed temperature versus preform height and experimental data at the end of heating time for the outside preform surface temperature. The average relative error is 1.7%.

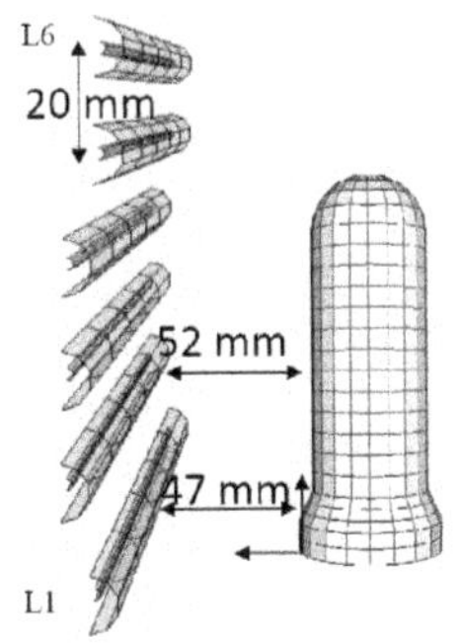

Figure 13.21. *Geometrical definition of oven*

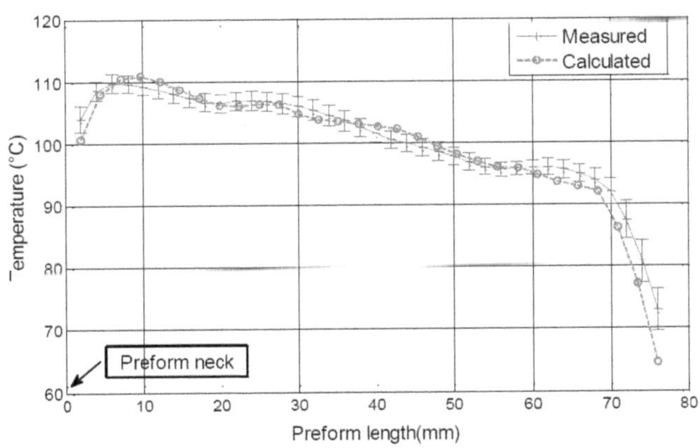

Figure 13.22. *Comparison between computed and experimental temperatures versus height. For a color version of the figure, see www.iste.co.uk/boyard/heat.zip*

Now, using the optimal temperature profile calculated in the previous step (Figure 13.17), we can perform an optimization procedure using the nonlinear constrained algorithm sequential quadratic programming (SQP) in order to calculate the best heating system parameters [BOR 07]. The objective was to obtain the desired temperature along the preform length, by modifying the process parameters related to the IR oven. In the case presented, seven optimization variables associated with the percentage of power of lamps of an industrial oven (Logoplast Company) have been used. The objective function f is simply defined as follows:

$$f(X) = \sum_{i=1}^{7} \left(T_{comp,i}(X) - T_{obj,i} \right)^2 \qquad [13.18]$$

where X represents the set of optimization variables, $T_{comp,i}(X)$ is the computed temperature for the ith lamp and $T_{obj,i}$ is the desired temperature value for the ith lamp. The temperature of each lamp can be related to the lamp power using the Stefan–Boltzmann law (section 13.2.2). Starting with the initial values referenced in Table 13.7, we obtain after four optimization iterations and a CPU time of 7 h 30 min on a Pentium M 1.8 Ghz 512 Mo, the optimized values.

	P_1 (%)	P_2 (%)	P_3 (%)	P_4 (%)	P_5 (%)	P_6 (%)	P_7 (%)	f (°C^2)
Initial value	35	35	35	35	35	35	35	0.134
Optimized value	39.1	35.9	17.2	19.2	23.4	29.3	33.1	0.016

Table 13.7. *Initial and optimized values of lamps power*

Figure 13.23 shows the comparison between the temperature profiles before and after optimization. The obtained profile fits well the targeted profile.

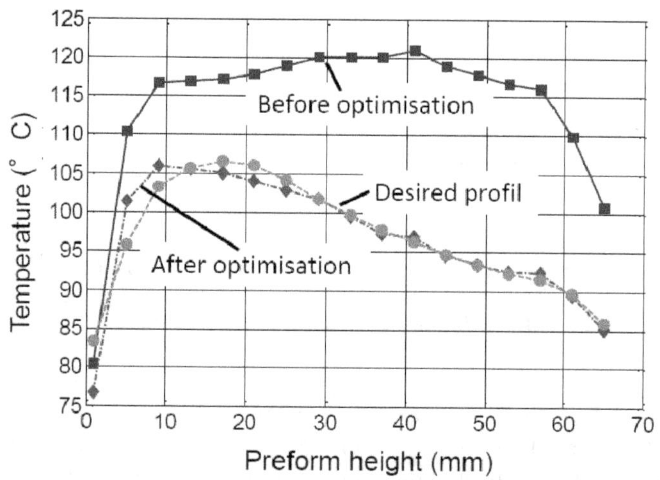

Figure 13.23. *Temperature profiles before and after optimization. For a color version of the figure, see www.iste.co.uk/boyard/heat.zip*

Other parameters such as heating time and cooling time can be added as optimization variables. We may also improve the objective function by accounting for *energy efficiency* for example.

a)

b)

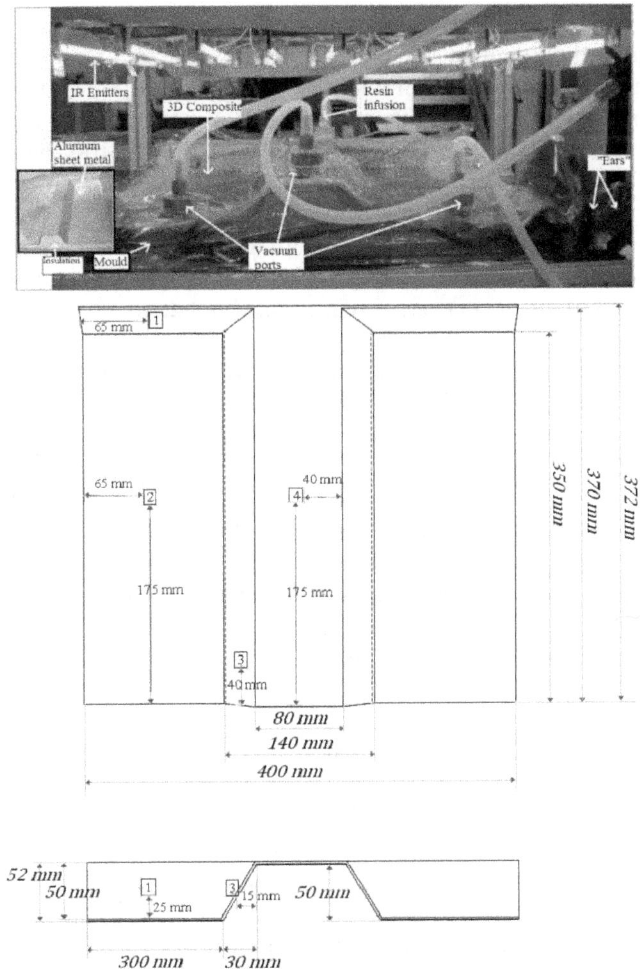

Figure 13.24. *Infrared composite curing process*

13.4.2. *Optimal infrared composite curing*

Epoxy resins have several applications in the aerospace and automobile industry. Because of their good adhesive properties, superior mechanical, chemical and thermal properties and resistance to fatigue and microcracking, they produce high-performance composites. In addition, rising energy costs have also promoted an interest in using infrared energy to manufacture these composites [KAR 00, CHE 02]. In the technology presented here, the composite is cured using an IR oven which includes halogen lamps (Figure 13.24(a)). The liquid resin infusion (LRI) process is used to manufacture the composite, whereby liquid resin (epoxy resin: RTM6) is infused through a fiber reinforcement (carbon reinforcement IMS60, non crimp fabric [45/90/-45/0]) previously in a one-sided mold (Figure 13.24(a)). The 3D geometry used to validate numerical simulations is a simplified car underbody and its dimensions are sketched in Figure 13.24(b). The manufacturing temperature for such composite is between 160 and 190°C.

The objective of this work was to compare the temperature measurements of the 3D composite geometry realized using an infrared camera and thin thermocouples (less than 100 µm in diameter) versus numerical simulations.

In order to measure surface temperatures, we used a microbolometer camera equipped with a lens of focal length $F = 18$ mm. Its resolution is 320 × 240 pixels. Its spectral range corresponds to the average infrared (7.3 – 13 µm). The image frequency is 60 Hz. The apparent emissivity of the composite in the spectral range of the camera $\varepsilon_{7-13} = 0.95$ has been calculated due to the apparent reflectivity R_{7-13} [SIE 02]. The apparent reflectivity has been deduced from spectral reflectivity measurements, performed due to an infrared spectrometer Bruker Vertex 70. It is to be noted that this spectrometer is equipped with two detectors enabling coverage of all the IR spectral ranges (0.8 to 25 µm). An important detail is that the resulting measurement includes all the different films used for resin infusion (recovery surface and demolding fabric) as well as carbon fibers.

In the previous papers [NAK 13, NAK 11], a FE thermal model based upon the COMSOL Multiphysics™ software was fully detailed, for the prediction of the infrared incident heat flux on the top surface of the composite during heating. Although we have applied the ray tracing method previously [NAK 10], it is not convenient to use it for an industrial application because of the computational expense when solving the radiation problems and analysis time resulting from coupling ray tracing method with the optimization algorithms. The method employed by the FE heat transfer solver to compute the radiation heat transfer between opaque diffuse surfaces is based upon radiosity which is defined as the total radiative flux leaving a surface (see section 13.3.1).

The composite was meshed into 8,248 tetrahedral elements. A very dense mesh is used for the volume material discretization along and near the edges highlighted in Figure 13.25(a), as higher temperature gradients will occur here, while a coarser mesh is used in the remaining volume. The resulting temperature distribution is represented in Figure 13.25(b). Due to the optimization procedure [NAK 10], the maximum temperature difference throughout the composite geometry is less than

5°C. Again, the positions and associated numbers of thermocouples and surfaces are sketched in Figure 13.25(b).

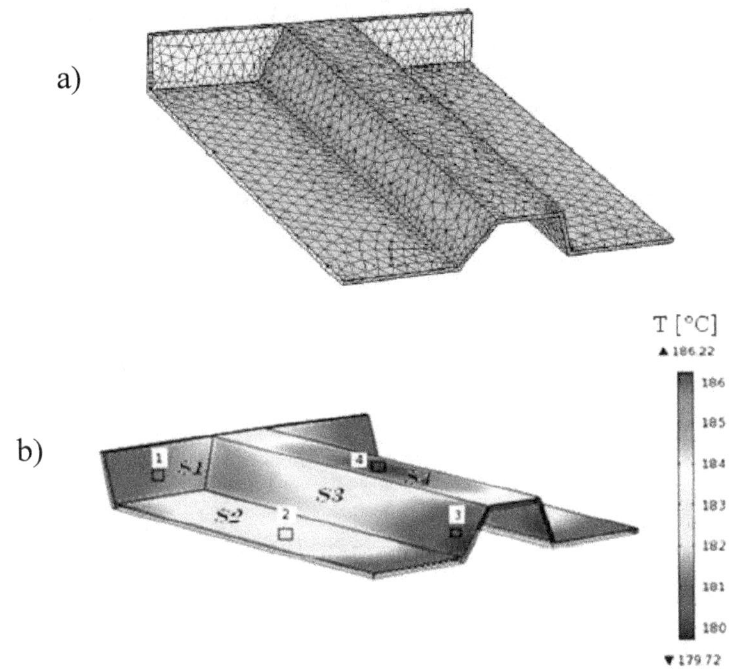

Figure 13.25. *3D mesh of the composite geometry and computed temperature distribution. For a color version of the figure, see www.iste.co.uk/boyard/heat.zip*

In Figure 13.26, measured temperatures issued from both thermocouples and an IR camera, and computed ones, are plotted versus time. The 3D numerical model is developed in the finite element software COMSOL Multiphysics™, where the heat balance equation is coupled with the cure kinetic model of the resin [SIE 02]. The computed radiative heat flux is exported to COMSOL Multiphysics™, and imposed as a boundary condition on the top surface of the composite. This numerical model allows the calculation of the temperature distribution in the composite during curing as well as the degree of curing. From a global point of view, the agreement is unexpectedly good considering all the input data involved in the numerical model. At position 1, the measured temperature using thermocouple (TC1) and computed one is in good agreement, especially during the temperature rise. However, a temperature difference of 5°C is calculated during the steady state.

At position 2, the surface temperature measured by IR camera is slightly higher than those measured by thermocouple (TC2). Nevertheless, the measured and computed temperatures exhibit again a good agreement.

At position 3, again the best agreement between experimental and numerical temperatures occurs during temperature rise (transient state). A temperature difference of 8°C is observed during the steady state, both with IR camera and thermocouple measurement. Indeed, the proximity of this thermocouple to a vacuum port leads to underestimate the measured temperature.

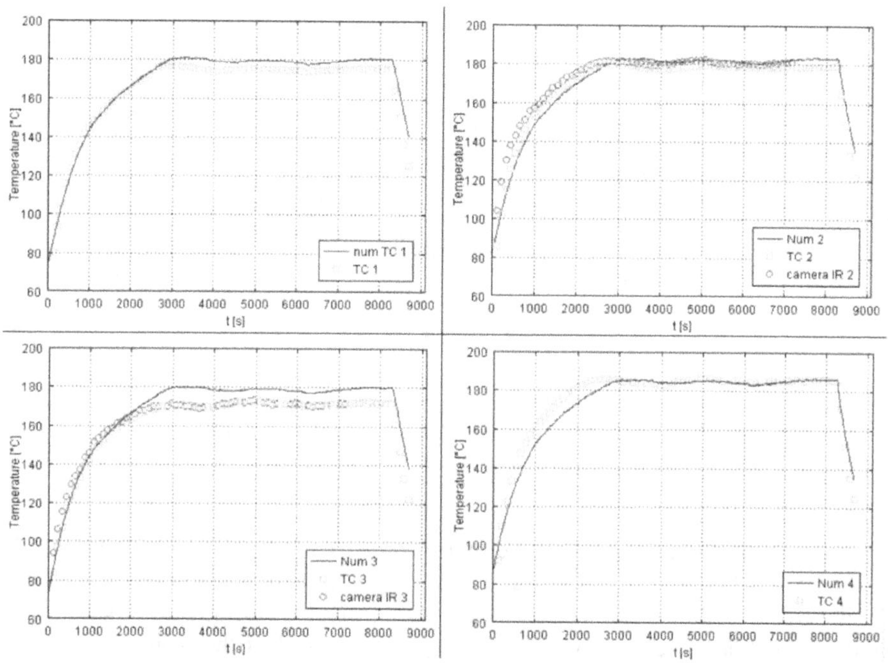

Figure 13.26. *Comparison between measured and computed temperatures. For a color version of the figure, see www.iste.co.uk/boyard/heat.zip*

Different temperature measurements due to an IR camera and thermocouples allows for validating computed temperatures. In order to simulate the curing process, 3D radiation numerical simulations also include a kinetic model based upon modified Kamal–Sourour equation [NAK 13, RYA 79]. For estimating the influence of IR radiation upon mechanical properties of the final composite piece, it would be important to achieve mechanical tests (DMA, 3-points bending, shear test, etc.). For example, comparisons may be performed between samples cured by conduction mode (or autoclave), and by infrared oven. In addition, careful examinations of SEM images would lead to better understand the microstructure of the composite piece obtained using the IR curing process.

13.5. Future work

Infrared heating of polymers is still a challenging and exciting research topic. Many questions still remain unanswered. An important one deals with the interaction of radiation and the microstructure of polymer. As mentioned previously (see section 13.3.2.2), when adding submicronic mineral fillers or dyes in PET [BIL 12], IR heating is affected. This means rays may be refracted and radiation diffusion may occur. This is also the case for thermoplastic composites [AKU 15] (for example, polycarbonate matrix and glass fibers reinforcement) or some polymers crystallizing during IR heating. The ray tracing method can be used to account for scattering but it can be time-consuming for industrial applications. Different physical models, based on ratio between the diameter D of spherical particles and IR wavelength λ (size parameter known as $\pi D/\lambda$), such as Rayleigh approximation and Mie theory [VAN 81] may be applied to account for interactions between IR radiation and microstructure as in [DEN 96].

Another interesting challenge deals with the measurement of the absorption coefficient for semi-transparent polymers (for example, PET). The common procedure uses a spectrometer in transmission mode. But measuring the absorptivity independently of the thickness samples requires careful cuts using microtome, meaning operator dependancy. In the case of specular reflection, we could take advantage of Kramers–Kronig [KIT 04] relation in order to assess to the absorption coefficient.

In addition, the characterization of heater lamps and more particularly halogen lamps, still remains a big challenge. Most laboratories are not equipped for measuring such high temperatures ($\cong$ 2500 K). Again, an IR camera may be a good way of evaluating the desired temperature. The experimental set-up consists of heating a thermoplastic sheet for which all the thermo-physical and thermo-optical properties are well known. Thus, when performing accurate numerical simulations of infrared heating, we can apply an inverse design approach in order to find the optimal temperature value. The objective function may compare, for example, experimental and numerical temperature profiles at the back surface of the polymer sheet. For this method, work has to be done on the influence of the boundary condition such as convection and environment temperature. Lastly, the experiments involving an infrared camera could also bring information on the irradiance map in W/m^2 provided by the halogen lamp on the thermoplastic sheet (profile, amplitude, etc.).

13.6. Acknowledgements

The authors warmly thank all the PhD and postdoctoral students for their indispensable help, exciting scientific discussions and apologize to those they forgot to mention:

Serge Monteix, Sylvia Andrieu, Cedric Champin, Maxime Bordival, Benoit Cosson, Sawsane Nakouzi, Johann Pancrace, Myriam Dauphin, Andre Chateau and Akue Asseko, among others.

Thanks also go to our colleague Jean-Paul Arcens, now retired, who helped us in many technical aspects during the past decade on the different problems described in this chapter.

Special thanks go to our colleague Gary Menary (Queen's University of Belfast) for reviewing this chapter.

A special thought for our colleague Bernard Plantamura (Sidel Company) who left us prematurely in 2015.

13.7. Bibliography

[ADE 02] ADEME, "Energies radiantes et leurs applications industrielles", available at: http://www.ademe.fr/energies-radiantes-applications- industrielles, 2002.

[AKU 15] AKUE ASSEKO A.-C., COSSON B., SCHMIDT F. *et al.*, "Laser transmission welding of composites-Part A: thermo-physical and optical characterization of materials", *Infrared Physics and Technology*, vol. 73, pp. 304–311, 2015.

[AND 05] ANDRIEU S., Étude expérimentale et numérique du chauffage infrarouge court sur des plaques polymères pour le thermoformag, PhD Thesis, ENSMP, 2005.

[BAR 94] BARDON J.P., JAVELAS R., Mesure de températures de surface sur des matériaux semi-transparents, Internal report, CNRS-Ecotech/Ademe-Arc-Metrologie, 1994.

[BIL 12] BILLON N., HAUDIN J.M., VALLOT C. *et al.*, "Stretch blow molding of mineral filled PET", *Key Engineering Materials*, vol. 504–506, pp. 1099–1104, 2012.

[BOR 07] BORDIVAL M., SCHMIDT F., LE MAOULT Y., "Numerical modelling and optimization of infrared heating system for the blow molding process", *Proceedings of 9th Esaform Conference on Material Forming*, 2007.

[BOR 09] BORDIVAL M., LE MAOULT Y., SCHMIDT F., "Optimization of preform temperature distribution for the stretch-blow moulding of PET bottles: infrared heating and blowing modeling", *Polymer Engineering and Science*, vol. 49, no. 4, pp. 783–793, 2009.

[CHE 02] CHERN B.-C., MOON T.J., HOWELL J.R., "On-line processing of unidirectional fiber composites using radiative heating: I. model", *Journal of Composite Materials*, vol. 1935, pp. 1905–1934, 2002.

[COH 85] COHEN M.F., GREENBERG D.P., "The hemicube: a radiosity approach for complex environments", *Computer Graphics*, vol. 19, no. 3, pp. 31–40, 1985.

[COS 11] COSSON B., SCHMIDT F., LE MAOULT Y. *et al.*, "Infrared heating stage simulation of semi-transparent media (PET) using ray tracing method", *International Journal of Material Forming*, vol. 4, pp. 1–10, 2011.

[DEN 96] DENIS A., DARGENT E., LEBAUDY P. *et al.*, "Dependence on the spectral scattering coefficient on cristallinity into semicrystalline polyester", *Journal of Applied Polymer Science*, vol. 62, pp. 1211–1218, 1996.

[FRA 02] FRANCK I., DAVID D.W., *Fundamentals of Heat and Mass Transfer*, 5th ed., John Wiley & Sons, 2002.

[HAI 94] HAIJI N., SPRUIELL J.E., "Radiation pyrometry on semi-transparent media with wavelength dependent absorption coefficient", *Polymer Engineering and Science*, vol. 34, no. 2, pp. 122–127, 1994.

[HOW 98] HOWELL J.R., "The Monte-Carlo method in radiative heat transfer", *Journal of Heat Transfer*, vol. 120, pp. 547–560, 1998.

[IEC 87] IEC, International electrotechnical vocabulary, Lighting, Internal Report, 1987.

[KAR 00] KARKANAS P.-I., PARTRIDGE I.-K., "Cure modelling and monitoring of epoxy/amine resin systems II. Network formation and chemoviscosity modeling", *Journal of applied Polymer Science*, vol. 77, pp. 2178–2188, 2000.

[KIT 04] KITTEL C., *Introduction to Solid State Physics*, 8th ed., Wiley, 2004.

[KNI 96] KNIGHTS M., "The truth about heaters", *Plastics Technology*, vol. 42, no. 5, pp. 38–46, 1996.

[LI 97] LI B., TAO W., LIU R., "Ray effect in ray tracing method for radiative heat transfer", *International Journal of Heat Mass Transfer*, vol. 40, pp. 3419–3426, 1997.

[MAO 05] LE MAOULT Y., Mesures infrarouges appliquées aux transferts thermiques et à la science des matériaux, confrontation de modèles aux expérimentations, Thesis, Mines Albi / INP Toulouse, 2005.

[MOD 03] MODEST M.F., *Radiative Heat Transfer*, 2nd ed., Acad. Press, New York, 2003.

[MON 01] MONTEIX S., Modélisation du chauffage convecto-radiatif de préformes en P.E.T. pour la réalisation de corps creux, PhD Thesis, ENSMP, 2001.

[MON 04] MONTEIX S., LE MAOULT Y., SCHMIDT F. *et al.*, "Quantitative infrared thermography applied to blow moulding process: measurement of a heat transfer coefficient", *Quantitative Infrared Thermography Journal*, vol. 1, no. 2, pp. 133–150, 2004.

[NAK 10] NAKOUZI S., PANCRACE J., SCHMIDT F. *et al.*, "Curing simulation of composites coupled with infrared heating", *International Journal of Material Forming*, vol. 3, no. 1, pp. 587–590, 2010.

[NAK 11] NAKOUZI S., PANCRACE J., SCHMIDT F. *et al.*, "Simulations of an infrared composite curing process", *Advanced Engineering Materials*, vol. 13, no. 7, pp. 604–608, 2011.

[NAK 12] NAKOUZI S., BERTHET F., DELAUNAY D. *et al.*, "Optimization of the incident IR heat flux upon a 3D geometry composite part (carbon/epoxy)", *Key Engineering Materials*, vol. 504–506, pp. 1085–1090, 2012.

[NIX 15] NIXON J., Analysis of the stretch blow moulding process and subsequent simulation development, PhD Thesis, Queen's University Belfast, 2015.

[NAK 13] NAKOUZI S., BERTHET F., LE MAOULT Y. *et al.*, "Simulations of an infrared composite curing process", *Key Engineering Materials*, vol. 554–557, pp. 1517–1522, 2013.

[ROS 04] ROSATO D.V., ROSATO A.V., DIMATTIA D.P., *Blow Molding Handbook*, 2nd ed., Hanser Verlag, 2004.

[RYA 79] RYAN M.E., DUTTA A., "Kinetics of epoxy cure: a rapid technique for kinetic parameter estimation", *Polymer*, vol. 20, pp. 203–206, 1979.

[SCH 03] SCHMIDT F., LE MAOULT Y., *et al.*, "Modelling of infrared heating of thermoplastic sheet used in thermoforming process", *Journal of Materials Processing Technology*, vol. 143, pp. 225–231, 2003.

[SIE 02] SIEGEL R., HOWELL J.R., *Thermal Radiation Heat Transfer*, 4th ed., Taylor and Francis-Hemisphere, Washington, 2002.

[THR 99] THRONE J.L., *Understanding Thermoforming*, Hanser Gardner Publications, 1999.

[VAN 81] VAN DE HULST H.C., *Light Scattering By Small Particles*, Dover, 1981.

[WIL 14] WILSON C., MCGRANAGHAN G., "Infrared heating comes of age", *Reinforced Plastics*, vol. 58, no. 2, pp. 43–47, 2014.

[WOL 89] WOLFE W.L., ZISSIS G.J., *The Infrared Handbook*, ERIM ed., 1989.

List of Authors

Jean-Luc BAILLEUL
Laboratoire de Thermocinétique de Nantes
CNRS, University of Nantes
France

Jean-Christophe BATSALE
TREFLE department
ENSAM
France

M'hamed BOUTAOUS
CNRS CETHIL
INSA Lyon
France

Nicolas BOYARD
Laboratoire de Thermocinétique de Nantes
CNRS, University of Nantes
France

Fabien CARA
TFX SA
Switzerland

Thierry COUPEZ
Institut de Calcul Intensif
Ecole Centrale de Nantes
France

Didier DELAUNAY
Laboratoire de Thermocinétique de Nantes
CNRS, University of Nantes
France

Hugues DIGONNET
Institut de Calcul Intensif
Ecole Centrale de Nantes
France

Frédéric JACQUEMIN
Institut de Recherche en Génie Civil et Mécanique
University of Nantes
France

Patrice LAURE
Laboratoire J.A. Dieudonné
University of Nice Sophia-Antipolis
France

Steven LE CORRE
Laboratorie de Thermocinétique Nantes
CNRS, University of Nantes
France

Yannick LE MAOULT
Institut Clément Ader
Université de Toulouse
Mines d'Albi
Campus Jarlard
France

Yasir NAWAB
Textile Composite Material Research Group
Faculty of Engineering and Technology
National Textile University
Faisalabad
Pakistan

Christophe PRADERE
TREFLE department
ENSAM
France

Michel QUINTARD
Institut de Mécanique des Fluides de Toulouse
CNRS, University of Toulouse
INPT, UPS
France

Gilles REGNIER
Laboratoire PIMM
Arts et Métiers ParisTech, CNRS
CNAM
France

Benoit ROUSSEAU
Laboratoire de Thermocinétique de Nantes
CNRS, University of Nantes
France

Fabrice SCHMIDT
Institut Clément Ader
Université de Toulouse
Mines d'Albi
Campus Jarlard
France

Luisa SILVA
Institut de Calcul Intensif
Ecole Centrale de Nantes
France

Vincent SOBOTKA
Laboratoire de Thermocinétique de Nantes
CNRS, University of Nantes
France

Mathieu ZINET
Ingénierie des Matériaux Polymères
CNRS, University Claude Bernard Lyon
France

Index

A, B, C

Arrhenius law, 275
automated tape placement (ATP), 380
Bailleul's model, 95–97
bidirectional reflectivity distribution function (BRDF), 370
bidirectional transmissivity distribution function (BTDF), 371,
Big Data, 325
blackbody, 365–370, 381
chemical shrinkage, 158, 160, 163–167, 170–172
composite forming simulation, 283
contactless, 311, 312, 326
contact measurement, 301

D, E, F

Darcy's law, 272-273, 282, 291
de Gennes, 243–245
differential scanning calorimeter (DSC), 40-41
discrete channels, 229–231
Drapcocot, 248
effective thermal conductivity, 176, 177, 180, 181–183, 194, 195
E-glass, 360, 362, 379
Far Infrared Range (FIR), 359, 374
finite difference method (FDM), 278
finite element method (FEM), 278, 282, 285, 291, 294
flash method, 44, 48-49, 54
Flow equations, 271
fluxmeter, 302, 305, 311
Fokker and Plank equation, 277
Fourier, 83
fractal, 239–241

G, H

gradient algorithm, 209, 216
guarded hot plate, 47, 54, 58
heat flux, 301, 302, 304–309, 311, 326
  sensor, 333–338, 344–348, 352, 355, 356
  types, 345–347

heat transfer, 175–196
modes, 334–335
Heaviside function, 280–282, 290, 293
hot disk, 310, 311
hot plane, 309–311
hot wire, 309, 310
method, 44, 49-50, 54
hygral, 158, 167

I, K L, M

IKV model, 34
inertia, 303, 304, 306, 309, 310, 312
injection molding, 122, 124, 125
interfacial adhesion, 249
irradiance, 363, 370, 372
Kirchhoff's law, 370, 374
light scattering techniques, 124
Lorentz model, 374,
mechanical analysis, 82
micro channel, 2
Mid Infrared Range (MIR), 374
Moldflow(R), 3, 10
Monte Carlo, 404, 407
multi-scale, 175, 195, 196

N, O, P

Navier–Stokes equations, 271, 273, 291
Near Infrared Range (NIR), 374
non-equilibrium, 175, 187, 189–191, 193–195
objective functions, 207, 216
optimal design, 211, 222
Peclet field estimation, 322
Peltier, 344, 346, 348
polyetheretherketone (PEEK) tape, 360
polyethylene (PE), 362
polymer chain, 241–243, 247, 263
polymer welding, 235
poly(methyl methacrylate) (PMMA), 362
polypropylene (PP) rods, 360
pore-scale, 175, 178,
porosity, 22
probes, 302, 308–310, 312, 326
PvT data, 30-38, 50

R, S

representative volume element (RVE), 61
residual stress, 157–159, 161, 167
resin transfer, 352, 353
molding (RTM), 203
rheometer, 140
segment model, 242
shish-kebab, 131–133, 137, 139
shock probe, 50, 54
Snell–Descartes law, 373,
specific volume, 29-37
squeezing, 239, 240, 257, 259, 260, 263

T, U, V, W

Tait model, 34
thermal dispersion, 184, 186, 187, 188, 190
thermal modeling, 215, 235, 253, 262
thermal sources, 193
thermocouple, 2, 12, 302–306, 310, 311
thermopile, 336
thermoset composite, 159, 161, 165

Thery, 337
transfer mode, 204, 205, 224, 231
transformation kinetics, 121, 123, 143, 146
transient plane source (TPS) method, 44, 51
TWINTEX, 360, 378–380
ultrasonic welding, 236, 237, 255, 259
viscoelasticity, 140, 148
Vitrex, 251
warpage, 157, 161, 165, 168

Other titles from

ISTE

in

Mechanical Engineering and Solid Mechanics

2016

DE SAXCÉ Géry
Galilean Mechanics and Thermodynamics of Continua

2015

KARLIČIĆ Danilo, MURMU Tony, ADHIKARI Sondipon, MCCARTHY Michael
Non-local Structural Mechanics

SAB Karam, LEBÉE Arthur
Homogenization of Heterogeneous Thin and Thick Plates

2014

ATANACKOVIC M. Teodor, PILIPOVIC Stevan, STANKOVIC Bogoljub, ZORICA Dusan
Fractional Calculus with Applications in Mechanics: Vibrations and Diffusion Processes

ATANACKOVIC M. Teodor, PILIPOVIC Stevan, STANKOVIC Bogoljub, ZORICA Dusan
Fractional Calculus with Applications in Mechanics: Wave Propagation, Impact and Variational Principles

CIBLAC Thierry, MOREL Jean-Claude
Sustainable Masonry: Stability and Behavior of Structures

ILANKO Sinniah, MONTERRUBIO Luis E., MOCHIDA Yusuke
The Rayleigh−Ritz Method for Structural Analysis

LALANNE Christian
Mechanical Vibration and Shock Analysis – 5-volume series – 3rd edition
Sinusoidal Vibration – volume 1
Mechanical Shock – volume 2
Random Vibration – volume 3
Fatigue Damage – volume 4
Specification Development – volume 5

LEMAIRE Maurice
Uncertainty and Mechanics

2013

ADHIKARI Sondipon
Structural Dynamic Analysis with Generalized Damping Models: Identification

ADHIKARI Sondipon
Structural Dynamic Analysis with Generalized Damping Models: Analysis

BAILLY Patrice
Materials and Structures under Shock and Impact

BASTIEN Jérôme, BERNARDIN Frédéric, LAMARQUE Claude-Henri
Non-smooth Deterministic or Stochastic Discrete Dynamical Systems: Applications to Models with Friction or Impact

EL HAMI Abdelkhalak, BOUCHAIB Radi
Uncertainty and Optimization in Structural Mechanics

KIRILLOV Oleg N., PELINOVSKY Dmitry E.
Nonlinear Physical Systems: Spectral Analysis, Stability and Bifurcations

LUONGO Angelo, ZULLI Daniele
Mathematical Models of Beams and Cables

SALENÇON Jean
Yield Design

2012

DAVIM J. Paulo
Mechanical Engineering Education

DUPEUX Michel, BRACCINI Muriel
Mechanics of Solid Interfaces

ELISHAKOFF Isaac *et al.*
Carbon Nanotubes and Nanosensors: Vibration, Buckling and Ballistic Impact

GRÉDIAC Michel, HILD François
Full-Field Measurements and Identification in Solid Mechanics

GROUS Ammar
Fracture Mechanics – 3-volume series
Analysis of Reliability and Quality Control – volume 1
Applied Reliability – volume 2
Applied Quality Control – volume 3

RECHO Naman
Fracture Mechanics and Crack Growth

2011

KRYSINSKI Tomasz, MALBURET François
Mechanical Instability

SOUSTELLE Michel
An Introduction to Chemical Kinetics

2010

BREITKOPF Piotr, FILOMENO COELHO Rajan
Multidisciplinary Design Optimization in Computational Mechanics

DAVIM J. Paulo
Biotribolgy

PAULTRE Patrick
Dynamics of Structures

SOUSTELLE Michel
Handbook of Heterogenous Kinetics

2009

BERLIOZ Alain, TROMPETTE Philippe
Solid Mechanics using the Finite Element Method

LEMAIRE Maurice
Structural Reliability

2007

GIRARD Alain, ROY Nicolas
Structural Dynamics in Industry

GUINEBRETIÈRE René
X-ray Diffraction by Polycrystalline Materials

KRYSINSKI Tomasz, MALBURET François
Mechanical Vibrations

KUNDU Tribikram
Advanced Ultrasonic Methods for Material and Structure Inspection

SIH George C. *et al.*
Particle and Continuum Aspects of Mesomechanics

www.ingramcontent.com/pod-product-compliance
Lightning Source LLC
LaVergne TN
LVHW020421301125
826312LV00002B/22
9781848217614